D1930180

The Minerals, Metals & Materials Series

Hojong Kim · Shafiq Alam
Neale R. Neelameggham
Harald Oosterhof · Takanari Ouchi
Xiaofei Guan
Editors

Rare Metal Technology 2017

TMS

Springer

Editors
Hojong Kim
Pennsylvania State University
University Park, PA
USA

Harald Oosterhof
Umicore
Olen
Belgium

Shafiq Alam
University of Saskatchewan
Saskatoon, SK
Canada

Takanari Ouchi
Massachusetts Institute of Technology
Cambridge, MA
USA

Neale R. Neelameggham
IND LLC
South Jordan, UT
USA

Xiaofei Guan
Watertown, MA
USA

ISSN 2367-1181 ISSN 2367-1696 (electronic)
The Minerals, Metals & Materials Series
ISBN 978-3-319-51084-2 ISBN 978-3-319-51085-9 (eBook)
DOI 10.1007/978-3-319-51085-9
TMS owns copyright; Springer has full administrative rights

Library of Congress Control Number: 2016960563

© The Minerals, Metals & Materials Society 2017
This work is subject to copyright. All rights are reserved by the Publisher, whether the whole or part of the material is concerned, specifically the rights of translation, reprinting, reuse of illustrations, recitation, broadcasting, reproduction on microfilms or in any other physical way, and transmission or information storage and retrieval, electronic adaptation, computer software, or by similar or dissimilar methodology now known or hereafter developed.
The use of general descriptive names, registered names, trademarks, service marks, etc. in this publication does not imply, even in the absence of a specific statement, that such names are exempt from the relevant protective laws and regulations and therefore free for general use.
The publisher, the authors and the editors are safe to assume that the advice and information in this book are believed to be true and accurate at the date of publication. Neither the publisher nor the authors or the editors give a warranty, express or implied, with respect to the material contained herein or for any errors or omissions that may have been made.

Printed on acid-free paper

This Springer imprint is published by Springer Nature
The registered company is Springer International Publishing AG
The registered company address is: Gewerbestrasse 11, 6330 Cham, Switzerland

Preface

Rare Metal Technology 2017 is the proceedings of the symposium on rare metal extraction and processing sponsored by the Hydrometallurgy and Electrometallurgy Committee of the TMS Extraction and Processing Division. The symposium has been organized to encompass the extraction of rare metals as well as rare extraction processing techniques used in metal production and mineral processing. This is the fourth symposium since 2014, which will be held in San Diego, California.

This symposium intends to cover research and developments in the extraction and processing of less common rare metals that are not covered by other TMS symposia. These elements include antimony, bismuth, barium, beryllium, boron, calcium, chromium, gallium, germanium, hafnium, indium, manganese, molybdenum, platinum group metals, rare earth metals, rhenium, scandium, selenium, sodium, strontium, tantalum, tellurium, and tungsten. These are rare metals of low tonnage sales compared to high tonnage metals such as iron, copper, nickel, lead, tin, zinc, or light metals such as aluminum, magnesium, or titanium and electronic metalloid silicon. Rare processing includes bio-metallurgy, hydrometallurgy, and electrometallurgy, as well as extraction of values from electric arc furnace (EAF) dusts, and less common waste streams not discussed in recycling symposia. Rare high-temperature processes included microwave heating, solar-thermal reaction synthesis, molten salt electrochemical processes, cold crucible synthesis of the rare metals, and the design of extraction equipment used in these processes as well as laboratory and pilot plant studies.

This volume covers extraction and processing techniques of various platinum group metals, rare earth elements as well as other less common metals such as arsenic, indium, antimony, molybdenum, chromium, titanium, and vanadium, including electrochemical processing, aqueous processing, biological separation, and microwave heating. The symposium is organized into the following sessions: (1) Rare-Earth Elements, (2) Platinum Group Metals, and (3) Base and Rare Metals (Co, Cr, Sn, Ti, Mo and V).

We acknowledge the efforts of the symposium organizers and proceedings editors: Hojong Kim, Shafiq Alam, Neale R. Neelameggham, Harald Oosterhof, Takanari Ouchi, and Xiaofei Guan. The support from TMS staff members Trudi

Dunlap and Patricia Warren is greatly appreciated in assembling and publishing the proceedings. We sincerely thank all the authors, speakers, and participants and look forward to continued collaboration in the advancement of science and technology in the area of rare metal extraction and processing.

Hojong Kim
Lead Organizer

Contents

Part I Rare Earth Elements I

The Economics of the Search Minerals Direct Extraction Process for Rare Earth Element Recovery 3
David Dreisinger and Greg Andrews

Recovery of Critical Rare Earth Elements for Green Energy Technologies 19
Jyothi Rajesh Kumar and Jin-Young Lee

Selective Reduction and Separation of Europium from Mixed Rare-Earth Oxides from Waste Fluorescent Lamp Phosphors 31
Mark L. Strauss, Brajendra Mishra and Gerard P. Martins

Application of Rare Earths for Higher Efficiencies in Energy Conversion 37
W.D. Judge, Z.W. Xiao and G.J. Kipouros

Microwave Treatment for Extraction of Rare Earth Elements from Phosphogypsum 47
Adrian Lambert, Jason Tam and Gisele Azimi

Selective Separation of Rare Earth Chlorides Utilizing Vapor Phase Extraction 55
Katelyn M. Lyons, Jerome P. Downey, Jannette L. Chorney and Katie J. Schumacher

Microstructure Observation of Oxidation of Nd-Magnet at High Temperatures 65
Muhamad Firdaus, M. Akbar Rhamdhani, Yvonne Durandet, W. John Rankin, Kathie Mcgregor and Nathan A.S. Webster

Part II Rare Earth Elements II and Platinum Group Metals

Electrochemical Behavior of Neodymium in Molten Chloride Salts 77
L. Diaz, P. Chamelot, M. Gibilaro, L. Massot and J. Serp

Novel Reactive Anode for Electrochemical Extraction of Rare Earth Metals from Rare Earth Oxides 87
Aida Abbasalizadeh, Seshadri Seetharaman, Prakash Venkatesan, Jilt Sietsma and Yongxiang Yang

Electrochemical Formation of Nd Alloys Using Liquid Metal Electrodes in Molten LiCl–KCl Systems 93
Hirokazu Konishi, Hideki Ono, Eiichi Takeuchi, Toshiyuki Nohira and Tetsuo Oishi

Challenges in the Electrolytic Refining of Silver—Influencing the Co-deposition Through Parameter Control 103
Ann-Kathrin Maurell-Lopez, Bernd Friedrich and Wolfgang Koch

Vapor Treatment for Alloying and Magnetizing Platinum Group Metals 119
Yu-ki Taninouchi and Toru H. Okabe

Biotechnological Recovery of Platinum Group Metals from Leachates of Spent Automotive Catalysts 129
Norizoh Saitoh, Toshiyuki Nomura and Yasuhiro Konishi

Recovering Palladium from Chloridizing Leaching Solution of Spent Pd/Al_2O_3 Catalyst by Sulfide Precipitation 137
Qian Li, Qiang Zou, Bin Xu, Yongbin Yang, Xuefei Rao, Long Hu and Tao Jiang

Mechanism of Intensifying Cyanide Leaching of Gold from a Calcine by the Pretreatment of Acid or Alkali Washing 147
Yan Zhang, Qian Li, Xiaoliang Liu, Yongbin Yang, Bin Xu, Tao Jiang and Hongwei Li

Part III Base and Rare Metals

Disclosure of the Kinetic Relations of Semidirect Cemented Carbide Leaching in Acid Media 159
Gregor Kücher, Stefan Luidold, Christoph Czettl and Christian Storf

A New Two-Stage Aluminothermic Reduction Process for Preparation of Ti/Ti-Al Alloys 167
Kun Zhao and Naixiang Feng

Study on Pre-reduction Mechanisms of Chromium Ore Pellets in SRC Process 177
Pei-xiao Liu, Yan-xiang Li and Han-jie Guo

Recovery of Valuable Metals from High-Content Arsenic Containing Copper Smelting Dust 183
Xuepeng Li and Dachun Liu

Sulfuric Acid Leaching of Mechanically Activated Vanadium–Bearing Converter Slag 193
Junyi Xiang, Qingyun Huang, Xuewei Lv and Chenguang Bai

Present Status and Development of Comprehensive Utilization of Vanadium-Titanium Magnetite 203
Shiju Zhang, Songli Liu, Wenhui Ma, Kuisong Zhu, Li Cao and Yongnian Dai

Review of TiO_2-Rich Materials Preparation for the Chlorination Process 211
Songli Liu, Li Cao, Kuisong Zhu, Shiju Zhang and Kui He

Part IV Poster Session

Adsorbents for Selective Recovery of Heavy Rare Earth Elements 219
Takeshi Ogata, Hirokazu Narita and Mikiya Tanaka

Behavior of Sec-Octylphenoxy Acetic Acid (CA-12) in Yttrium Recovery from High Concentrated Heavy Rare Earths Mixture 225
Corradino Sposato, Alessandro Blasi, Assunta Romanelli, Giacobbe Braccio and Massimo Morgana

Preparation of Molybdenum Powder from Molybdenite Concentrate Through Vacuum Decomposition-Acid Leaching Combination Process 235
Chongfang Yang, Yuezhen Zhou, Dachun Liu, Wenlong Jiang, Fansong Liu and Zewei Liu

Pressure Leaching Behavior of Molybdenum-Nickel Sulfide from Black Shale 247
Zhigan Deng, Xingbin Li, Chang Wei, Cunxiong Li, Gang Fan and Minting Li

Selective Recovery of Scandium From Sulfating Roasting Red Mud By Water Leaching 255
Zhaobo Liu, Hongxu Li and Zihan Zhao

Study for Preparation of Industrial Ammonium Molybdate from Low Grade Molybdenum Concentrate 265
Qingwei Qin, Zhenwei Liu, Tiejun Chen, Zili Huang, Jianhong Yang and Wei Han

Study of a Synergistic Solvent Extracting System to Separate Yttrium and Heavy Rare Earths: A Deep Investigation on System Behavior 277
Alessandro Blasi, Corradino Sposato, Assunta Romanelli, Giacobbe Braccio and Massimo Morgana

The Recovery of Bismuth from Bismuthinite Concentrate Through Membrane Electrolysis 285
Si-yao Peng, Jian-guang Yang, Jian-ying Yang and Lei Jie

Author Index 303

Subject Index 305

About the Editors

Hojong Kim is Assistant Professor of Materials Science and Engineering, and Norris B. McFarlane Faculty Professor at the Pennsylvania State University. Dr. Kim received his B.S. from Seoul National University in South Korea in 2000 and Ph.D. degree at Massachusetts Institute of Technology (MIT) in 2004 both in Materials Science and Engineering. His doctoral research sought to identify the corrosion mechanisms of constructional alloys in high temperature and high pressure steam environments under Prof. Ronals Latanision in the Uhlig Corrosion Laboratory at MIT. After graduate research, Dr. Kim worked as a senior research scientist at Samsung-Corning Precision Glass Co. Ltd. and as a project lead to improve the process yield for thin film transistor liquid crystal display (TFT-LCD) glass melting processes by engineering high temperature materials of refractory ceramics (alumina and fused zirconia), molybdenum, and platinum alloys. After five years of industrial experience, Dr. Kim returned to MIT as a postdoctoral associate and later as a research scientist to contribute to the growing need for sustainable technology. He conducted research on high temperature electrochemical processes, including molten oxide electrolysis for carbon-free iron and steel production with focus on developing inert anode materials in molten slags as well as molten salt liquid metal batteries for large-scale energy storage.

Dr. Kim's current research interests embrace the development of environment-friendly electrochemical processes for resource extraction/recycling, corrosion-resistant materials, as well as energy storage systems. He was awarded The Minerals, Metals & Materials Society (TMS) Young Leaders Professional Development Award in 2013 from the Extraction and Processing Division in recognition of his achievement in the high temperature materials and electrochemical processing. In 2015, he was chosen to receive a Doctoral New Investigator Award from the American Chemical Society Petroleum Research Fund for his innovative research and education program. He is currently leading efforts to separate alkali/alkaline-earth elements from the used nuclear fuel recycling processes in the molten salt electrolytes. Dr. Kim is the lead organizer of the present Rare Metal Extraction & Processing Symposium at the TMS 2017 Annual Meeting and is the vice-chair of the Hydrometallurgy and Electrometallurgy Committee of the Extraction and Processing Division.

Shafiq Alam is Associate Professor at the University of Saskatchewan, Canada. In 1998, he received his Ph.D. degree in Chemical Engineering from Saga University, Japan. From 1999 to 2001, he was appointed as a post-doctoral research fellow at the University of British Columbia and the University of Toronto, Canada.

Dr. Alam has extensive experience in industrial operations, management, engineering, design, consulting, teaching, research, and professional services. Before joining the University of Saskatchewan in 2014, he was an assistant/associate professor at Memorial University of Newfoundland for about seven years. Prior to starting his career in academia, he worked with many different companies, such as, Shell, Process Research ORTECH Inc., Fluor Canada Ltd., and the National Institute of Advanced Industrial Science and Technology (AIST), Japan. Dr. Alam is highly experienced in the area of mineral processing and extractive metallurgy, and he possesses two patents and has more than 140 publications. He is the co-editor of four books and an associate editor of the *International Journal of Mining, Materials and Metallurgical Engineering* (*IJMMME*). He is the winner of the 2014 TMS Extraction & Processing Division's Technology Award.

Dr. Alam is a registered professional engineer and has worked on projects with many different mining companies including, Falconbridge, INCO (Vale), Barrick, Hatch, Phelps Dodge, Rambler, Anaconda, etc. He is an Executive Committee Member of the Hydrometallurgy Section of the Canadian Institute of Mining, Metallurgy and Petroleum (CIM) and currently, he holds the office of secretary (2013–2017). Dr. Alam is also the chair of the Hydrometallurgy and Electrometallurgy Committee of the Extraction & Processing Division (EPD) of The Minerals, Metals & Materials Society (TMS) for the period of 2015–2017. He is a co-organizer of many symposia at the international conferences through CIM and TMS. Dr. Alam is also involved in organizing the TMS 2017 Symposium on Energy, Environment and Materials Process Engineering, which is an EPD Symposium in Honor of Ramana Reddy focusing on applications of process engineering principles in materials processing, energy and environmental technologies.

Neale R. Neelameggham is 'The Guru' at IND LLC, involved in international consulting in the field of metals and associated chemicals (boron, magnesium, titanium, and lithium and rare earth elements), thiometallurgy, energy technologies, soil biochemical reactor design, etc. He was a visiting expert at Beihang University of Aeronautics and Astronautics, Beijing, China. He was a plenary speaker at the Light Metal Symposium in South Africa—on low carbon dioxide emission processes for magnesium.

Dr. Neelameggham has more than 38 years of expertise in magnesium production and was involved in process development of its startup company NL Magnesium through to the present US Magnesium LLC, UT until 2011. Neelameggham and Brian Davis authored the ICE-JNME award winning (2016) paper—"21st Century Global Anthropogenic Warming Convective Model"—which notes that constrained air mass warming is independent of the energy conversion source—fossil or renewable energy. He is presently developing Agricoal™ and agricoalture to improve arid soils.

Dr. Neelameggham holds 16 patents and patent applications, and has published several technical papers. He has served on the Magnesium Committee of Light Metals Division (LMD) of TMS since its inception in 2000, chaired it in 2005, and in 2007 he was made a permanent co-organizer for the Magnesium Symposium. He has been a member of the Reactive Metals Committee, Recycling Committee, Titanium Committee, and Program Committee Representative of LMD and LMD council.

Dr. Neelameggham was the inaugural chair, when in 2008, LMD and EPD (Extraction & Processing Division) created the Energy Committee, and has been a co-editor of the energy technology symposium proceedings through the present. He received the LMD Distinguished Service Award in 2010. While he was the chair of Hydro and Electrometallurgy Committee he initiated the rare metal technology symposium in 2014. He is co-editor for the 2017 proceedings for the symposia on magnesium technology, energy technology, rare metal technology, and solar cell silicon.

Harald Oosterhof graduated as a chemical engineer from Twente University (TU) in the Netherlands in 1994. In the same year he assumed a position as researcher at TU Delft. His research on antisolvent crystallization of well-soluble salts was rewarded with two patents and several publications. After receiving his Ph.D. from Delft University in 1999, he assumed the position of project manager at Umicore, a global materials and technology group that is based in Belgium. During his first assignment as Project Leader Hydrometallurgy, he focused on the refining of cobalt, nickel, and germanium. Since 2011, Dr. Oosterhof has worked as a scientist in the Recycling and Extraction Technology group at Umicore's Central R&D department. His main competence areas are special metals hydrometallurgy, recycling and refining of rare earth metals, base metal hydrometallurgy, and recycling of spent rechargeable batteries and superalloys. In his current job, Dr. Oosterhof is frequently involved in business development of scarce metals recycling and he is heading a team of hydrometallurgical specialists.

Takanari Ouchi is a research scientist in Materials Processing Center at Massachusetts Institute of Technology (MIT). He received his Ph.D. in Nano-Science and Nano-Engineering from Waseda University on 2011. He developed electrochemical deposition processes to fabricate metal nanostructures with both well-controlled crystallinity and uniformity at the single nanometer scale and demonstrated the applicability of these processes to fabrication of bit patterned magnetic recording media for future hard disk drives. After completing his doctoral degree, Dr. Ouchi joined MIT, where he has developed liquid metal batteries, which is in principle a bi-directional electrolysis (electro-refining) cell, to apply for the grid-scale energy storage. He led the systematic investigation of electrochemical properties of liquid metal electrodes in molten salt electrolytes and developed novel lithium, calcium, and sodium liquid metal batteries. Dr. Ouchi has been an author of 16 peer reviewed papers and conference proceedings and of 47 talks at conferences. He has constantly contributed to create the vibrant field of metal extraction by working as a member of Hydrometallurgy and Electrometallurgy Committee at The Minerals, Metals & Materials Society (TMS), organizing technical symposiums at TMS, and soliciting papers as a guest editor of *JOM*. He has earned several awards and honors, such as TMS Extraction & Processing Division (EPD) Young Leaders Professional Development Award in 2015 based on his reputation in the electrochemical metal extraction processing.

Xiaofei Guan received his B.S. degree in Applied Physics from Nankai University, China, in 2009, and Ph.D. degree in Materials Science and Engineering from Boston University in 2013. His Ph.D. research in the group of Prof. Uday Pal was on resource recovery and recycling, and electrolytic production of energy intensive metals from minerals. In 2014, he joined the Ramanathan Research Group at Harvard University as a postdoctoral fellow, where he led research on energy conversion and storage devices including solid oxide fuel cells, protonic ceramic fuel cells, and hydride-air batteries. He is currently a postdoctoral fellow in the Clarke Research Group and the Girguis Laboratory at

Harvard University, where his research is centered on growth and characterization of semiconductor thin films for solar energy applications, a joint project at the interface between materials science, microbiology, and electrochemistry.

Dr. Guan received the Outstanding Ph.D. Dissertation Award in Materials Science and Engineering from Boston University in 2014, and the Young Leaders Professional Development Award from the TMS Extraction & Processing Division in 2015. He also serves as an advisor and guest editor in the Recycling and Environmental Technology Committee for *JOM*.

Part I
Rare Earth Elements I

The Economics of the Search Minerals Direct Extraction Process for Rare Earth Element Recovery

David Dreisinger and Greg Andrews

Abstract The Foxtrot deposit of Search Minerals contains a range of rare earth element containing minerals including allanite, fergusonite and bastnasite. The Search Minerals Direct Extraction Process for treating these minerals involves coarse crushing of the ore to-6 mesh (3.45 mm), acid treatment at 200 °C in a novel reactor configuration, water leaching, various purification steps to reject iron, aluminum and uranium/thorium (present in small amounts) and finally precipitation of rare earths as a oxalate and calcination to form a mixed rare earth oxide product for refining. The economics of the process are presented.

Keywords Rare earths · Extraction · Mixed rare earth oxide · Foxtrot · Capital and operating cost

Introduction

Search Minerals Inc. (Search) is exploring and developing a number of deposits for rare earth element (REE) recovery in Labrador, Canada. The Port Hope Simpson District is in the southeast of Labrador and is highly prospective for heavy and light rare earth elements. The Foxtrot deposit sits within the Port Hope Simpson REE District, which is 70 km long by up to 10 km wide. The infrastructure available at Foxtrot is excellent; a deep-water port, air strip, and road and power infrastructure are pre-existing at St. Lewis. The three communities of Port Hope Simpson, St. Lewis and Mary's Harbour are in close proximity to the site.

A Preliminary Economic Assessment has been developed for Foxtrot. Roscoe Postle Associates (RPA) has determined a 43–101 compliant resource for the Foxtrot deposit [1]. The resource comprises 7.39 Mt of Indicated Resource and 1.96 Mt of Inferred Resource. An NSR cut-off of $165/t was used for material that could be accessed by open pit mining and $260/t was used for material to be mined

D. Dreisinger (✉) · G. Andrews
Search Minerals, #211, 901 West 3rd Street, North Vancouver, BC V7P 3P9, Canada
e-mail: david_dreisinger@yahoo.com

© The Minerals, Metals & Materials Society 2017

H. Kim et al. (eds.), *Rare Metal Technology 2017*,
The Minerals, Metals & Materials Series,
DOI 10.1007/978-3-319-51085-9_1

underground. The resource is summarized in Table 1. The deposit is rich in the critical rare earth elements Pr, Nd, and Dy.

Search Minerals have developed a direct extraction process for recovery of rare earth elements from the Foxtrot mineralization [2–4]. This process was conceived to meet the challenges posed in early work on Foxtrot metallurgy. Namely, the new direct extraction process was designed to (1) eliminate complex and costly beneficiation processes (e.g. gravity, flotation and magnetic separation), (2) reduce the use of reagents (e.g. acid, base, and oxalic acid) and (3) improve the control and rejection of key impurities including minor amounts of thorium contained in the Foxtrot mineralization.

The direct extraction process treats crushed material (eliminating costly grinding and beneficiation) using a low temperature acid extraction process. The acid treated material is then water leached to extract rare earths. The leachate is then purified in a series of steps followed by rare earth oxalate precipitation and calcination. The mixed rare earth oxide calcine is then available for refining to individual rare earth elements and final products.

The direct extraction process has now been the subject of an engineering design and cost study. The details of the process and associated project capital and operating costs are outlined below.

The Direct Extraction Process

The direct treatment of Foxtrot mineralization was investigated through a series of studies on acid baking/water leaching, solution purification, RE precipitation, RE re-dissolution and purification to remove thorium and finally RE precipitation with oxalic acid and calcination to make a mixed REO. The goal again was simplification, cost reduction and production of a premium product. The general flowsheet for the treatment scheme is shown in Fig. 1.

The testing of the direct extraction process has been reported in Dreisinger et al. [3, 4]. A summary of the key steps follows.

Table 1 Foxtrot resource [1]

Classification	Cut-off	Tonnage	Pr	Nd	Dy	LREE	HREE	TREE
	$NSR	000s	ppm	ppm	ppm	(%)	(%)	(%)
Open pit								
Indicated	$165	4129	372	1393	177	0.69	0.17	0.86
Inferred	$165	228	368	1378	179	0.68	0.17	0.85
Underground								
Indicated	$260	3263	429	1602	209	0.78	0.19	0.97
Inferred	$260	1730	430	1602	201	0.80	0.19	0.99
Total indicated		**7392**	**397**	**1485**	**191**	**0.73**	**0.18**	**0.91**
Total inferred		**1958**	**423**	**1576**	**199**	**0.79**	**0.18**	**0.97**

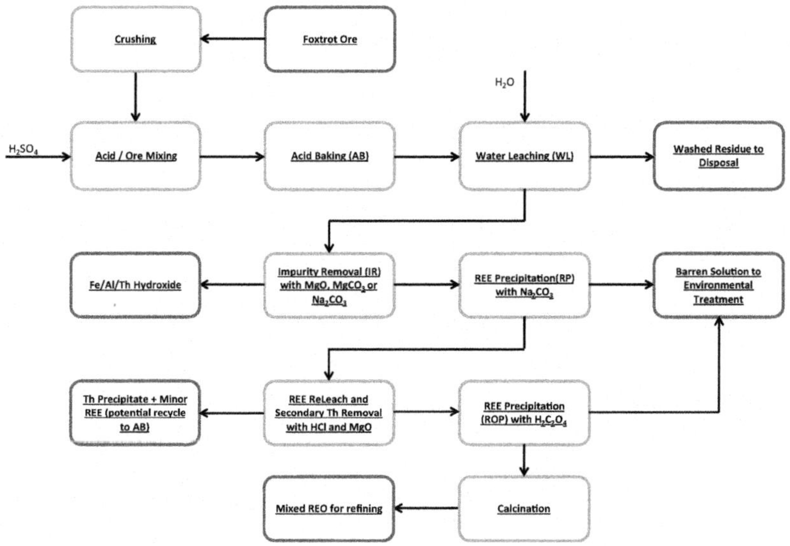

Fig. 1 Conceptual flowsheet for testing of the new Foxtrot process

Acid Treatment/Water Leach

The material is crushed to a nominal 6 mesh particle size and treated at 200 °C with minimum acid followed by water leaching. The average extraction (Table 2) was 78% for the series La-Er. The extractions of Tm–Lu were lower. The radioactive elements Th and U were extracted but the major gangue elements (Si, Al, Fe, Na, K) were only weakly extracted. Some Mg, Ca, Ti, P and Mn were also extracted.

Impurity Removal by Oxidation and Precipitation

The initial removal of impurities was tested by pH adjustment to pH 3.75 with $MgCO_3$ and oxidation (for Fe removal). More than 90% of the iron was eliminated along with 88.4% of the thorium. There was also significant rejection of Si, Al, Ti and P. The losses of REEs ranged from 0.74 to 3.6% from La to Lu.

Bulk Rare Earth Precipitation

The purified solution was treated with a soda ash solution (Na_2CO_3) at pH 7.25 to precipitate the REEs into a mixed carbonate product for further purification. The precipitation of REEs approaches 100% (Table 3). The co-precipitation of Th, U, Fe, Al is similarly very high. The mixed REE carbonate precipitate may be further

Table 2 Bulk acid bake—water leach results

Rare earth extraction					Impurity extraction				
Element	Analysis (mg/L or g/t)			Extn	Element	Analysis (mg/L, % or g/t)			Extn
	Feed	PLS	Residue	(%)		Feed	PLS	Residue	(%)
La	1720	144	392	76.7	Th	109	9.35	34.6	70.4
Ce	3720	321	805	78.1	U	22.4	1.2	13.5	44.4
Pr	437	39.7	90.8	79.6	Si	31.32	288	32.44	0.8
Nd	1610	148	330	80.0	Al	3.99	212	3.97	4.6
Sm	297	27.9	63.8	79.6	Fe	7.83	496	7.69	5.5
Eu	15.5	1.51	3.6	79.1	Mg	0.12	43.5	0.07	36.7
Gd	244	22.6	56.1	78.4	Ca	1.45	642	0.96	40.0
Tb	37.3	3.59	8.5	79.2	Na	2.13	47	2.14	2.6
Dy	223	20.8	54.6	77.5	K	3.36	384	3.44	9.2
Ho	43.7	4.09	11.7	76.0	Ti	0.27	4.6	0.28	1.5
Y	1090	107	288	77.0	P	0.01	5	0.02	33.5
Er	122	11.2	36	73.8	Mn	0.23	82.3	0.19	27.6
Tm	17.2	1.49	5.8	69.9					
Yb	111	8.56	41.8	64.9					
Lu	15.8	1.02	7.3	55.9					

Conditions: acid bake at 6 mesh, 200 °C, 2 and 24 h water leach at 90 °C with 600 rpm mixing intensity

Table 3 REE Prec. results

Rare earth precipitation					Impurity precipitation				
Element	Analysis (mg/L or g/t)			Pptn	Element	Analysis (mg/L, % or g/t)			Pptn
	Feed Soln	Filtrate	Ppte	(%)		Feed Soln	Filtrate	Ppte	(%)
La	134	0.24	57,700	99.8	Th	0.84	0.04	422	95.7
Ce	315	0.39	129,000	99.9	U	1.07	0.13	467	88.4
Pr	36.3	0.06	15,600	99.8	Al	83.6	0.05	3.67	99.4
Nd	136	0.18	58,800	99.9	Fe	39.3	0.2	1.81	99.5
Sm	26.1	0.05	11,000	99.8	Mg	1550	1560	0.505	0.7
Eu	1.39	0.03	584	97.6	Ca	616	583	1.95	6.6
Gd	23	0.04	10,900	99.8	Na	191	1050	0.125	0.3
Tb	3.62	0.03	1640	99.1	K	298	289	0.03	2.2
Dy	20.8	0.05	9270	99.7	P	5	5	0.003	1.2
Ho	4.08	0.02	1800	99.5	Mn	77.4	68.8	0.131	3.9
Y	107	0.3	42,100	99.7					
Er	11.1	0.04	4880	99.6					
Tm	1.46	0.04	636	97.1					
Yb	7.92	0.03	3470	99.6					
Lu	0.93	0.03	409	96.6					

Conditions: pH 7.25 with Na_2CO_3 addition for 3 h at 25 °C [4]

refined by a re-leach, oxalate precipitation and calcination method to form a mixed REO for refining. The overall recovery of REEs from the original mineralized material to mixed carbonate precipitate has been calculated and summarized in Table 4.

The mixed REE carbonate precipitate was re-dissolved in HCl solution followed by pH adjustment for purification and oxalic acid precipitation. The REE oxalate was then calcined to make a mixed REO product. The analysis of the mixed REO product from the testing is summarized in Table 5. This product was generated in a continuous kiln test campaign of the acid treatment followed by batch water leach and purification (as outlined above).

The final product analyzed 98.9% REO+Y with <0.01% S, < 0.01% F, 2 g/t, Th and 97 g/t U. Other key impurity values were 93 g/t Si, 0.15% Ca, 44 g/t P, 465 g/t Mn, 34 g/t Cr, 23 g/t Sr, and 56 g/t V. The critical and highly values rare earths (Pr, Nd, Eu, Tb, Dy, Y) accounted for 37.95% of the Mixed REO oxide.

The direct extraction process has been demonstrated to be technically feasible. The Foxtrot mineralization can be crushed, acid treated, and leached with the leachate purified to recover a mixed rare earth oxide for refining. At the present time

Table 4 Overall recovery (%) of rare earth elements from Foxtrot material to mixed carbonate precipitate

La	Ce	Pr	Nd	Sm	Eu	Gd	Tb	Dy	Ho	Y	Er	Tm	Yb	Lu
75.9	77.3	78.9	79.3	78.9	78.4	77.6	78.4	76.7	75.3	76.3	73.1	69.2	64.3	55.4

Table 5 Chemical analysis of the mixed REO calcine product from the kiln testing

Rare earth		Impurity	
Element	Analysis (%)	Element	Analysis (g/t)
La	12.0	Th	2.1
Ce	31.4	U	96.6
Pr	3.87	Si	93.5
Nd	15.4	Al	<53
Sm	2.53	Fe	69.9
Eu	0.147	Mg	92
Gd	2.13	Ca	1572
Tb	0.314	Na	<300
Dy	2.00	K	<83
Ho	0.413	Ti	<60
Y	9.8	P	43.6
Er	1.07	Mn	465
Tm	0.156	Zn	<200
Yb	0.705	F	290
Lu	0.0887	Ctot	<100
TREE	82.0	S	<100
LREE	65.2		
HREE	16.8		
TREO	98.9		

Search Minerals are embarking on a \$1.9M CDN pilot plant program to continue the "scaling up of the process" as part of the drive to commercialization. This pilot plant program will confirm the process parameters in large-scale equipment while providing product for refinery evaluation.

Engineering and Cost Studies

The engineering and cost studies have proceeded in two steps. First, SNC Lavalin (Australia) has prepared a concept study focusing on the processing of rare earth mineralization to a mixed rare earth oxide product [5]. Second, Roscoe Postle Associates (RPA) have completed a Preliminary Economic Assessment on the Foxtrot Deposit at a processing rate of 1000 tpd from underground and open pit resources [1]. The results of the SNC Lavalin study were used as the basis for the processing plant cost in the PEA produced by RPA. The SNC Lavalin study costs were updated by RPA to take into account changes in currency exchange rates and the cost of raw materials (e.g. steel and other metal prices).

SNC Lavalin Engineering and Cost Study [5]

The SNC Lavalin engineering and cost study comprised the following items.

- Testwork Review
- Process Design Criteria
- Preliminary METSIM® Model
- Preliminary Mass Balance
- Block Flowsheet
- Major Equipment List
- Operating Cost Estimate
- Capital Cost Estimate
- Study Report

The project battery limits were defined as:

- Reclaim from the ROM stockpile (pre-crushed to 100% passing 30 mm).
- Acid storage tank inlet at the port.
- Reagents as delivered to site (except sulphuric acid) and product dispatch from site.
- Boundaries of the residue storage facility.
- Water supply from a client identified year round water source, including an allowance for pumping and a water supply pipeline.
- Power supply 'over the fence' from a utility supplier, including an allowance for a transmission line and incoming transformers and distributions.

- Liquid effluent disposal (barren solution) to sea requirements after a long pipeline to the port.

The following items were excluded from the scope of work:

- Resource definition and mining.
- Testwork and testwork management.
- Engineering development and studies prior to project commitment.
- Project definition and scheduling.
- Study of social and environmental issues.
- Definition of or obtaining project approvals and permits.
- Financial modeling.
- Power delivery outside of the battery limits.
- Water delivery outside of the battery limits.
- Logistics of reagents and product outside of the battery limits.
- Port logistics.

The testwork review highlighted the results presented in this paper as a basis for design. The SNC Lavalin review concluded that a simple direct acid bake and water leaching process has been developed for treatment of Foxtrot mineralization that involves crushing to 6 mesh particle size, application of 100 kg/t of H_2SO_4 to the material at 200 °C for 2 h followed by a 24 h water leach to produce a weakly acidic product leach solution. Favourable material handling characteristics of the acid/feed mixture and calcine as well as easy filtration characteristics of the water leach residue have led to the development of a much more operable acid bake circuit than other "typical" REE acid bake circuits. The review also highlighted areas where future testing and process development would be most useful including optimization of crush size for balance between leaching and solids handling, reduced impurity levels (e.g. U) in the final concentrate, determining solid-liquid separation design parameters, determining oxidant requirements in impurity removal and a general reduction in reagent requirements. The current pilot plant program is designed to address these areas.

The conceptual block flow diagram is shown in Fig. 2.

Feed Preparation

The purpose of feed preparation is to mine and crush the mineralization to a size of 100% passing 3.45 mm, while minimizing the fines generated. This sized feed is approaching the minimum size possible for a dry crushing circuit. This is achieved by:

1. Open cut mining with in pit crushing to below 30 mm and delivered to the run of mine (ROM) stockpile at the processing facility.
2. Further size reduction by high pressure grinding rollers (HPGR, in closed circuit with a vibrating screen) during day shift will ensure an appropriately sized feed

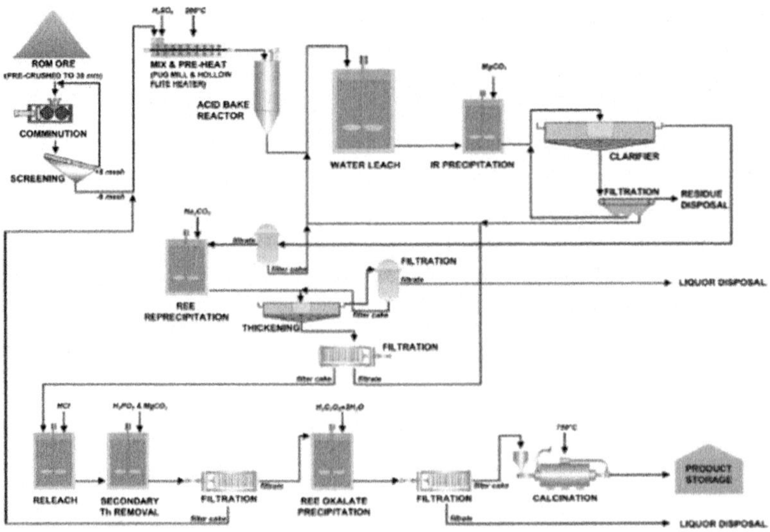

Fig. 2 Conceptual block flow diagram [5]

for the continuous operation of the acid mixing and acid bake section of the facility.

3. This area is supported by a small dust collection system for control of dusts, recognizing that this is the highest risk factor when handling low radiation mineralization. Dust collected is mixed back with the feed to the acid mixing process.

Mixing and Acid Baking

The purpose of the mixing stage is to intimately mix the crushed ore with concentrated (93%) sulphuric acid at a rate of 100 kg of acid (100% equivalent) for every tonne of ore. This mixture must then be heated and maintained at 200 °C for 90 min. This section of the plant operates continuously 24 h a day. This is achieved by:

1. A pug mixer is used to intimately mix the acid with the ore. A recycled filter cake from the later Secondary Th Removal stage is also added to pug mill to minimize REE losses.
2. A Holo-Flite® preheater to raise the temperature of the mixture to 200 °C. The Holo-Flite® heater is an indirect heat exchanger utilising hollow screws and jacketed walls for heating of bulk solids and pastes. The screws provide movement of the solids similar to a screw conveyor while also improving the heat transfer by further mixing of the solids. The action of the Holo-Flite® system (screw conveyor movement) is expected to mimic the “rabbling” used in

the laboratory testwork for ensuring excellent acid/ore contact through the heating system.
3. The heated mixture is then transferred to a refractory lined bin to provide the 90 min at temperature for the acid to react.
4. The hot acidic gases evolved from the Holo-Flite® and refractory lined bin are contained and scrubbed in a caustic scrubber.

Water Leaching, Impurity Removal and REE Precipitation

The purpose of the water leaching stage is to provide the time, agitation and temperature to leach the metals converted to soluble form by the acid bake. The slurry is then neutralized with $MgCO_3$ slurry to reduce the free acid sufficiently to substantially remove impurities by precipitation onto the leach residue. The leach residue is then separated from the leach solution and the rare earths in that solution are precipitated by further neutralization with Na_2CO_3 solution. The water leaching, impurity removal, and REE precipitation are achieved by:

1. Hot acid baked ore is mixed with hot water in the 3 large gravity overflow Water Leach Tanks providing the 24 h residence time and intense mixing required for the Water Leach.
2. Wash filtrates from downstream in this area are recycled to the water leach to maximize the washing of products, minimizing the dilution of the leach liquor and minimizing the overall water consumption and disposal. These flows are supplemented by additional raw water supply.
3. The liquid feeds to the water leach are heated by a combination of heat recovery from the REE precipitation overflow liquor prior to disposal and steam heating.
4. The last of the Water Leach Tanks overflows to the first of the 2 smaller Impurity Removal Tanks where the controlled addition of $MgCO_3$ solution neutralizes the free acid to achieve a solution pH of 3.85. As the pH is raised to this level and over the 2-h residence time, the majority of the Fe, Al and Th impurities are precipitated onto the leach residue.
5. The relatively coarse leach residue is thickened as feed for a vacuum belt filter. On the vacuum belt filter the solids are thoroughly washed to maximize the recovery of REE. The resultant solids are transferred to a bin and dumped into a dump truck for dry disposal to a lined dry tailings storage facility assumed to be 2.5 km from the processing facility.
6. The overflow from the thickener is filtered with a precoated polishing filter to ensure none of the ore fines or precipitated impurities are contaminating the solution and thus maximizing the purity of the REE Precipitation product.
7. The solids free solution is mixed with a Na_2CO_3 solution to preferentially precipitate RE carbonates from the liquor.
8. The precipitated solids are thickened and two thirds of these solids are recycled as seed for the precipitation process. The remaining third are filtered and washed

on a diaphragm filter. The resultant solids are transferred to a belt conveyor to one of the two HCl Releach tanks for further processing.

9. The overflow from the thickener is filtered with a precoated polishing filter to ensure none of precipitated product is lost and the liquor is then disposed to sea via a 2.5 km pipeline.

Releach and Precipitation

The purpose of this step is to further purify the RE carbonate product by releaching in HCl, a second Th removal step and selectively re-precipitating as RE oxalates. This section of the processing facility operates 24 h a day, but processes the solids as an 8 h batch from the previous continuous sections. This is achieved by:

1. An 8 h batch of REE Precipitation filter cake is allowed to accumulate in one of the two HCl Releach Tanks. The cake is constantly repulped by continually adding a minimal level of 33% HCl. At change of shift, the REE Precipitation Filter cake is diverted to the other HCl Releach Tank that has been emptied on the previous shift.
2. Additional 33% HCl is added to the batch now being processed to reach an excess of 10 g/L acid and the tank is agitated under these conditions for an hour as the HCl Releach step.
3. After the one hour releach, the Chloride Tails Filter cake from a previous batch is added to reduce the free acid levels and releach some of the value metals.
4. A small quantity of H_3PO_4 is added to this tank to assist in lowering the pH required to remove Th (and consequently reducing REE losses). Then sufficient $MgCO_3$ slurry is added to further neutralize the solution to achieve a solution pH of 3.8. As the pH is raised to this level and over the 2 h residence time, the majority of the Fe, Al and Th impurities are precipitated onto the releach residue.
5. The resultant slurry is filtered through a precoated diaphragm filter. The solids are washed, air dried and dumped into a transportable hopper. Once full, the hopper is rolled and hoisted into a position for a metered return to the Acid Bake pug mixer. The solution is collected in the RE Oxalate Precipitation Tank.
6. Solids from the previous batch are repulped into the high strength RE chloride solution, and then an oxalic acid solution is mixed into the solution to preferentially precipitate RE oxalates from the liquor.
7. The precipitated solids are filtered and two thirds of these solids are reserved and used as seed for the next batch. The remaining third are washed and air dried on the diaphragm filter and dumped into a calcination hopper for further processing.
8. The filtrate is collected in the Chloride Tails Tank, further neutralized with a $MgCO_3$ slurry and filtered with the same filter used to filter the second Th removal step. The filter cake from this filter is used to neutralize the free acid

from the HCl Releach. The filtrate is pumped directly into the RE Precipitation overflow stream and disposed to sea via a 2.5 km pipeline.

9. The majority of the tanks in this section of the plant are covered and vented to a HCl scrubbing system.

Calcination

The RE oxalate precipitate is calcined at 750 °C to break down to the RE oxides, cooled, transferred to 1 t bulk bags, weighed and sampled for quality control and then transferred in bags to a shipping container ready for export. This is achieved by:

1. A calcination hopper loosely filled with oxalate filter cake is wheeled into a 450 kW oven and a pre-programmed heating/cooling cycle is activated to heat the contents to 750 °C and hold them at this temperature for 4 h and then allow the contents to cool.
2. Once the hopper is cool, the hopper can be removed from the oven, hoisted into a packaging platform and used to give a metered and measured feed into 1 m^3 bulk bags.
3. Packaging is largely a manual operation of hanging the bag, initiating feed to near desired weight, jogging the feeder to achieve desired weight, sampling for analysis and custody transfer and affixing the relevant identifying documentation. The bag can then be closed, unhooked from the packaging frame and taken by forklift to a shipping container at the loading dock.
4. Once the container is fully loaded with product and sealed it can be loaded onto a truck for export.

Reagents

Reagent use in the REE extraction process is central to ore treatment. Reagents are provided as follows:

1. 93% Sulphuric acid is delivered in 10,000 DWT shipments and unloaded to insulated acid tanks at the port. The level in a small storage tank at the processing facility is maintained by pumping through an insulated and electrically traced pipeline from the port. The acid from this tank is dosed directly into the process.
2. Diesel fuel for heating and steam generation is delivered by standard fuel tankers to the onsite fuel tanks.
3. 33% HCl acid is delivered in 20 t rubber lined steel isotainers. These are unloaded at port and progressively emptied into 2 storage tanks at the processing facility from a top unloading gantry while the isotainers are on the back of trucks. The unloading facility and storage tanks are vented through the process facility HCl scrubbing system.

Table 6 Reagent consumption for 1000 t/d of ore treatment

Reagent	Annual consumption (t/a)
Sulphuric acid	36,500
Magnesium carbonate	20,400
Filter aid	540
Flocculant	4
Sodium carbonate	9200
Hydrochloric acid	9800
Phosphoric acid	420
Sodium hydroxide	120
Oxalic acid	2000
Hydrogen peroxide (allowance)	0.2
Diesel (litres)	2,280,000
Product shipping	2940
Process plant consumables	0.8

4. Smaller scale liquid reagents are stored inside in the bulk box packaging that they arrive in until they need to be diluted for use in the process and then a batch at the correct concentration is prepared and transferred to the dosing tank.
5. Stocks of solid reagents are stored undercover, but outside the building and are brought inside prior to reagent preparation for thawing (if required). Bulk bags of these reagents will be transferred by forklift to the appropriate mixing tank after which the bag is hooked to a local hoist, lowered onto a bag breaker above the tank and mixed to the appropriate concentration. This is then transferred to the corresponding dosing tank. It is expected that 40–50 bags per day will need to be processed in this manner.

A provision in the capital cost estimate has been included for laboratory services and mobile equipment for servicing the plant.

The annual consumption of reagents is shown in Table 6. A total of 72 people are employed at the processing plant.

Preliminary Economic Assessment

The preliminary economic assessment of the Foxtrot project was completed by RPA and incorporated the SNC Lavalin engineering and cost study results [1]. The basis for the PEA and some of the main results were as follows:

- 1000 tonnes per day processing rate
- Mine life: 14 years: 8 years open pit, 6 years underground
- $353 Canadian per tonne processing facility average unit revenue, net of pay factors and third party separation charges
- $238 Canadian per tonne processing facility average unit operating cost

- Feed grade-weighted average REE recovery of 76.8%
- Total Life-of-Mine production of 44,129 tonnes of TREO, or 3300 tonnes per year at peak production
- Life-of-Mine production includes 7.095 million kg of neodymium oxide (Nd_2O_3), and 0.836 million kg of dysprosium oxide (Dy_2O_3)
- Revenue is recognized at the time of production
- Metallurgical process produces a dry stackable residue, for ease of disposal

The capital cost of the project is summarized in Table 7. The initial capital for the first 8 years of operation is estimated at $152.2 Million Canadian including a $32.7 Million Canadian contingency.

The operating costs are summarized in Table 8. The operating cost for the open pit operation averages $220.99 Canadian per tonne treated. The underground costs increase to $259.28 Canadian per tonne due to the higher cost of underground mining.

The economic indicators and net present value of the deposit are summarized in Table 9. The pre-tax NPV of $93 Million Canadian, the 22.2% IRR and the short payback of 3.5 years demonstrate the attractiveness of continuing to work on the development of the Foxtrot deposit into Canada's first REE mine. The total revenue expected over the mine life was calculated as $1.713 Billion Canadian.

Table 7 Capital cost summary

Area	Capital ($Million CDN)
OP and surface infrastructure	19.5
Processing	72.0
Indirects/owners	28.1
Contingency	32.7
Total initial capital	**152.2**
Sustaining capital	8.8
Underground capital (Year 8)	56.7
Reclamation and closure	14.0
Total capital cost	**231.7**

Table 8 Unit operating cost summary

Area	Unit	OP	UG
OP mining by contractor	CDN$/t processed	55.11	
UG mining by owner	CDN$/t processed		87.91
Crushing	CDN$/t processed	5.00	5.00
Processing—concentration	CDN$/t processed	141.35	141.35
G&A	CDN$/t processed	19.52	25.02
Total operating costs	**CDN$/t processed**	**220.98**	**259.28**

OP Open Pit, *UG* Underground
Note OP mining by contractor based on $5.50/t moved and $4.50/t moved for ore and waste, respectively

Table 9 Economic and financial results of the PEA[a] [1]

	Pre-tax	After-tax
Net present value (NPV) (10% discount rate)	\$93M	\$ 48M
Internal rate of return (IRR)	22.2%	16.7%
Payback period	3.5 years	4.4 years
Undiscounted cash flow	\$327M	\$226M

[a]It is important to note that the PEA is preliminary in nature. It includes inferred mineral resources, which are considered too speculative geologically to have the economic considerations applied to them that would enable their categorization as mineral reserves. There is no certainty that the PEA forecast will be realized. Mineral resources that are not mineral reserves do not have demonstrated economic viability

Conclusions

Search Minerals has rapidly advanced the exploration and development of the Port Hope—Simpson Rare Earth Element District. The Foxtrot deposit contains a variety of rare earth minerals and is rich in heavy rare earth elements. A simple direct leaching process has been developed for treatment of Foxtrot mineralization that represents a breakthrough in the development of this deposit. The product is a high purity mixed rare earth oxide product suitable for refining through to separate rare earth products. The process is simple, uses readily available reagents and technologies and is easily scaled to larger size as required.

A Preliminary Economic Assessment of the Foxtrot deposit was completed by Roscoe Postle Associates [1] and confirmed the economic attractiveness of the development. The total project cost for the mine, processing plant and local infrastructure was estimated at \$152 Million Canadian initial capital cost including a \$32.7 Million Canadian in contingency. The mine life was estimated at 14 years. The project operating cost was estimated at \$220.99 CDN/t for open pit operation (years 1–8) and \$259.28 CDN/t for underground operation (years 9–14). The pretax NPV of the project was calculated at \$93 Million Canadian with an IRR of 22.2%.

The PEA confirms the technical and financial feasibility of the Foxtrot deposit at the current state of mineral exploration and technical development. Search Minerals will continue to explore and develop the deposit (and other targets within the Port Hope—Simpson District) toward the goal of advancing to REE production.

Acknowledgements Support was received from the Research & Development Corporation ("**RDC**") of Newfoundland and Labrador and from the Atlantic Canada Opportunities Agency ("**ACOA**") for this development.

References

1. Roscoe Postle Associates, *Technical Report on the Foxtrot Project in Labrador, Newfoundland and Labrador Canada* (Report NI 43–101) (Roscoe Postle Associates Inc., Toronto, ON, 2016), posted on http://www.sedar.com
2. D.B. Dreisinger, J.D. Clucas, N. Verbaan, T. Grammatikopoulos, M. Aghamirian, C. Forstner, The processing of REE's from search minerals foxtrot resource, in *Proceedings of Rare Earths 2012*, ed. by J.R. Goode, G. Moldoveanu, M.S. Reyat (CIM Metsoc, Montreal, 2012), pp. 81–94
3. D.B. Dreisinger, N. Verbaan, M. Johnson, The processing of REE's from search minerals foxtrot resource—an update, in *COM 2014—Conference of Metallurgists Proceedings* (2014), 13 p
4. D.B. Dreisinger, N. Verbaan, M. Johnson, The search minerals direction extraction process for rare earth element recovery, in *Rare Metal Technology 2016*, ed. by S. Alam, H. Kim, N.R. Neelameggham, T. Ouchi, H. Oosterhof (The Minerals, Metals & Materials Society, 2016), pp. 3–16
5. SNC Lavalin, Foxtrot REE project conceptual study (2015, June)

Recovery of Critical Rare Earth Elements for Green Energy Technologies

Jyothi Rajesh Kumar and Jin-Young Lee

Abstract 21st century is electronic revolution in human lives as well as energy is a one of the most significant task. Rare earth elements (REEs) occupies key role in our daily life as well as high-tech industrial applications. REEs are one of the energy critical elements, for sustainable growth, it depends on their utilization and re-use, reduce, recycling policy employments. The rapid growth of population and their needs will raise the demand of certain REEs. Global wide REE deposits mainly located at China, from 2010 onwards China made policy to control exports of REE's to foreign countries. REEs are utilized in many modern electrical and electronic devices such as smart phones, computers, LED lights etc. Recovery of the REEs from secondary resources is one of the best solution and alternative option, it needs co-ordination between nations as well as sustainable environmental regulations and implications and certain education to undeveloped countries. As per the Department of Energy (DoE), US the following rare earth elements (REEs) were under critical situation: Dy, Y, Nd, Tb, Eu, in another two namely Ce and Te under near to critical matrix. It shows that REEs scarcity for industry as well as sustainable developments is growing fast. As well as REEs are used in clean energy and defense technologies. As well as REEs predicted shortfall (million US $) are as follows: Y-85, Dy-22, Er-12 and Tb-7 etc. Green energy technologies should have high efficiency, clean and renewable and carbon-di-oxide emissions under control.

J. Rajesh Kumar · J.-Y. Lee (✉)
Convergence Research Center for Development of Mineral Resources (DMR),
Korea Institute of Geoscience and Mineral Resources (KIGAM),
Daejeon 305-350, Korea
e-mail: jinlee@kigam.re.kr

J. Rajesh Kumar
e-mail: rkumarphd@kigam.re.kr; rajeshkumarphd@rediffmail.com

© The Minerals, Metals & Materials Society 2017

H. Kim et al. (eds.), *Rare Metal Technology 2017*,
The Minerals, Metals & Materials Series,
DOI 10.1007/978-3-319-51085-9_2

Introduction

The fields of the rare earths recovery technologies are fascinating. Significant research and development work continues globally to explore and establish ways and means to put the rare earths to use, precious metals to recover and individually and collectively, in the service of civilization. The demand for rare-earths (REs) in the international markets increases with increasing developments in the advanced and high-tech technologies. The rare earths have an ever growing variety of applications in the modern technology. They provide many an industry with crucial materials and they provide many a customer with benefits. From these beginnings and over many years, industrial applications of rare earths have developed in metallurgy, magnets, ceramics and electronics, chemical, optical, medical and nuclear technologies.

A chronological account of the Chemistry and metallurgy especially hydrometallurgy (aqueous processing of the metal ions) of the REs arranges into three eras or ages. The period prior to 1950 may be called the Dark Age. The next two decades were the Age of Enlightenment. The period after early 1970s may be considered as Golden Age. In the first three decades of this golden era a number of remarkable advances and discoveries were made in the field of rare metals and these have left the future of the rare metals will be glorious and full of excitement, be it in Science, Technology, or in commercial utilization [1].

The REs have an ever growing variety of applications in the modern technology (Fig. 1). They provide many an industry with crucial materials and they provide many a customer with benefits. From these beginnings and over many years, industrial applications of rare metals have developed in metallurgy, magnets, ceramics and electronics, chemical, optical, medical and nuclear technologies. RE primary products are mainly used as raw materials for high-purity individual RE chemicals, and in making of petroleum and environment protection catalysts, misch metal and polishing powders.

Although the demand for REs will further increase in future, the problem of balancing between medium and heavy rare metals with light ones will continue to create trouble as before. The rare earth ore contains various metals of a nearly constant composition. Since RE cost level is likely to go up when, as a result of increasing demand for some specific rare metals the remaining ones are unutilized,

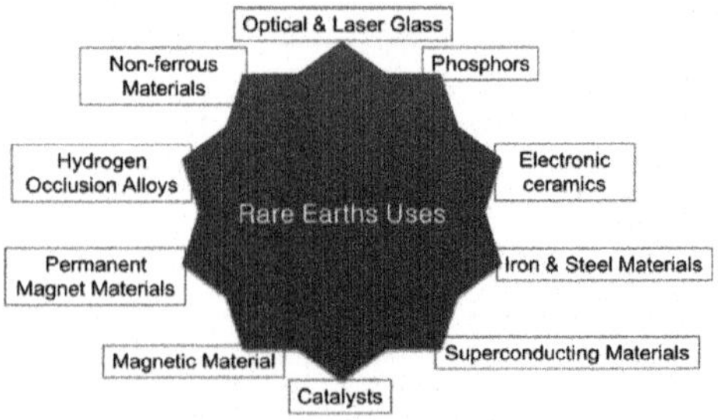

Fig. 1 Rare earths applications and uses in various sectors

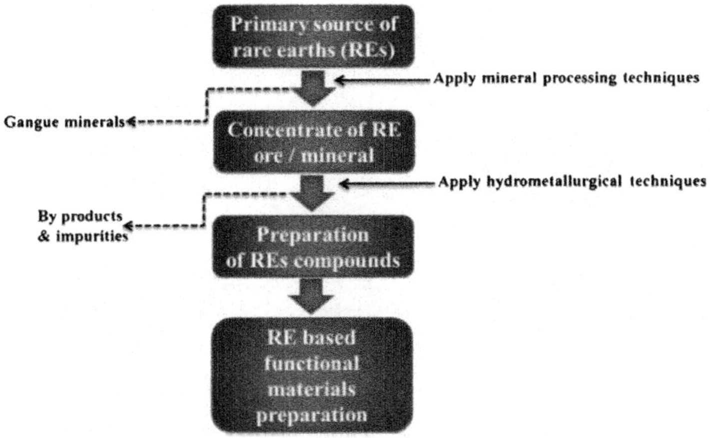

Fig. 2 Mineral processing and hydrometallurgical processing of the REs from primary sources to prepare functional materials

the rare earth manufactures will find it difficult to maintain stability in the supply of large quantities. There is a need to develop a demand for utilized rare metals or preferably a balanced demand for the available quantities of various rare metals.

There is yet another problem of distinguishing between the resources producing countries and resources consuming countries. Except America which possesses a vast reserve, major rare earth consuming countries like South Korea, Europe, Japan and others have to totally depend on imports, as mentioned earlier. As they don't have natural resources, it is difficult for them to ensure either a stable price or a stable volume of rare earth supply. Therefore, it might be necessary to promote mutual exchange of information and market exploration between the producing countries and to establish international collaboration at every stage from the starting material to final product. For REs processing from primary (or) secondary resources hydrometallurgy is one of the economical and environmentally friendly subject areas. The general flowsheet of the primary sources of REs processing was presented in Fig. 2.

Solvent extraction (liquid-liquid extraction) is widely used method for the separation of REs. In liquid-liquid extraction process, organophosphorus extractants have been commonly employed to achieve high separation efficiency between REs. The role of liquid-liquid extraction in metal recovery was presented as Fig. 3. The present paper deals the all possible area of hydrometallurgy to recover the REs. The general flowsheet of the LLE processing was presented as Fig. 4.

China having majority of the RE reserves about 36%, Russia other CIS (Common wealth of Independent States) nations having 19%, USA having 13% and Australia has 5% of the RE reserves. A global market receives the REs from various sources currently such as bastnasite, xenotime, RE laterite, ion adsorption

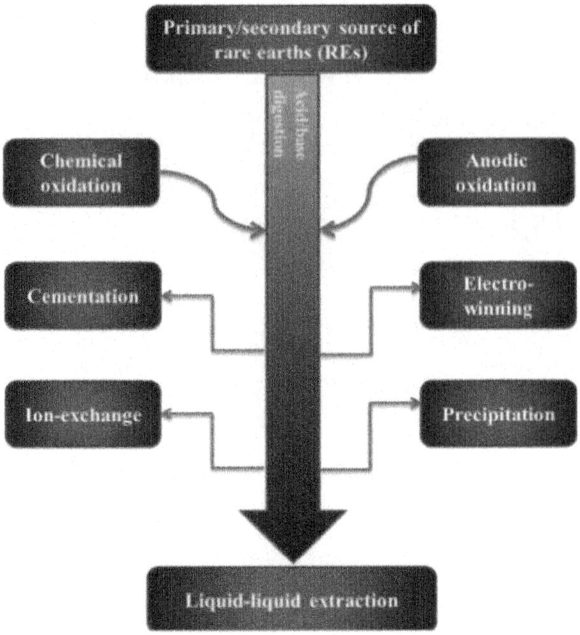

Fig. 3 The role of liquid-liquid extraction in metallurgical processing of REs from primary (or) secondary resources

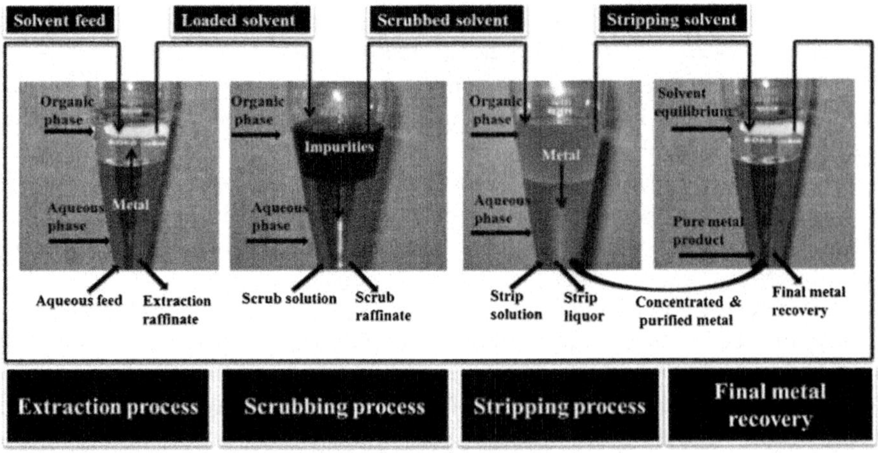

Fig. 4 General metal recovery flowsheet through liquid-liquid (solvent) extraction technique

clays and loparite etc. From year 2010 China reduces export quotas up to −40%, this is the main cause to elevation the criticality of demand REs. In new millennium, advanced and cleaner technologies were prime priorities in developed/devolving nations. South Korean government made goal on this subject area of

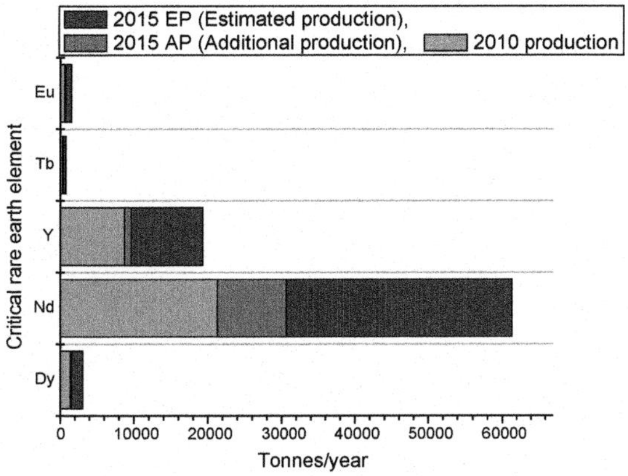

Fig. 5 Critical REEs supply and production comparisons in the years 2010 and 2015 [2]

research is make sure a reliable supply of materials critical to Korean mainstay industries. And develop the research and development policy for recycling of end-use products and designing for recyclability [2]. The critical rare earth elements supply and production was compared with year 2010 and 2015, the data was presented in Fig. 5.

Critical Rare Earths Processing by Hydrometallurgical Routes

Light rare earths processing by using tri-butyl-phosphate is well known route for single LREs separation and extraction (Fig. 6). The other organo-phosphorous based extractant such as TOPS 99 (equivalent to di-ethyl-hexyl-phosphoric acid). This process was developed by two stages, the beginning stage cerium was removed by precipitation methodology by using NaOCl mixed with NaOH, and purity of Ce is 78%. Cerium free raffinate solutions processed for other LREs (Fig. 7). Another organo-phosphorous based extractant such as PC-88A was saponified with NaOH then utilized for mixed rare earth chloride processing (Fig. 8).

The other critical REs such as Y, Dy and Tb were proceed by using various extractants through liquid-liquid extraction process. The first extraction cascade proceed by Versatic acid, the extraction raffinate solutions were generated LRE (light rare earths) rich concentration and loaded organic further processed by scrubbing and final LO contains HREs (Heavy rare earths). The scrubbed raffinate solutions were proceed for yttrium oxide by using TBP. The final extraction cascade was proceed for Tb and Dy by using saponified PC-88A (Fig. 9).

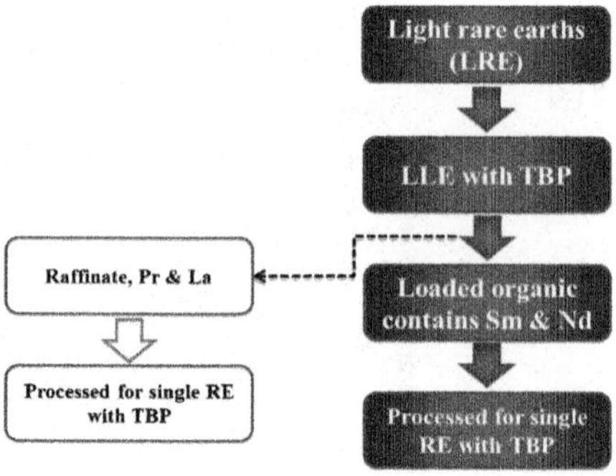

Fig. 6 Solvent extraction (liquid-liquid extraction (LLE)) of LREs from aqueous solutions by using TBP as an extractant system (this method developed by Indian Rare Earths Ltd (IREL, India)) [1]

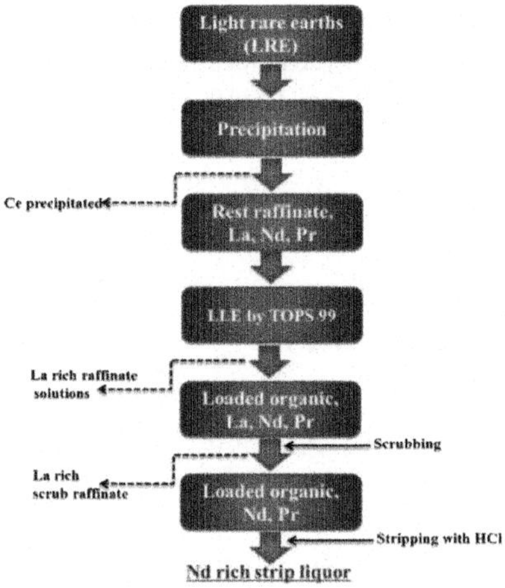

Fig. 7 LREs processed by hydrometallurgical methods [3]

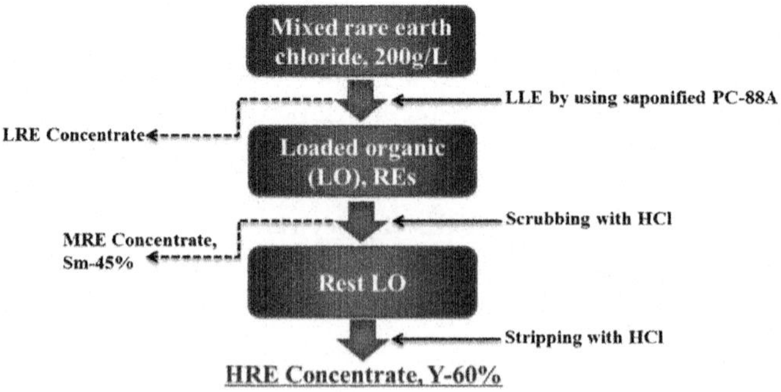

Fig. 8 REs processed by liquid-liquid extraction (LLE) by saponified PC-88A [1, 4]

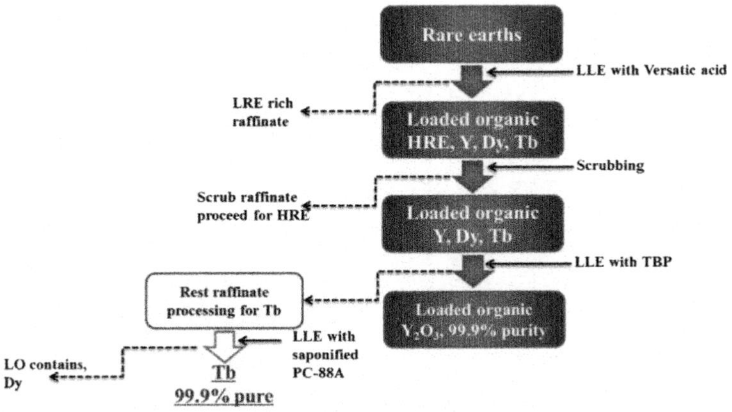

Fig. 9 Simultaneous extraction of Y, Tb and Dy by using various extractants [1, 4]

Critical Rare Earth Elements Processing by Iono-Metallurgical Routes

The media for metal recovery process is most important in metallurgy. Recent times, new class of solvents are applying in metal recovery, those are ionic liquids (molten (or) around ambient temperatures) [5, 6]. By using hydrometallurgical methods people were facing two major problems, one is waste water generation and further treatment, the other one is involving toxic chemicals such as cyanide, concentrated mineral acids, and caustic soda etc. [5]. Ionic liquids (ILs) were used in iono-metallurgy subject; these were involving low temperature processing for metals, this property will avoid the aqueous chemistry [5]. ILs was composed solely of ions, the liquid temperature below 100 °C. Various ionic liquids utilized for critical REs extraction processing, this summary data was presented in Table 1.

Rare Earths Processing by Using Organo-Phosphorus Extractants

Acidity ranged 0.5–5 mol/L was studied by using TOPS 99 (an equivalent to di-ethyl-hexyl-phosphoric acid) as an extractant system and H_3PO_4 solutions. The obtained data was presented as Fig. 10. Based on obtained data lower acidic conditions were more promising to critical REs such as Y, Dy, Tb and Nd. Concern separation factors data was also included in the Fig. 10. The other P based extractant PC 88A was utilized for REs separation at lower acidic ranges; the

Table 1 Critical rare earths extraction processing by using ILs

Name of the IL	Media	Remarks	References
Neodymium extraction			
Tri-hexyl(tetradecyl)phosphonium benzoate [T66614][BA] Tri-hexyl(tetradecyl)phosphonium bis (trifluoromethylsulfonyl)imide [T66614][TFSA]	Chloride	The extraction efficiency was reported up to 98% for 2–52 g/L of feed Nd solutions The loaded Nd metal was back extracted (stripping) by using HCl (or) HNO_3	[7]
1-Hexyl-3-methylimidazolium bis (trifluoromethylsulfonyl)imide [C6mim][NTf2]	pH 1–6 studied range	The possible extraction of neodymium was proposed as: $Nd(NO_3)_3 \cdot [Ph_3PC_3P(OEt)_2][PF_6]$ (IL)	[8]
Trihexyl(tetradecyl)phosphonium chloride (Cyphos® IL 101)	Nitrate	Nd and Dy were separated by using EDTA and 99.6% pure Nd_2O_3 was recovered by this method	[9]
Choline bis(trifluoromethylsulfonyl)imide, [Chol][Tf2N (IL) combined with choline hexafluoroacetylacetonate, [Chol][hfac] (extractant)	pH 2–6 studied range	The extracted neodymium compound was proposed as [Chol][Nd(hfac)4]. This compound crystal studies were examined	[10]
Trioctylmethylammonium dioctyl diglycolamate, [A336][DGA] (IL) mixed with trioctylmethylammonium nitrate, $[A336][NO_3]$ (diluent)	pH 1–6.8 studied range	100% of Nd extraction was achieved by this IL, this extraction process was exothermic in nature was observed	[11]
1-Hexyl-3-methylimidazolium bis (trifluoromethylsulfonyl)imide, $[C_6mim][NTf_2]$, 1-hexyl-3-methylpyrrolidinium bis (trifluoromethylsulfonyl)imide, $[C_6mpyr][NTf_2]$, and tributylmethylammonium bis(trifluoromethylsulfonyl) imide, $[N_{1444}][NTf_2]$	pH 0.5–6.0 studied range	Initial pH 2 is optimized condition for maximum Nd extraction. By using 1 mol/L nitric acid loaded Nd can recover by with four stages	[12]
Dysprosium extraction			
Tetraoctylphosphonium oleate [P8888][oleate]	pH 0.0–8.0 studied range	At pH 6.5 100% Dy extraction was reached. The trend of extraction efficiencies compare with other REs as follows: Yb > Er > Dy > Sm > Nd > La	[13]

(continued)

Table 1 (continued)

Name of the IL	Media	Remarks	References
TBP and triethyl-pentyl-phosphonium bis (trifluoromethyl-sulfonyl)amide ([P2225][TFSA])	–	The complex formed as $[RE(TBP)_m^{3+}(TFSA^-)_{x+y}]_{IL}$ reported, here RE = Pr (or) Nd (or) Dy	[14]
Yttrium extraction			
[1-Alkyl-3-(1-carboxylpropyl)im][PF6]	Nitric acid	This IL was saponified then used for Y extraction. The stripping efficiency reported about 95%. The cation-exchange mechanism was proposed	[15]
Europium and terbium extraction			
Trifluoromethyl-1,3-diketonate-functionalised IL	Neutral pH	99% of the Eu and Tb were extracted	[16]

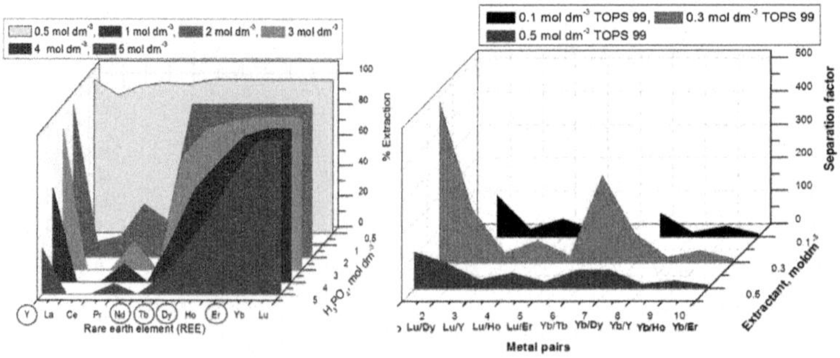

Fig. 10 Acid effect on rare earths using TOPS 99 as an extractant (adopted/reproduced from our own paper) [17]

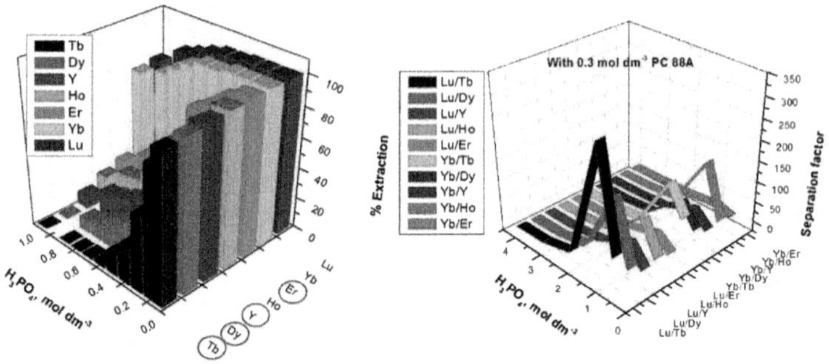

Fig. 11 Acid effect on rare earths using PC 88A as an extractant (adopted/reproduced from our own paper) [17]

obtained data was presented in Fig. 11. And it make known that, below 0.2 mol/L H3PO4 condition is better for critical REs.

Acknowledgements This research was supported by the Convergence Research Project (CRC-15-06-KIGAM) funded by the National Research Council of Science and Technology (NST).

References

1. N. Krishnamurthy, C.K. Gupta, *Extractive Metallurgy of Rare Earths*, 2nd edn. (CRC Press, 2016)
2. Critical materials strategy, DOE, USA, 2010
3. S. Radhika, B.N. Kumar, M.L. Kantam, B.R. Reddy, Solvent extraction and separation of rare-earths from phosphoric acid solutions with TOPS 99. Hydrometallurgy **110**, 50–55 (2011)
4. N.V. Thakur, Mineral Process. Extractive Metal. Rev. **21**(1–5), 277–306 (2000)

5. A.J. Hunt, *Element recovery and sustainability* (RSC Publishing, 2013) (ISBN: 978-1-84973-616-9)
6. A.P. Abbott, G. Frisch, S.J. Gurman, A.R. Hillman, J. Hartley, F. Holyoak, K.S. Ryder, Ionometallurgy: designer redox properties for metal processing. Chem. Commun. **47**, 10031–10033 (2011)
7. M. Panigrahi, M. Grabda, D. Kozak, A. Dorai, E. Shibata, J. Kawamura, T. Nakamura, Liquid–liquid extraction of neodymium ions from aqueous solutions of $NdCl_3$ by phosphonium-based ionic liquids. Sep. Puri. Tech. **171**, 263–269 (2016)
8. W. Junping, Z. Jieying, F. Da, K. Xinchun, S. Yunyu, Z. Lingling, L. Hongze, J. Rare Earths **34**, 83–90 (2016)
9. S. Riano, K. Binnemans, Extraction and separation of neodymium and dysprosium from used NdFeB magnets: an application of ionic liquids in solvent extraction towards the recycling of magnets. Green Chem. **17**, 2931–2942 (2015)
10. B. Onghena, J. Jacobs, L.V. Meervelt, K. Binnemans, Homogeneous liquid–liquid extraction of neodymium(III) by choline hexa fluoroacetylacetonate in the ionic liquid choline bis (trifluoromethylsulfonyl)imide. Dalton Trans. **43**, 11566–11578 (2014)
11. A. Rout, K. Binnemans, Solvent extraction of neodymium(III) by functionalized ionic liquid trioctylmethylammonium dioctyl diglycolamate in fluorine-free ionic liquid diluent. Ind. Eng. Chem. Res. **53**, 6500–6508 (2014)
12. A. Rout, J. Kotlarska, W. Dehaen, K. Binnemans, Liquid–liquid extraction of neodymium(III) by dialkylphosphate ionic liquids from acidic medium: the importance of the ionic liquid cation. Phys. Chem. Chem. Phys. **15**, 16533 (2013)
13. D. Parmentier, T.V. Hoogerstraete, S.J. Metz, K. Binnemans, M.C. Kroon, Selective extraction of metals from chloride solutions with the tetraoctylphosphonium oleate ionic liquid. Ind. Eng. Chem. Res. **54**, 5149–5158 (2015)
14. M. Matsumiya, Y. Kikuchi, T. Yamada, S. Kawakami, Extraction of rare earth ions by tri-*n*-butylphosphate/phosphonium ionic liquids and the feasibility of recovery by direct electrodeposition. Sep. Puri. Tech. **130**, 91–101 (2014)
15. W. Wei, L. Yu, X. Aimei, Y. Hualing, C. Hongmin, C. Ji, Solvent extraction of yttrium by task-specific ionic liquids bearing carboxylic group. Chin. J. Chem. Eng. **20**(1), 40–46 (2012)
16. J.H. Olivier, F. Camerel, R. Ziessel, Lanthanide ion extraction by trifluoromethyl-1,3-diketonate-functionalised ionic liquids adsorbed on silica. Chem. Eur. J. **17**, 9113–9122 (2011)
17. B.R. Reddy, J.R. Kumar, Rare Earths Extraction, Separation, and Recovery from Phosphoric Acid Media. Sol. Ext. Ion Exch. **34**, 226–240 (2016)

Selective Reduction and Separation of Europium from Mixed Rare-Earth Oxides from Waste Fluorescent Lamp Phosphors

Mark L. Strauss, Brajendra Mishra and Gerard P. Martins

Abstract Europium is a critical material required for LED, florescent lamp, cell phone screen, and flat panel display production. This process recycling europium from waste lamp phosphors is an innovative method to supply europium for high technology applications. Waste phosphor powder from recycled lamps is retorted, sieved, and leached to produce a europium/yttrium leach solution. The separation of europium and yttrium from the pregnant leach solution is conducted by selectively reducing Eu(III) to Eu(II) via zinc powder and precipitating europium (II) sulfate from solution using sulfuric acid as the precipitating agent. After one stage of selective reduction and precipitation, the purity and recovery of europium (II) sulfate was greater than 95% and 80%, respectively.

Keywords Waste Fluorescent Lamp · Phosphor Dust · Rare Earths · Recycling · Europium

Introduction

Recycled phosphor dust has the potential to supply a significant portion of US clean energy demand for europium. According to the DOE 2011 Critical Materials Strategy Report [1], the demand will increase from 200 to 220 tons per year in 2020. In 2007, 8000 tons per year of phosphor dust were discarded into landfills. Based a conservation calculation of concentration of 0.3% europium in the discarded dust, there is a potential to supply 240 tons per year. However, the concentration of europium is the dust will increase due to fewer halophosphate

M.L. Strauss (✉) · B. Mishra
Worcester Polytechnic Institute, 100 Institute Road, Worcester,
MA 01609, USA
e-mail: mlstrauss@wpi.edu

G.P. Martins
Colorado School of Mines (Kroll Institute of Extractive Metallurgy),
1500 Illinois Street, Golden, CO 80401, USA

© The Minerals, Metals & Materials Society 2017
H. Kim et al. (eds.), *Rare Metal Technology 2017*,
The Minerals, Metals & Materials Series,
DOI 10.1007/978-3-319-51085-9_3

based T12 lamps being recycled and a greater ratio of newer T8 lamps being recycled.

Red phosphor is phosphor used to create the color red in fluorescent lamps. It is composed of Y_2O_3:Eu or yttrium oxide doped with europium. Based on the previous work presented by Eduafo et al. [2], europium and yttrium are leached at the same time into solution. Therefore, the goal of this research was to develop a process to separate europium from yttrium, in order develop a saleable europium oxide product. However, the product of this work is europium (II) sulfate—which can be converted to europium oxide in a few steps.

Originally, Molycorp [3], developed a process to recovery 99.9% pure europium oxide form Eu-Sm-Gd concentrates separated from monazite. More recently, Preston et al. [4], Morais et al. [5], and Rabie et al. [6] have used zinc metal and sulfuric acid to separate and purity europium from samarium and gadolinium concentrates. The paper proposes using the same method for europium and yttrium concentrates, as a product of waste lamp phosphor leaching. The equation below demonstrates the equation for converting Eu(III) to Eu(II) via selective reduction.

$$2Eu^{3+} + Zn = 2Eu^{2+} + Zn^{2+} \tag{1}$$

The equilibrium constant, K_c, can be used to describe the thermodynamics of a system. In the equation below the activities of species are replaced with their actual concentrations because there is no simple method to measure the activities of concentrated species in high ionic strength solutions.

$$aA + bB \Leftrightarrow cC + dD$$

$$K_c = \frac{[C]^c[D]^d}{[A]^a[B]^b}$$

In addition, the formation of europium (II) sulfate precipitate, and intermediate in the europium separation experiments is demonstrated in Eq. 2.

HSC 5.11 was used to explore the possible reactions for the oxalic precipitation work and selective reduction and precipitation of europium (II) sulfate.

$$Eu^{2+} + SO_4^{2-} = EuSO_4 \quad K_c = 5.94 \times 10^8 \ (T = 25\,°C) \tag{2}$$

In all likelihood, europium (II) is part of chloride complex in a system with excess chloride ions, so the following equation better describes the actual conditions.

$$EuCl^{+} + SO_4^{2-} = EuSO_4 + Cl^{-} \quad K_c = 3.01 \times 10^8 \ (T = 25\,°C) \tag{3}$$

Some of the possible by products of the reaction include europium (III) sulfate and yttrium (III) sulfate.

$$YCl^{2+} + 1.5SO_4^{-2} + 4H_2O = 0.5Y_2(SO_4)_3 * 8H_2O + Cl^- \; K_c = 4.90 \times 10^{-1} \; (T = 25°C) \quad (4)$$

$$EuCl^{2+} + 1.5SO_4^{-2} + 4H_2O = 0.5Eu_2(SO_4)_3 * 8H_2O + Cl^- \; K_c = 1.31 \times 10^{5} \; (T = 25°C) \quad (5)$$

Similarly, the process uses zinc metal to reduce Eu^{III} to Eu^{II} via Zn^0. Next, the reduced solution is precipitated by the addition of sulfuric acid. As a result, europium (II) sulfate is precipitated and is separated via vacuum filtration. The product was quantified by X-Ray Fluorescence and SEM.

Experimental

Materials

Yttrium and europium oxide (99.99% pure) and zinc metal (99.8% 20–30 mesh) were provided by Alfa Aesar. Sodium hydroxide (Sigma Aldrich, USA) is dissolved in deionized water. Yttrium and europium were dissolved in solution using analytical grade (Sigma Aldrich, USA) hydrochloric acid and deionized water. The pH was adjusted with NaOH (Sigma Aldrich, USA) dissolved in deionized water. ACS grade 18 M sulfuric acid (Sigma Aldrich, USA) was diluted with deionized water.

Analysis

X-Ray Fluorescence (XRF) (Thermo Fisher Scientific) and scanning electron microscopy electron dispersive spectroscopy (SEM-EDS) were used for quantification and identification.

Procedure

A 13:1 ratio of yttrium oxide to europium oxide was dissolved in concentrated 12 M hydrochloric acid. The alkalinity of the solution was adjusted with a dilute sodium hydroxide solution to pH 3. The solution was added to magnetically-stirred vessel. 4% Hydrogen was bubbled through the system. Next, 5 times the stoichiometric amount of zinc was added to the mixing vessel. After two hours, a the reduced europium solution and dilute solution of 1 M sulfuric acid, degassed with 4% hydrogen, were mixed together in a cylindrical vessel, bubbled with 4%

Table 1 This table enumerates the results of the europium (II) sulfate precipitation experiments

Experiment	Dry weight	$EuSO_4$ purity (%)	Y impurity (%)	$EuSO_4$ recovery (%)
1	1.26	95.00	1.84	85.50
2	1.25	95.33	1.46	85.12
3	1.29	95.30	1.73	87.81
4	1.35	96.00	1.43	92.57
5	1.35	95.80	1.30	92.38
6	1.23	96.80	0.74	85.05
Mean	1.29	95.71	1.42	88.07
Standard deviation	0.05	0.65	0.39	3.56

hydrogen. After two hours, the solution was filtered by vacuum filtration with Millipore 47 mm filter in a Pall vacuum filtration setup. The precipitate was washed with a 0.001 M sulfuric acid.

Results and Discussion

Analysis of Precipitate

The precipitate was analyzed by XRF. Based on the in 6 experiments listed in Table 1.

Based on the inputs of the system, the expected dry weight of the precipitate was 1.40 g of $EuSO_4$. However, the total weight was less than that value. The differences between the expected weight and the actual weight was likely to do incomplete conversion of Eu(III) to Eu(II). The incomplete version of the system may have been due to competitive reactions or Eu(II) reoxidizing due to incomplete isolation from oxygen during transferring or precipitation. The purity was determined from XRF by adding the values for sulfur and europium as shown in Table 2. This method is not quantitative in nature. The recovery was determined by multiplying the dry weight over the expected value multiplied by the europium sulfate purity.

In addition, SEM-EDS was run on the precipitate as shown in Fig. 1 and Table 3.

SEM-EDS is able to identify the major impurities such as chlorine, but cannot quantify yttrium. The nickel, manganese and aluminum values listed in the table are either due to misidentification or a mistaken introduction of impurities in the in the SEM.

Table 2 The table demonstrates the elements abundances determined via XRF for run #6

Element	Wt%	Est. error
Eu	85.39	0.18
Sx	11.43	0.16
Cl	0.971	0.048
Y	0.743	0.037
Na	0.579	0.029
K	0.102	0.0051
Al	0.095	0.0084
Sb	0.0927	0.026
I	0.0734	0.036
Hf	0.0633	0.021
Cs	0.0587	0.05
Ce	0.0522	0.016
Th	0.0492	0.033
Zn	0.0448	0.0072
Pr	0.0433	0.018
Ru	0.0418	0.015
Nd	0.0394	0.01
Mn	0.0322	0.014
Mg	0.032	0.024
Pt	0.185	0.016
Zr	0.018	0.014
Rb	0.0118	0.012
Ga	0.01	0.006
V	0.0055	0.0027

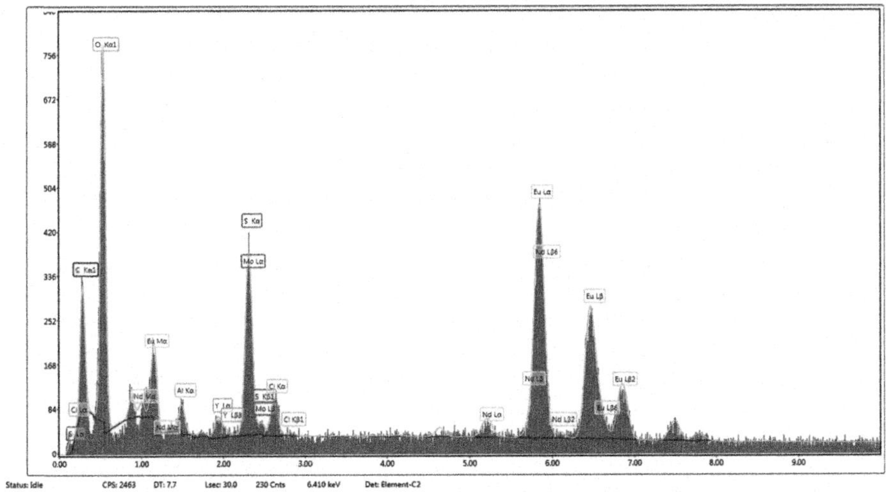

Fig. 1 This SEM-EDS spectrum was indicated for europium (II) sulfate

Table 3 The table shows the quant for europium (II) from SEM-EDS

Element	Weight %	Atomic %	Error %
C K	8.94	23.41	14.96
O K	27.24	53.57	9.32
AL K	0.85	0.99	25.97
S K	10.13	9.95	6.68
Cl K	0.95	0.84	20.98
Eu L	50.48	10.45	5.51
Mn K	0.73	0.42	49.75
Ni K	0.69	0.37	58.91

Conclusions

The selective reduction and precipitation is an interesting method to purify yttrium and europium rich solutions. Europium (II) sulfate purities greater than 95% and recoveries greater than 80% are easily attained.

Further work will be conducted using actual solutions with yttrium and europium attained from waste phosphor dust rather than pure powders. In addition, more attention will be paid to minimize disturbances from possible sources of oxygen so as to maximize recovery.

References

1. M. Mckittrick, U.S. Department of Energy Critical Materials Strategy, vol. 1. 2011
2. P.M. Eduafo, M. Strauss, Experimental investigation of recycling rare earth metals from waste fluorescent lamp phosphors. Rare Met. Technol. **2015**, 253–259 (2015)
3. C.K. Gupta, *Extractive Metallurgy of Rare Earths*. 2005
4. J.S. Preston, A.C. Preez, *The Separation of Europium from a Middle Rare Earth Concentrate bv Combined Chemical Reduction, Precipitation ind Solvent-Extraction Methods* (1996), pp. 93–101
5. C.A. Morais, V.S.T. Ciminelli, Recovery of europium by chemical reduction of a commercial solution of europium and gadolinium chlorides. Hydrometallurgy **60**(3), 247–253 (2001)
6. K.A. Rabie, S.A. Sayed, T.A. Lasheen, I.E. Salama, Europium separation from a middle rare earths concentrate derived from Egyptian black sand monazite. Hydrometallurgy **86**(3–4), 121–130 (2007)

Application of Rare Earths for Higher Efficiencies in Energy Conversion

W.D. Judge, Z.W. Xiao and G.J. Kipouros

Abstract Rare earths have taken a prominent position in the Hi-Tech field after the discovery of the NdFeB permanent magnets. Their use in the magnets sparked miniaturization of the communication devices such as computers and mobile phones and are being considered in many more applications. Potential applications are energy conversion devices that require strong motors and photovoltaics where rare earths can be used to improve the efficiencies of silicon. The presentation is summarizing the role of rare earth elements in every category of the energy conversion devices and the sustainability of these operations given the restrictions in availability of resources, limitation of production methods and environmental consequences.

Keywords Rare earth · Energy conversion · Renewable energy · Magnet · Photovoltaic · Sustainability

Introduction

Over the last few decades, the world's energy needs have increased steadily and are predicted to continue to increase over the next three decades by up to 48% [1]. These energy demands may be largely met through a number of fossil-fuel-based energy sources such as coal, oil, and natural gas. However, concerns over air

W.D. Judge · Z.W. Xiao · G.J. Kipouros (✉)
Chemical Engineering, College of Engineering, University of Saskatchewan, 57 Campus Drive, Saskatoon, SK S7N 5A9, Canada
e-mail: georges.kipouros@usask.ca

W.D. Judge
e-mail: will.judge@dal.ca

Z.W. Xiao
Faculty of Metallurgical and Energy Engineering, Kunming University of Science and Technology, Kunming 650093, China

© The Minerals, Metals & Materials Society 2017

H. Kim et al. (eds.), *Rare Metal Technology 2017*,
The Minerals, Metals & Materials Series,
DOI 10.1007/978-3-319-51085-9_4

pollution and climate change, or 'global warming', has also favoured the development of alternative 'low-carbon' technologies such as nuclear power as well as the so-called 'renewable' energy sources to meet future energy demands.

In the industrial and transportation sectors, the use of fossil-fuel-based energy sources are more difficult to avoid as the present infrastructure is heavily invested in this technology and, in many cases, practical alternatives do not exist [2]. The electric power sector, however, is highly dynamic and one of the fasting growing sectors, presenting an opportunity to implement low-carbon technologies, especially in nations with developing electric power systems. The success of low-carbon initiatives, however, is somewhat dependent upon the performance of these technologies, many of which are still under advancement.

To generate electrical power always requires some type of energy conversion process, Fig. 1, characterized by an efficiency which the fundamental laws of thermodynamics dictate must be less than unity—i.e. there is always an energy loss associated with energy conversion. In many cases, particularly with the developing low-carbon technologies, there is room for improvements in energy conversion efficiencies for electric power generation which would serve to increase energy outputs and reduce environmental impacts of power generation [3].

In this work, the applications of the rare earth elements to improve efficiencies in energy conversion are presented. The energy supply chain and the role of renewable energy sources are first reviewed with respect to energy conversion processes before the most prominent applications of rare earth elements in energy conversion are outlined. The sustainability of these applications are then discussed given the supply, production, and environmental challenges surrounding the rare earth industry.

Energy Supply Chain

To efficiently provide electricity *en masse* requires a well-developed 'energy supply chain' to deliver electricity from power generating stations to end users. In the basic energy supply chain, shown in Fig. 2, power stations utilize 'primary energy

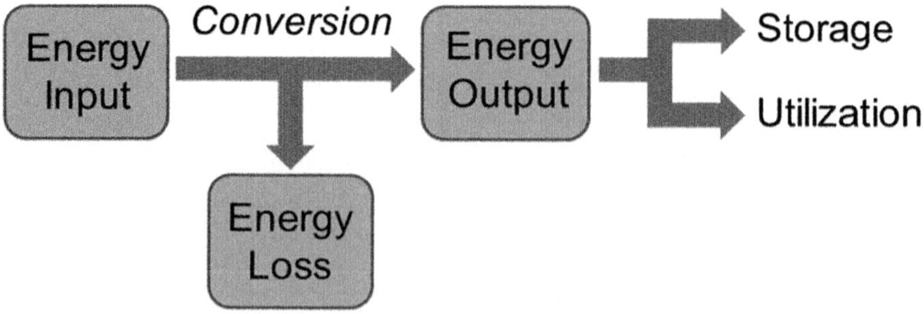

Fig. 1 The general energy conversion process

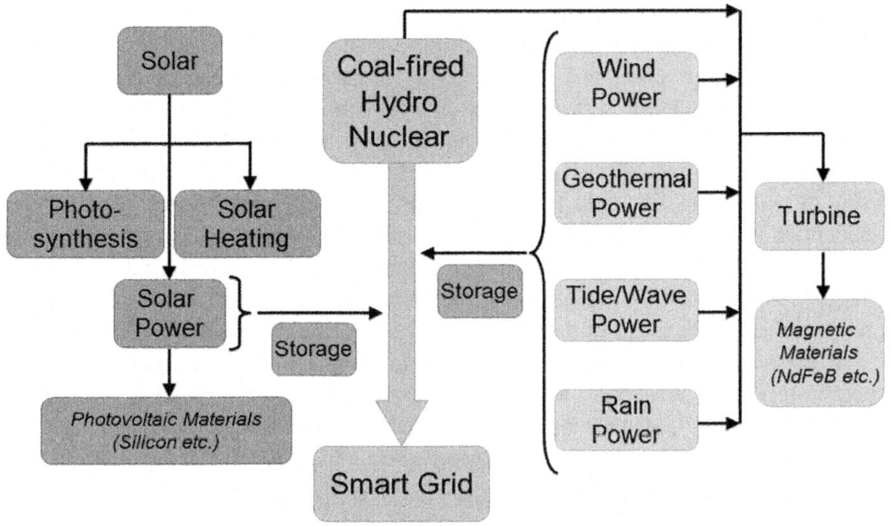

Fig. 2 The basic energy supply chain and the role of renewable energy sources

sources' such as coal, natural gas, or natural uranium, as well as renewable sources such as solar, wind, or geothermal energy, and convert them into 'secondary energy sources' or 'energy carriers'—the most common being electricity [4]. Electricity generated from power stations is delivered to end users through a so-called 'smart grid', comprising of a series of transformers, substations, transmission lines, distribution lines, etc., which detects and reacts to regional changes in usage to provide more reliable and optimal power allocation.

Interactive smart grids are a necessity as virtually all electricity delivered to end users is generated only moments prior to delivery—there are currently limited implemented solutions to grid-scale energy storage. Perhaps the most promising technology for grid-scale energy storage is the liquid metal battery which operates at elevated temperature with two liquid metal electrodes and a fused salt electrolyte [5]. As shown in Fig. 2, energy storage is a key component enabling the advancement of renewable energy sources as the power generated from these sources is not always consistent nor predictable.

As a secondary energy source, electricity is preferred as it is fairly easily and quickly transported over long distances and, upon delivery to the end user, may be converted to other forms of energy, including thermal, mechanical, chemical, etc., easily and typically with quite high efficiency. The conversion of energy from primary sources to electricity, however, is generally limited in efficiency and may be a targeted area for improvements [6].

Coal-fired, natural gas fuelled, or nuclear power stations which use primary energy sources to harvest thermal energy typically utilize a steam or gas turbine in conjunction with an alternator, or electrical generator, to convert thermal energy to mechanical and, subsequently, electrical energy [7]. For hydroelectric power stations, water turbines capture the energy from falling water and, in conjunction with an electrical generator, convert this to electrical energy.

Renewable sources of energy which exploit geothermal energy or the kinetic energy of wind, ocean waves, tides, etc. utilize the same basic technological principles to generate electricity—conversion of primary energy sources using turbines coupled with electrical generators [7]. In the case of solar energy, light from the sun may be collected and converted to thermal energy used to heat residential homes or may be directly converted to electrical energy using solar panels. Plants and other organisms utilize solar energy by converting it to chemical energy through photosynthesis, however, the development of a practical artificial photosynthesis process to utilize solar energy still remains a challenge for scientists!

At the heart of these energy conversion technologies are materials that require specific properties which is where rare earth elements find application. The unique magnetic, electric, and spectroscopic properties of rare earth elements may improve the efficiencies of energy conversion processes which involve magnetic and photovoltaic materials, Fig. 2.

Applications of Rare Earths

Permanent Magnets

The first rare earth containing permanent magnets to garner attention by industry were based on the samarium-cobalt compound $SmCo_5$, however, the high price of samarium limited their applications. In the early 1980s, several research groups, but notably General Motors Research Laboratories, independently discovered the neodymium-iron-boron permanent magnet material, $Nd_2Fe_{14}B$, which revolutionized electronics [8]. At General Motors, permanent magnets were commercially made by powder metallurgy. The compound was melt spun to produce powder which was aligned and compacted in a magnetic field in order to magnetize all particles in the same direction. The compacts were sintered and then remagnetized.

In the neodymium-iron-boron permanent magnets, the role of iron is to provide most of the magnetization and to ensure a high 'Curie temperature'—that is the temperature at which permanent magnetism is lost. The role of neodymium is to provide magnetocrystalline anisotropy, while boron stabilizes the compound and ensures an amorphous structure [9]. To raise the Curie temperature further, neodymium-iron-boron permanent magnets now contain additions of 2–4 weight % of the heavy rare earth elements terbium or dysprosium [10]. The result is a permanent magnetic material whose energy density is ten times greater than that of ferrite permanent magnets.

Rare earth permanent magnets provide more strength, less volume, and less weight which significantly improves the efficiency of turbines for energy conversion. Their most prominent use for energy conversion has been in wind turbines where it is estimated 15% of all rare earth permanent magnet material is used [11].

Typically, wind turbines contain about 150–200 kg of neodymium and 20–30 kg of dysprosium per megawatt of generating capacity [10]. This means most on- or offshore wind turbines require over 500 kg of rare earth elements per turbine!

Photovoltaics

Photovoltaics were discovered as early as the late 19th century (e.g. selenium), however, the first photovoltaics capable of actually powering electrical equipment were developed at Bell Laboratories in the early 1950s. These pioneering photovoltaics were made from silicon and had efficiencies of only 6% which initially limited their application primarily to the space program [12]. Today, silicon remains the dominant photovoltaic material and accounts for 99% of the photovoltaic market, with the next largest share, cadmium telluride (CdTe), accounting for less than 1% of the market [13].

The majority of silicon photovoltaics utilize either polycrystalline or single crystal silicon, however, amorphous silicon thin films have been playing an increasing role. For photovoltaic applications, metallurgical grade silicon (~98% pure) is unacceptable and 'solar grade' silicon having 6 N purity (i.e. 99.9999% pure) must be produced using the energy intensive and expensive Siemens process involving distillation and chemical vapour deposition. To produce single crystal silicon requires yet another tedious and expensive step, Czochralski crystal pulling. For photovoltaics, silicon wafers 0.2–0.5 mm thick are extracted from crystal rods using a wire sawing technique [12]. The high production costs of solar grade silicon are a major limiting factor in the application of photovoltaics.

The photovoltaic effect is a phenomenon where potential difference is generated in a material upon exposure to light which, in solar cells, is used to induce electric current through a circuit. From a solid state physics perspective, silicon is a far from ideal photovoltaic material, however, is very prevalent as the technology supporting silicon production is well-developed with high quality materials being produced on a large scale for the semiconductor industry [12]. The efficiencies of commercial polycrystalline silicon photovoltaics are generally in the range of 14–19% with the bandgap of silicon allowing it to capture about 48% of the available power from the sun [14]. The performance of silicon photovoltaics, however, may be improved at reasonable cost by applying layers containing low concentrations rare earth elements.

In silicon photovoltaics, rare earths element doped layers act as so-called 'wavelength converters' to capture additional wavelengths of light through either 'up-conversion' of the near-infrared spectral range or 'down-conversion' of the ultraviolet spectral range to wavelengths which silicon can utilize [15]. Up- or down-conversion materials are typically ceramics doped with small amounts of various rare earth elements, particularly the heavy rare earth elements, with up-conversion layers applied to the underside of the silicon photovoltaic and down-conversion layers applied to the topside of the silicon photovoltaic [16].

Of the total solar spectrum, approximately 35% is available for up-conversion and 15% is available for down-conversion—clearly the rare earth elements have the potential to drastically improve the efficiency of silicon photovoltaics [15].

Sustainability

Although rare earth elements are frequently hailed as the future of clean energy production, which they certainly do play an especially critical role, sustainability in their production is often overlooked where it is important to consider the resource restrictions, technology limitations, and environmental impacts of producing rare earth elements.

China is the dominant producer of rare earth elements and accounted for up to 97% of the total world production in the 2000s [17], but this was not always the case. The transition of Chinese dominance in the marketplace began in the 1980s as China began exporting rare earth concentrates, and eventually, in the 1990s, became sweeping as the extraction and supply chain of China developed more towards end-use, exporting products like rare earth magnets and metals [18]. During this time, competitors were unable to contend with Chinese prices, largely due to higher environmental and labour standards, but also related to China's convenient Bayan Obo deposit, where rare earths are mined as a by-product of iron ore, and ion adsorption clay, where expensive mining and beneficiation steps are avoided.

Since 2010, China has imposed rare earth production and export quotas, export taxes, and environmental policies which drastically increased prices and put severe pressure on western markets relying on rare earths leading to the so-called 'rare earth crisis' [17]. The resulting price increases has made it economical for competitors outside of China to mine rare earth elements, however most of these mines are producing only the light rare earth elements (La-Sm) due to the nature of their deposits [17]. The supply of the heavy rare earth elements (Eu–Lu, Y), which are critical to clean energy, is expected to continue to fall short of demand [17]. If sustainability in the rare earth industry is to be realized, supply reliability and rare earth recycling must be addressed.

The production of rare earth metals is cumbersome and difficult due to their low concentrations in ore bodies and the high reactivity of the metals. In China's Bayan Obo deposit, rare earth containing minerals are a by-product of the main iron ore, magnetite and hematite, which are removed by magnetic separation and the rare earth containing minerals further concentrated by a flotation process [19]. Chemical beneficiation occurs by treatment of the flotation concentrate with hot sulfuric acid and eventually a mixed rare earth product is precipitated. In the case of the ion adsorption clays, no physical beneficiation is conducted and the ore is directly leached with sulfuric acid solution. Individual rare earths are then separated from the mixed rare earth product by solvent extraction to yield individual rare earth oxides.

On the industrial scale, reduction to rare earth metals is mostly conducted through molten salt electrolysis of rare earth oxides or chlorides—some operating conditions are shown in Tables 1 and 2. Process conditions are typically chosen to minimize operating temperature, while still ensuring a liquid metal electrolysis product. The rare earth elements lanthanum through neodymium are electrowon as pure metals at temperatures above their melting points, Table 2. The high-melting point rare earth metals are electrowon as a ferroalloy, allowing electrolysis to be conducted at lower temperatures, Table 2. Improved procedures and efficiencies in extractive metallurgy would serve to increase the sustainability of the rare earth industry.

Until recently, rare earth mining in China has essentially proceeded unimpeded for maximum output with little to no regard for environmental impact. As a result,

Table 1 Economic and technical conditions for the production of rare earth metals by molten salt electrolysis

Metal	Method	Electrolyte	Current (A)	Current efficiency (%)	Yield (%)	Power consumption (kWh/kg)
La	Chloride	$LaCl_3$-KCl	800	70	90	20
	Oxide	LaF_3-LiF-BaF_2-La2O_3	1000	79	95	11
Ce	Chloride	$CeCl_3$-KCl	800	65	90	22
	Oxide	CeF_3-LiF-BaF_2-CeO_2	1300	74	95	11
Pr	Chloride	$PrCl_3$-KCl	800	60	85	24
	Oxide	PrF_3-LiF-BaF_2-Pr_6O_{11}	1028	70	95	12
Misch metal	Chloride	$REECl_3$-KCl	10,000	40	90	25
	Oxide	$REEF_3$-LiF-BaF_2-REE_2O_3	20,000	80–85	93	–

Translated from Ref. [20]

Table 2 Cell materials and operating conditions for the production of rare earth metals and alloys by molten salt electrolysis of rare earth oxides

	La	Pr	Nd	Dy–Fe alloy	Gd–Fe alloy
Electrolyte (weight %)	LiF:LaF_3 (85:15)	LiF:PrF_3 (90:10)	LiF:NdF_3 (90:10)	LiF:DyF_3 (90:10)	LiF:GdF_3 (90:10)
Metal collection device	W or Mo crucible			Fe crucible	
Cathode	W bar			Fe bar	
Anode	Graphite				
Electrode distance (cm)	10–15	10–15	10–15	10–14	10–14
Temperature (°C)	950–1000	1000–1050	1050–1080	1050–1100	1050–1100
Current (A)	4000–6000, 10,000, 25,000			3000–3500	
Cell voltage (V)	8–10			10–12	

Translated from Ref. [20]

there has been extreme environmental consequences which now showcase the hidden dilemmas behind the rare earth elements and clean energy. Rare earth mining and mineral processing consumes a large amount of water, chemicals, and energy, and generates a significant amount of waste which must be properly treated for sustainability of the rare earth industry.

Rare earth minerals are frequently associated with or contain thorium, uranium, sulfides, and fluorides which eventually end up in tailings ponds or off gases. At Bayan Obo, the tailings reservoir is approximately 12 km long, 11 km^2 in area, and contains about 150 million tons of tailings [19]—this is approximately 100 times the size of the Hungarian alumina factory pond which collapsed in 2010 releasing toxic red sludge! Groundwater contamination and radioactive dust emissions have been documented in the local environment.

It is estimated that to produce 1 ton of rare earth elements generates 1 ton of radioactive containing tailings [19]. This does not include other waste products generated in rare earth metal production. In the chemical beneficiation of rare earth ores, sulfuric acid produces off gases of hydrogen fluoride, carbon dioxide, and sulfur oxides, as well as generating acidic waste water. Ion adsorption clays treated by pond or heap leaching leave behind leachant that damages the environment. The electrolytic production of rare earth metals and alloys produces chloride or fluoride containing waste water and gas. Clearly the utmost care must be taken to ensure these processes are conducted in an environmentally friendly and sustainable manner.

Conclusions

As the worlds energy demands continue to increase, so do concerns over fossil-fuel driven climate change. A move towards low-carbon technologies for energy supply is dependent upon the performance of these technologies which always involve some form of energy conversion in the energy supply chain to end users.

Rare earth elements have found applications in improving efficiencies of energy conversion technology, especially for renewable energy sources. Here, the most prominent applications of rare earths are in permanent magnets for turbines and as up- or down-conversion materials for photovoltaics. These technologies become more efficient due to the unique electric, magnetic, and spectroscopic properties of the rare earth elements.

The sustainability of the rare earth industry is an important consideration when evaluating their potential in clean energy technologies. Current situations involving resource restrictions, technology limitations, and environmental impacts of rare earth production are challenges which must be addressed and overcome to ensure sustainability.

Acknowledgements The authors would like to acknowledge the financial support of the Natural Sciences and Engineering Research Council (NSERC) of Canada. One of the authors (Z. W. Xiao) also acknowledges the financial support from the China Higher Education Council during his Sabbatical Leave in our laboratory.

References

1. International Energy Outlook 2016, *U.S. Energy Information Administration, U.S. Department of Energy* Report No. DOE/EIA-0484. (2016)
2. H. Lund, Renewable energy strategies for sustainable development. Energy **32**(6), 912–919 (2007)
3. B. Sorensen, *Renewable Energy: Physics, Engineering, Environmental Impacts* (Economics & Planning. Elsevier, Burlington, MA, 2011)
4. H.M. Wee, W.H. Yang, C.W. Chou, M.V. Padilan, Renewable energy supply chains, performance, application barriers, and strategies for further development. Renew. Sustain. Energy Rev. **16**(8), 5451–5465 (2012)
5. H. Kim, D.A. Boysen, J.M. Newhouse, B.L. Spatocco, B. Chung, P.J. Burke, D.J. Bradwell, K. Jiang, A.A. Tomaszowska, K. Wang, W. Wei, L.A. Ortiz, S.A. Barriga, S.M. Poizeau, D.R. Sadoway, Liquid metal batteries: past, present, and future. Chem. Rev. **113**(3), 2075–2099 (2013)
6. I. Dincer, Renewable energy and sustainable development: a crucial review. Renew. Sustain. Energy Rev. **4**(2), 157–175 (2000)
7. N. Armaroli, V. Balzani, The future of energy supply: challenges and opportunities. Angew. Chem. Int. Ed. **46**(1–2), 52–66 (2007)
8. A.L. Robinson, Powerful new magnet material found. Science **223**(4639), 920–922 (1984)
9. J.M.D. Coey, Rare-earth magnets. Endeavor **19**(4), 146–151 (1995)
10. G.P. Hatch, Dynamics in the global market for rare earths. Elements **8**(5), 341–346 (2012)
11. X. Du, T.E. Graedel, Global rare earth in-use stocks in NdFeB permanent magnets. J. Ind. Ecol. **15**(6), 836–843 (2011)
12. A. Goetzberger, C. Hebling, Photovoltaic materials, past, present, future. Sol. Energy Mater. Sol. Cells **62**(1–2), 1–19 (2000)
13. A. Goetzberger, C. Hebling, H.W. Schock, Photovoltaic materials, history, status and outlook. Mater. Sci. Eng. R **40**(1), 1–46 (2003)
14. B. Parida, S. Iniyan, R. Goic, A review of solar photovoltaic technologies. Renew. Sustain. Energy Rev. **15**(3), 1625–1636 (2011)
15. S.V. Eliseeva, J.G. Bunzli, Rare earths: jewels for functional materials of the future. New J. Chem. **35**(6), 1165–1176 (2011)
16. D. Verma, T.O. Saetre, O.M. Midtgard, in *Review on Up/Down Conversion Materials for Solar Cell Application*. Photovoltaic Specialists Conference (PVSC). (2012), pp. 2608–2613
17. K.A. Gschneidner Jr., The rare earth crisis-the supply/demand situation for 2010–2015. Mater. Matters **6**(2), 32–37 (2011)
18. K.A. Gschneidner Jr., Replacing the rare earth intellectual capital. Magn. Business and Technol. **10**(1), 6 (2011)
19. N. Krishnamurthy, C.K. Gupta, *Extractive Metallurgy of Rare Earths* (CRC Press, Boca Raton, 2016)
20. S. Pang, S. Yan, Z. Li, D. Chen, L. Xu, B. Zhao, Development on molten salt electrolytic methods and technology for preparing rare earth metals and Alloys in China. Chinese J. Rare Metals. **35**(3), 440–444 (2011). [Translated by Z.W. Xiao]

Microwave Treatment for Extraction of Rare Earth Elements from Phosphogypsum

Adrian Lambert, Jason Tam and Gisele Azimi

Abstract Many advanced technologies in modern society require the use of rare earth elements (REEs). Since these technologies are dominating the world, the demand for REEs is increasing fast. Therefore, finding new sources for them is highly of interest. One of the secondary sources for REEs is phosphogypsum (PG) that is a by-product generated by phosphoric acid production. This research builds upon our previous studies investigating the hydrometallurgical recovery of REEs from PG. Here, we investigate the effect of microwaving the PG sample before leaching in acid. Microwave radiation results in the dielectric heating of water molecules in the PG crystals and vaporization, causing the formation of breaks and pores in these particles as the vapor escapes. The lixivant would then be able to penetrate and diffuse further into the PG particles, bringing more REEs into solution. Our results show that REEs leaching efficiency increases when microwave treatment is used.

Keywords Rare earth elements · Recovery · Phosphogypsum · Microwave treatment

Introduction

Rare earth elements (REEs) are a group of 17 elements including 15 lanthanides, scandium and yttrium. REEs have unique magnetic, spectroscopic, catalytic, and hydrophobic properties. Because of that they are essential for the production of many high-tech products, such as portable electronics, aeronautical components,

---

A. Lambert · G. Azimi (✉)
Laboratory for Strategic Materials, Department of Chemical Engineering and Applied Chemistry, University of Toronto, Toronto, ON M5S 3E5, Canada
e-mail: g.azimi@utoronto.ca

J. Tam · G. Azimi
Department of Materials Science and Engineering, University of Toronto, Toronto, ON M5S 3E4, Canada

© The Minerals, Metals & Materials Society 2017

H. Kim et al. (eds.), *Rare Metal Technology 2017*,
The Minerals, Metals & Materials Series,
DOI 10.1007/978-3-319-51085-9_5

and renewable energy. Despite their name, REEs are not rare, and even the most rare ones are 200 times more abundant than gold [1]. The problem with REEs is that they do not occur in high concentrations that are suitable for economic extractions. Also, they are mainly concentrated in a few minable resources across the world (97% in China). Also, they co-occur and their similar physicochemical properties make their individual separation and purification challenging [2].

The global demand for REEs was about 136,000 tonnes in 2013 [3] and currently it is about 200,000 tonnes (47% increase) [4]. Therefore, some of REEs are expected to be in short supply in the next 15 years [5]. Currently, 95% of the world's production of REEs takes place in China [6]. The availability of REEs is undergoing a temporary decline mainly because of quotas imposed by the Chinese government on export and actions taken against illegal mining operations. The reduction in availability coupled with increasing demand has led to increased prices for REEs over the past few years. For those reasons, REEs have become a strategically important class of materials, and the importing countries, including Canada, United States, and European Union members have accelerated efforts to find other routes to satisfy their demands for REEs [7, 8].

Apatite, which is a mineral used for production of phosphate fertilizer, contains about 0.04–1.57 wt% REEs [9, 10]. It has been reported that 65–85% of REEs in apatite are precipitated with the process by-product, called phosphogypsum (PG) [10]. For every tonne of phosphoric acid produced, about 5 tonnes of PG is produced. Global estimated production of PG is about 100–280 million tonnes annually [11]. In most cases, it is wet stacked adjacent to fertilizer production facilities, spanning hundreds of hectares in area and a few meters in height, occupying vast amounts of land [12]. Although the concentration of REEs in PG is low, its volume is large; therefore it is a suitable secondary source for REEs.

In a recent study, we investigated the recovery of REEs from PG, obtained from the Agrium's fertilizer production plant in Redwater, Alberta, Canada. We performed hydrometallurgical acid leaching experiments and showed that hydrochloric acid is the most efficient leachant in this process. The optimum operating conditions were determined to be 1.5 M concentration, 80 °C, solid-to-liquid ratio of 1/15, and 20 min residence time.

In the current study, we examined the effect of microwaving the PG sample before leaching in acid. Theoretically, the dielectric heating of water molecules in the PG crystals by microwave radiation would vaporize them, causing the formation of breaks and pores in these particles as the vapor escapes. The leachant would then be able to penetrate and diffuse further into the PG particles, bringing more REEs into solution. This innovative process would provide an effective and highly efficient way for recovering REEs from PG.

Results and Discussions

Phosphogypsum Characterization

Agua Regia Digestion and Inductively-Coupled Plasma (ICP-OES). The PG sample was obtained from Agrium Inc. To determine the amount of REEs present in the PG sample, we performed aqua regia digestion (Ethos EZ Microwave Digestion System) followed by Inductively-Coupled Plasma (Agilent 720-ES ICP-OES). Table 1 presents the concentration of REEs in the sample. The total content of REEs in the feed is 342 ppm. As can be seen, Y, La, Ce, Nd, and Sm have higher concentration than the rest of the REEs, so for extraction purposes, we only focused on these 5 elements.

Microwave Treatment and Leaching Experiments

To conduct microwave treatment experiments, a multi-power microwave oven (Panasonic, NN-ST775S) was used. A known amount of PG sample was placed inside an alumina crucible, which was placed inside the microwave oven for treatment. PG samples were treated at 1200 W power for 15 min. Microwave oven was purged continuously with 0.5 L/min of nitrogen gas. After each treatment, solid samples were collected and characterized by Transmission Electron Microscopy (TEM), Scanning Transmission Electron Microscopy-Energy-Dispersive X-ray Spectroscopy (STEM-EDS) (Hitachi HF-3300), Scanning Electron Microscopy (SEM) (Hitachi SU8230), and X-ray diffraction (XRD) (Philips PW1830 diffractometer). The microwave treated samples were used as the feed for the leaching experiments, using 1.5 HCl, S/L of 1/15 at 80 °C for 60 min as the leachant. Solution samples were taken using syringes and filtered by nylon syringe filters from VWR. They were then diluted and analyzed by ICP-OES (Agilent 720-ES).

Table 1 The concentration of REEs in the PG sample

Element	ppm	Element	ppm	Element	ppm
Y	116	Sm	22	Er	8
La	58	Eu	3	Tb	1
Ce	55	Gd	10	Yb	6
Pr	6	Dy	8	Lu	1
Nd	46	Ho	2		

Transmission Electron Microscopy (TEM), Scanning Transmission Electron Microscopy-Energy-Dispersive X-ray Spectroscopy (STEM-EDS)

To study the nature and structure of the PG sample after microwave treatment, the sample was characterized using TEM and STEM/EDS. As shown is Fig. 1b, c, this analysis confirms the presence of REEs and a cluster of Ce and La is observed in the particle.

Scanning Electron Microscopy (SEM) and X-ray Diffraction (XRD) Results

SEM analysis of as-received and microwave treated PG samples was performed to observe the morphology change of the powders due to treatment. As can be seen in Fig. 2, SEM images show that after microwave treatment, cracks and pores appear on the particles. Furthermore, XRD results indicate that the crystal structure of the particles changes. Initially there is 79% calcium sulfate dihydrate ($CaSO_4.2H_2O$) present. After 15 min treatment, there is no dihydrate left, and particle consists of hemihydrate ($CaSO_4.0.5H_2O$) and anhydrite ($CaSO_4$) phases that have less water molecules.

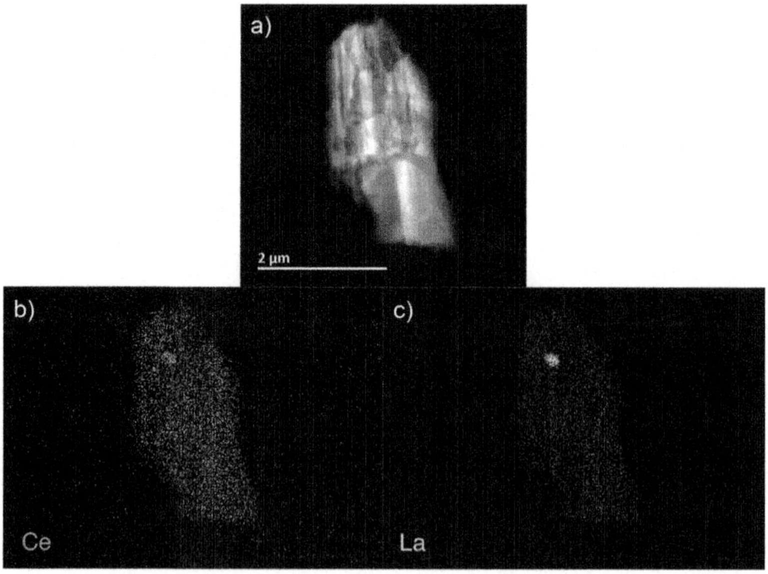

Fig. 1 **a** TEM image of microwaved treated PG. **b** STEM/EDS elemental map of Ce. **c** STEM/EDS elemental map of La

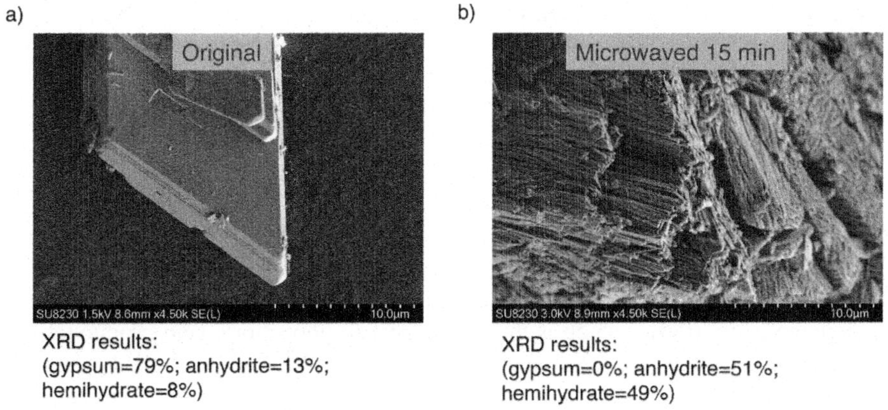

Fig. 2 **a** SEM image and XRD results of original PG with no treatment. **b** SEM image and XRD results of microwaved PG

Leaching Results

Leaching experiments were performed using both original PG with no treatment and microwaved PG. As can be seen in Fig. 3, microwave treatment increases the average leaching efficiency of Ce, La, Nd, Y, and Sm by 30%. The reason behind this result is that microwave radiation removes water molecules from the PG crystal

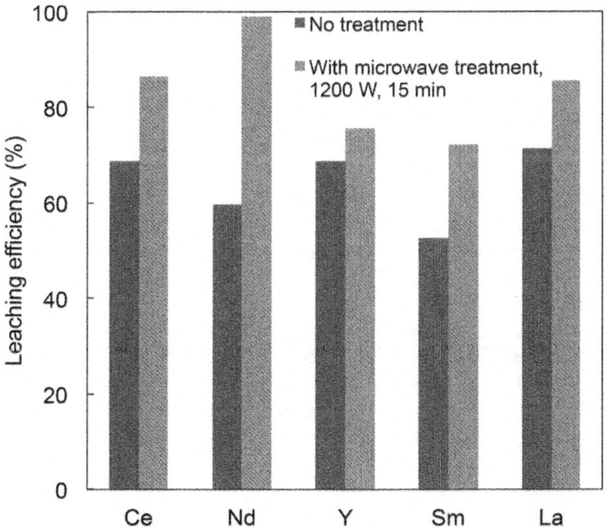

Fig. 3 Comparison of leaching efficiency of Ce, La, Nd, Sm, and Y from original PG with no treatment and microwave treated sample. Leachant: 1.5 M HCl, 80 °C, S/L = 1/15

(confirmed by XRD analysis, shown in Fig. 2); therefore it creates breaks and pores in these particles (shown in SEM images of Fig. 2). Therefore acid molecules can diffuse into the PG particles, extracting more REEs into solution.

Conclusions

In this research, we investigated the effect of microwave treatment on the hydrometallurgical recovery of REEs (mainly Ce, La, Nd, Sm, and Y) from phosphogypsum (PG) , which is the by-product of phosphoric acid production that is used for fertilizer applications. We obtained the PG sample from Agrium Inc. located in Alberta, Canada. Leaching experiments were performed using 1.5 M HCl at 80 °C with a solid-to-liquid ratio of 1/15. The results show that microwave treatment removes water molecules from PG crystal structure; creating cracks and pores on the particles. As a result, leachant molecules can diffuse into the particles and bring more REEs into the solution. Microwave treatment increases the average leaching efficiency of REEs by 30%. This innovative process results in a significant improvement of the recovery of REEs from PG and can potentially be implemented at large scale.

Acknowledgements The authors would like to acknowledge the financial support provided by Agrium Inc. and Natural Sciences and Engineering Research Council of Canada (NSERC) for this project. Also, Agrium Inc. is gratefully acknowledged for providing us with phosphogypsum samples and technical support.

References

1. N.E. Topp, The chemistry of the rare earth elements, in *Separating rare earth elements*, ed. by D.J. Fray (New York, NY: American Elsevier Publishing Company, 1965). Science **289**, 2295–2296 (2000)
2. M. Humphries, Rare earth elements: the global supply chain, in Congressional Research Service Report, R41347 (2013), p. 3
3. B. Mishra, A. Anderson, Extraction and recovery of rare-earth metals: challenges in processing. Paper presented in: ERES 2014, 1st European rare earth resources conference, Greece (2014), pp. 19–25
4. S. Al-Thyabat, P. Zhang, In-line extraction of REE from Dihydrate (DH) and HemiDihydrate (HDH) wet processes. Hydrometallurgy **153**, 30–37 (2015)
5. X. Wang, Y. Lei, J. Ge, S. Wu, Production forecast of China's rare earths based on the Generalized Weng model and policy recommendations. Resour. Policy **43**, 11–18 (2015)
6. U.S. Department of Energy, *Critical Materials Strategy* (2011), pp. 113–120
7. Report on critical raw materials for the EU. Report of the Working Group on defining critical raw materials (2014)
8. B.I. Pålsson, O. Martinsson, C. Wanhainen, A. Fredriksson, Unlocking rare earth elements from European apatite-iron ores. Paper presented in: ERES2014, 1st European rare earth resources conference, Milos Island, Greece (2014), pp. 211–220

9. F. Habashi, The recovery of the lanthanides from phosphate rock. J. Chem. Tech. Biothechnol. **35A**, 5–14 (1985)
10. H. Tayibi, M. Choura, F.A. López, F.J. Alguacil, A. López-Delado, Environmental impact and management of phosphogypsum. J. Environ. Manage. **90**, 2377–2386 (2009)
11. S. Enamorado et al., Transfer of Cd, Pb, Ra and U from phosphogypsum amended soils to tomato plants. Water Air Soil Pollut. **203**, 65–77 (2009)
12. M. Walawalkar, C. Nichol, G. Azimi, Process investigation of the acid leaching of rare earth elements from phosphogypsum using HCl, HNO_3, and H_2SO_4. Hydrometallurgy. **166**, 195–204 (2016). doi:10.1016/j.hydromet.2016.06.008

Selective Separation of Rare Earth Chlorides Utilizing Vapor Phase Extraction

Katelyn M. Lyons, Jerome P. Downey, Jannette L. Chorney and Katie J. Schumacher

Abstract The Metallurgical and Materials Engineering Department at Montana Tech is investigating a new method of extracting and refining rare earth elements (REEs) from mineral ores and concentrates. The relative stabilities of various REE compounds at elevated temperatures were evaluated using thermogravimetric and differential thermal analyses (TGA/DTA). The results, in combination with thermodynamic analyses, revealed that vapor phase extraction and selective condensation is a potentially viable separation method for rare earth halides. Selective vaporization and condensation experiments were performed on selected rare earth chlorides. A series of close-coupled tube furnaces provided a temperature gradient ranging from 1150 to 400 °C. Within the condensation regions, a series of one-inch-diameter ceramic tube sections were packed with stainless steel (316L) wool to create high surface area for condensate collection. The ceramic tube sections and stainless steel wool were leached in 18 MΩ water. Analysis of the leachate samples revealed that selective separation had occurred but oxychlorides were detected in the non-volatile matter.

Keywords Rare earth elements · Vapor phase extraction · Thermodynamics

Introduction

In the past four decades, technological and environmental applications of rare earth elements have grown significantly in diversity and importance. REEs are critical for a number of key defense systems and advanced materials as well as for emerging technology in cellphones, alternative energy, and medical devices such as MRI's and PET scanners, and batteries for hybrid vehicles. Currently, the worldwide production of rare earth oxides is dominated by China, which may cause the

K.M. Lyons (✉) · J.P. Downey · J.L. Chorney · K.J. Schumacher
Montana Tech of the University of Montana, 1300 West Park Street, Butte, MT 59701, USA
e-mail: klyons1@mtech.edu

© The Minerals, Metals & Materials Society 2017

H. Kim et al. (eds.), *Rare Metal Technology 2017*, The Minerals, Metals & Materials Series,
DOI 10.1007/978-3-319-51085-9_6

United States to face supply uncertainty in the future and lose its longstanding leadership in many areas of rare earth technology. Due to continued advances in material development, the demand for REE has increased and is expected to continue to increase [1].

The term "rare" is a historic misnomer for rare earth elements because some rare earth oxides or carbonates are similar in crustal abundance to that that of copper, zinc, and other common industrial metals. The difference is that REEs have little tendency to become concentrated in ore deposits that are economically viable to mine. Also, due to their similar physical and chemical properties, REEs are difficult to separate from one another. Current separation methods are hydrometallurgically intensive and produce large amounts of waste, including contaminated wastewater. Rare earth resources could greatly benefit from a process that could effectively separate the REEs while lowering waste generation [1].

The objective of this research is to evaluate an alternative method for separating and recovering rare earth chlorides (RECl). Recent research demonstrated that rare earth oxides can be successfully converted to rare earth chlorides using ammonium chloride (NH_4Cl) as the chloritizing agent [2, 3]. The conversion proceeds according to the general net chemical reaction shown in Eq. (1).

$$RE_2O_{3(s)} + 6NH_4Cl_{(s)} \rightarrow 2RECl_{3(s)} + 6NH_{3(g)} + H_2O_{(g)} \quad (1)$$

Rare earth chlorides are extremely soluble in water, which makes them easy to recover by hydrometallurgical processes. A disadvantage of hydrometallurgical processing is the difficulty in separating individual rare earth elements, often requiring multiple solvent extraction stages. By exploiting differences in rare earth chloride vapor pressure, it should be possible to selectively vaporize and condense certain rare earth chlorides at their characteristic temperatures.

Experimental

The individual vaporization temperatures for each RECl were determined by thermodynamic modeling and thermogravimetric analysis/differential thermal analysis (TGA/DTA). The TGA/DTA plots for dysprosium chloride and europium chloride are illustrated in Figs. 1 and 2.

HSC Chemistry 7.0 (Outotec) [4] software was used to model multiple reactions with the same set of species within the system. HSC uses a free energy minimization algorithm that compares multiple chemical reactions and their free energies to predict stable species at equilibrium. Figures 3 and 4 show the predicted stable phases generated by HSC.

Utilizing the information in Figs. 3 and 4, the vaporization temperature for dysprosium chloride ($DyCl_3$) and europium chloride ($EuCl_3$) are determined to be 880 and 1870 °C, respectively. The objective of this research is to selectively

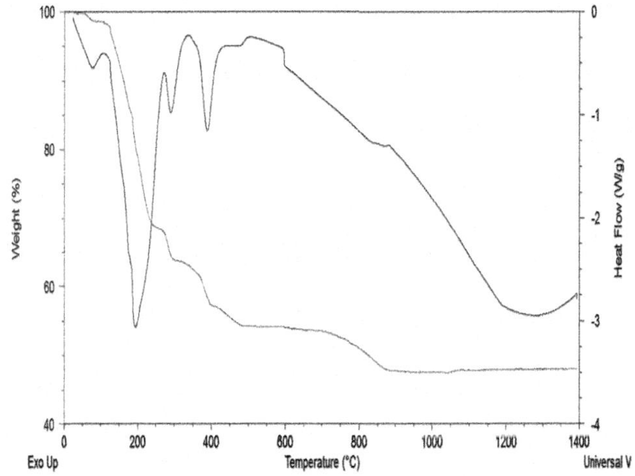

Fig. 1 TGA/DTA for dysprosium chloride

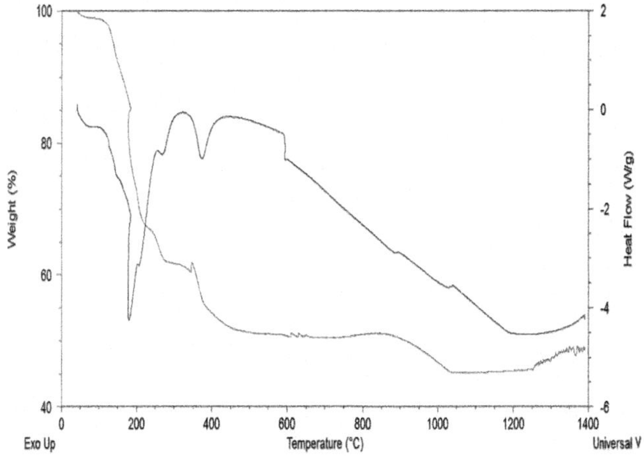

Fig. 2 TGA/DTA for europium chloride

isolate rare earth chlorides through differences in their vapor pressures. If the rare earth chlorides can be elevated to a high enough temperature to convert them to a vapor phase, they could be selectively collected as condensate upon cooling. When the equilibrium vapor pressure exceeds the partial pressure, a gas is formed. StabCal [5] was used to evaluate thermodynamic data generated in HSC and compare vaporization temperature to the partial pressure of the RECl (Fig. 5).

Based on thermodynamic analysis, it was determined that europium chloride and dysprosium chloride would be the initial focus of separation because they have the greatest difference in partial pressure and vaporization temperature.

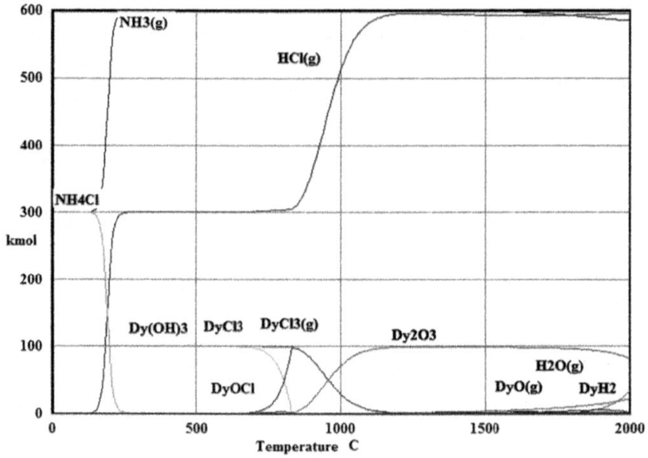

Fig. 3 Thermodynamic model for dysprosium species generated using HSC Chemistry 7.0

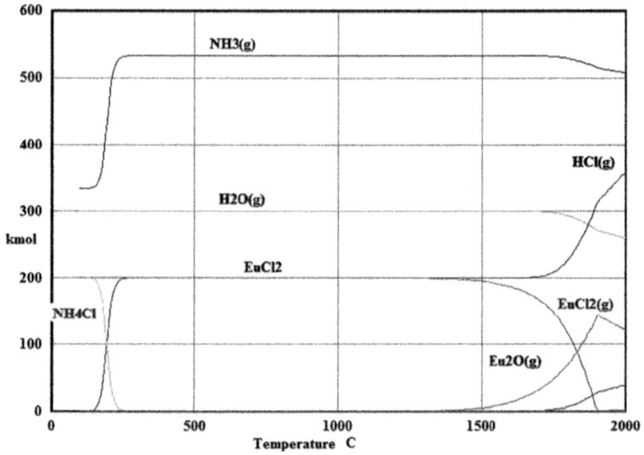

Fig. 4 Thermodynamic model for europium species generated using HSC Chemistry 7.0

A 3-zone MTI OTF-1500X tube furnace was used for proof of concept experimentation. Experiments were run with dysprosium chloride and europium chloride under an inert argon atmosphere varying time. The heating profile of the furnace was set to peak at 1150 °C and gradually decline along the length of the tube. From the thermodynamic data analyzed, this temperature profile would allow for the dysprosium chloride to vaporize, transport with the argon carrier gas, and condense in the cooler regions of the furnace. Europium chloride, which has a very high vaporization temperature, would remain in the boat. Within the condensation regions, a series of one-inch-diameter stainless steel washers were spaced one inch

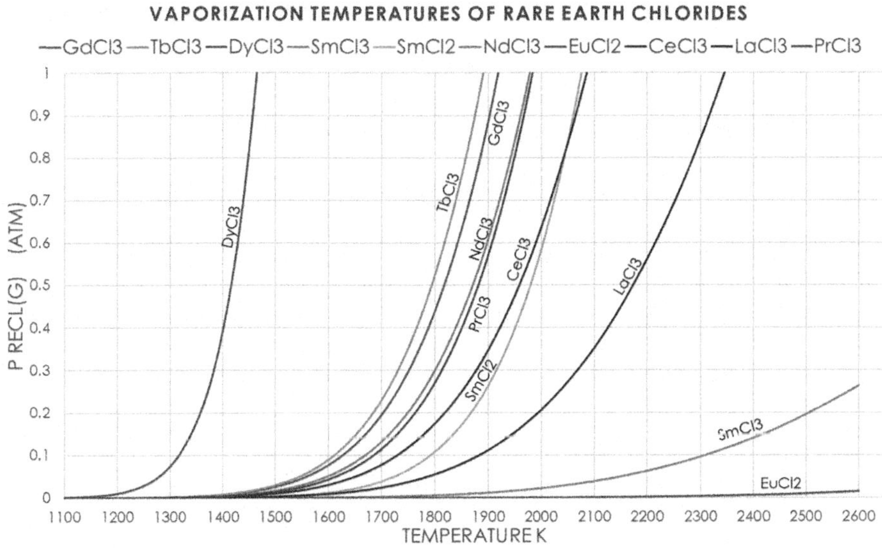

Fig. 5 Vaporization curve for select RECls

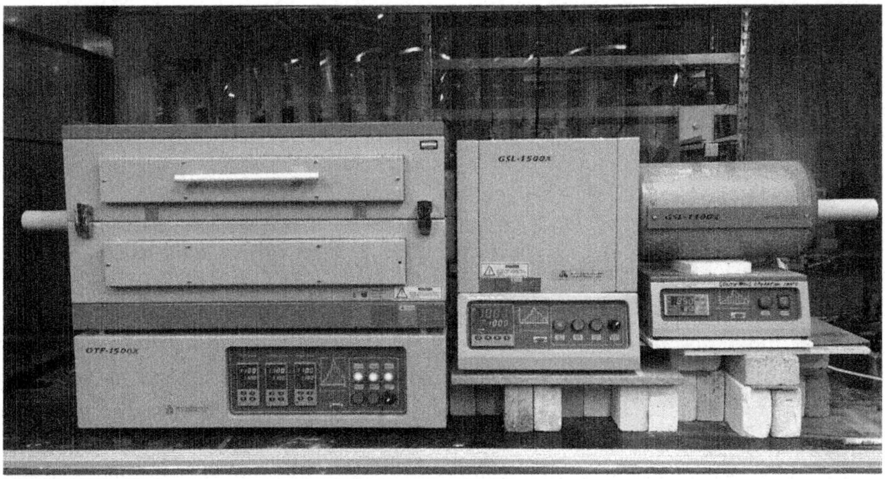

Fig. 6 MTI furnace train consisting of MTI OTF-1500X, OSL-1500X and GSL-1000X tube furnaces

apart from each other to create high surface area for condensate collection. The stainless steel washers were leached in 18 MΩ water for two hours.

Similar experiments were conducted in a furnace train comprised of three MTI furnaces: OTF-1500X, OSL-1500X and GSL-1000X, shown in Fig. 6. The furnace train allowed for larger vaporization and condensation regions. Within the

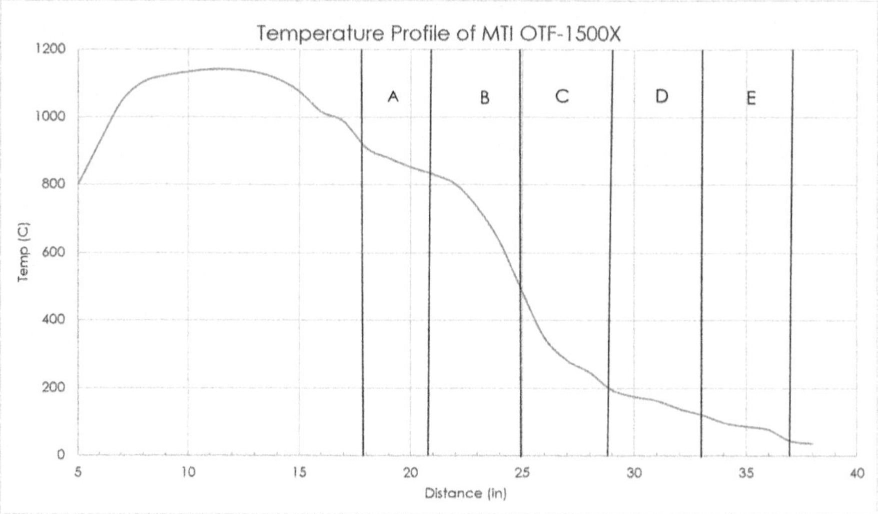

Fig. 7 Temperature profile of MTI OTF-1500X tube furnace used for experimentation

condensation regions, a series of one-inch-diameter ceramic tube sections were packed with stainless steel (316L) wool to create high surface area for condensate collection. Following each experiment, the ceramic tube sections and stainless steel wool were leached in 18 MΩ water for two hours. Analysis of the leachate samples revealed that selective separation had occurred but oxychlorides were detected in the non-volatile matter.

Results and Discussion

Figure 7 illustrates the temperature gradient of the MTI OTF-1500X three zone tube furnace. Letters *A–E* represent approximately three to five inch sections in the furnace where the condensate was recovered. Figures 8 and 9 show the results of the rare earth chloride recovered from the proof of concept experiments. *Dy/Eu* represent the leachate of the non-volatile material remaining in the boat. The data are summarized in Table 1.

By analyzing the data in Table 1, it can be determined that, for experiments less than 8 h, dysprosium is at least ten times the concentration of europium at a temperature less than 800 °C. According to thermodynamic analysis previously discussed, dysprosium chloride should condense anywhere below a temperature of 880 °C. In the two-hour experiment, dysprosium concentration was 1600 times that of europium in the coldest region in the furnace, verifying potential for separation. The low selectivity ratios for the eight-hour experiment are likely due to the limited

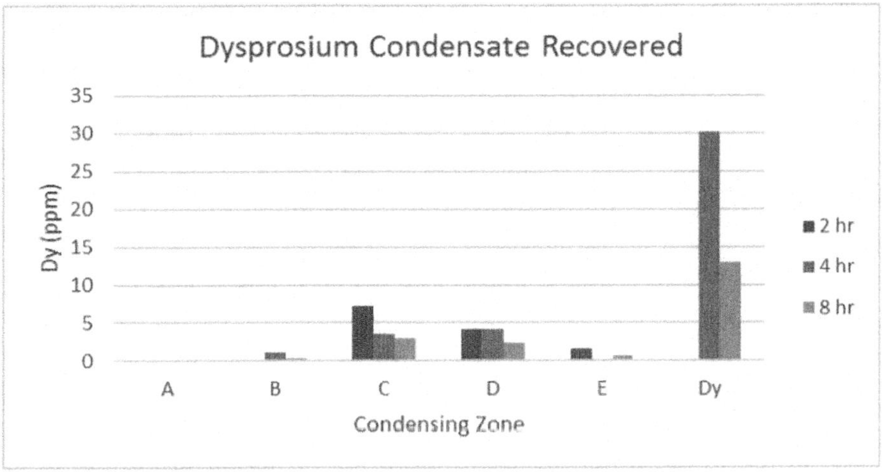

Fig. 8 Plot of dysprosium chloride condensate recovered from leach sections. *A–E* represent distance in furnace and *Dy* represent non-volatile matter remaining in boat

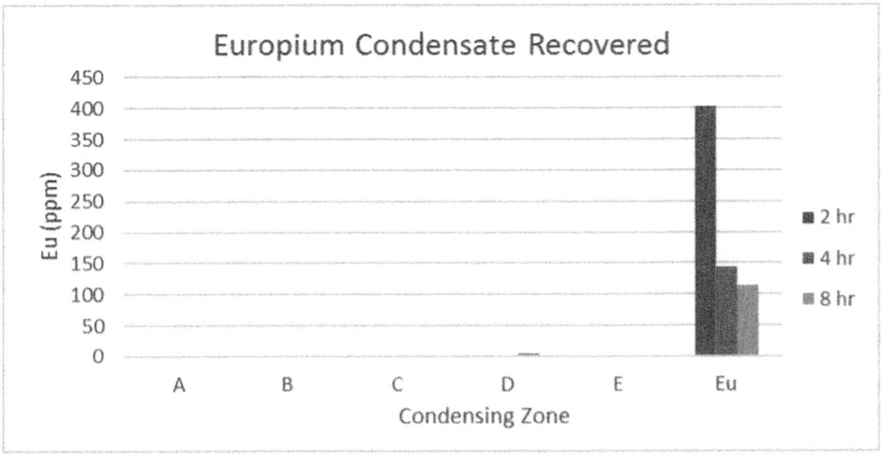

Fig. 9 Plot of europium chloride condensate recovered from leach sections. *A–E* represent distance in furnace and *Eu* represent non-volatile matter remaining in boat

number of experiments and potential for europium chloride to become entrained in the carrier gas or attached to the waters of hydration.

The images in Figs. 10 and 11 show the dysprosium chloride and europium chloride boats preceding and following vaporization, respectively.

X-ray diffraction (XRD) analysis of the non-volatile matter remaining in the boat indicated the presence of rare earth oxychloride (REOCl). The remaining mass loss is likely waters of hydration in the rare earth chloride. Reagent grade rare earth chlorides are extremely hydrated so it is difficult to account for water loss.

Table 1 Selectivity ratios of recovered condensate according to temperature

Sample	Temperature range (°C)	2 h	4 h	8 h
		Selectivity ratio (Dy:Eu)	Selectivity ratio (Dy:Eu)	Selectivity ratio (Dy:Eu)
A	910–830	0.0	0.0	0.0
B	800–480	0.0	35.5	2.8
C	350–200	47.6	43.7	2.1
D	140–100	11.5	36.2	0.5
E	<100	1590.0	–	0.6
Dy:Eu	–	0.0	0.2	0.1

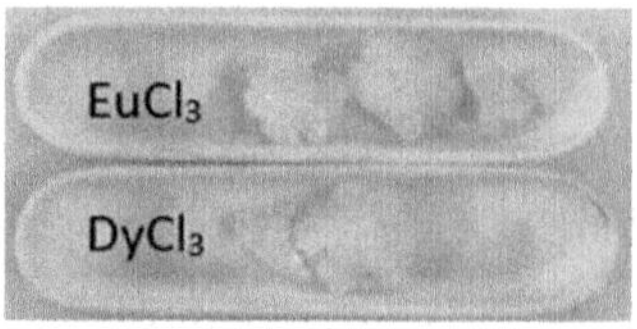

Fig. 10 Raw material placed in boat prior to vaporization

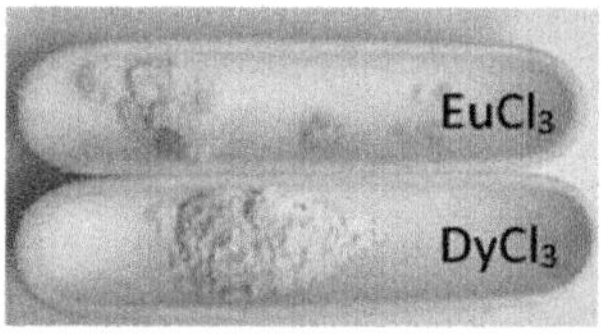

Fig. 11 Non-volatile matter remaining in boat post vaporization

Conclusions

Vapor phase extraction is a thermodynamically viable means of selectively separating rare earth chlorides. Leach solution analysis revealed that dysprosium is up to 1590 times more likely than europium to be recovered as condensate. The europium detected downstream is expected to be a result of a small amount of europium chloride powder becoming entrained in the argon purge gas or becoming attached to the waters of hydration. Despite the controlled atmosphere, hydration has been a chronic problem due to the hygroscopic nature or the RECls. Thermodynamic modeling has been further assessed and experimentation is currently underway to evaluate ways to prevent oxychloride formation. Future experiments will also examine the degree of selectivity that can be obtained with other combinations of RECls as well as with a rare earth bearing ore.

Acknowledgements Research was sponsored by the Army Research Laboratory and was accomplished under Cooperative Agreement Number **W911NF-15-2-0020**. The views and conclusions contained in this document are those of the authors and should not be interpreted as representing the official policies, either expressed or implied, of the Army Research Laboratory or the U.S. Government. The U.S. Government is authorized to reproduce and distribute reprints for Government purposes notwithstanding any copyright notation herein.

References

1. U.S. Department of the Interior and U.S. Geological Survey, "Rare Earth Elements—Critical Resources for High Technology" 17 May 2005. [Online]. Available: http://pubs.usgs.gov/fs/2002/fs087-02/. Accessed 6 September 2016
2. D.W. Gaede, Chlorination and selective vaporization of rare earth elements, M.S. Thesis, Montana Tech, Butte, MT, 2016
3. D.W. Gaede, B.D. Ruffier, J.P. Downey, L.G. Twidwell, J.L. Chorney, R.J. Foy, K.M. Lyons, Chlorination of rare earth element oxides, Drying, Roasting, and Calcining of Minerals, TMS, 11–18 (2015)
4. Outotec Research, HSC 7.0, Outotec Research, Antti Roine, 2014
5. H. Huang, W32-STABCAL, Butte, 2016

Microstructure Observation of Oxidation of Nd-Magnet at High Temperatures

Muhamad Firdaus, M. Akbar Rhamdhani, Yvonne Durandet, W. John Rankin, Kathie Mcgregor and Nathan A.S. Webster

Abstract There is a growing interest in recycling/recovery of rare earth elements from permanent magnets. A number of processing techniques are currently being developed but highly sensitive to the oxidation state of rare earth in the magnetic waste. This study investigated the microstructural changes of thermal oxidation of an Nd-based magnet and the behaviour of its oxides under high temperature recycling/recovery process. XRD analyses were carried out on a powdered sample (~ 10 μm) heated to 1273 K. SEM-EDS analysis was conducted on the heated bulk samples to provide detailed metallographic information. Metallographic analysis revealed multiple oxidation zones where the outer scale did not effectively inhibit further diffusion of oxygen. The thickness of this scale was found to be grown quite rapidly at temperatures higher than 973 K. The results indicated that the micro-mechanism of oxidation at higher temperature are more complex than at temperature below 773 K.

Keywords NdFeB magnet · Microstructure · Oxidation

Introduction

A number of processing techniques and strategies are currently being developed to recycle or recover the rare earth elements (REE) such as Nd, Dy and Pr from rare earth permanent magnets (REPM). Different processing routes such as hydrometallurgical, pyrometallurgical, physical/mechanical separation can be applied depending on the stage at which the recycled or end of life (EoL) products come into the material flow [1, 2]. All of these processing routes require significant

M. Firdaus (✉) · M. Akbar Rhamdhani · Y. Durandet
Department of Mechanical and Product Design, Swinburne University of Technology, Melbourne, VIC 3122, Australia
e-mail: mfirdaus@swin.edu.au

M. Firdaus · W. John Rankin · K. Mcgregor · N.A.S. Webster
CSIRO Mineral Resources, Private Bag 10, Clayton South, VIC 3169, Australia

© The Minerals, Metals & Materials Society 2017
H. Kim et al. (eds.), *Rare Metal Technology 2017*,
The Minerals, Metals & Materials Series,
DOI 10.1007/978-3-319-51085-9_7

understanding of the oxidation process of REPM at high temperature including its melting state in order for processes to be viable. There is currently a lack of information on the formation mechanisms and behaviour of the rare earth oxides in the waste REPM at high temperature. Oxidation observations were often done to support kinetics of reaction for specific techniques (e.g. in hydrometallurgy for leaching kinetics) without any attempt to understand the mechanisms. In addition, most investigations relied on the information from low temperature systems.

The corrosion and oxidation behaviour of REPM have been extensively studied particularly the degradation of magnetic properties at temperatures up to 873 K. The first report by Blank and Adler showed that oxidation between 673 and 873 K resulted in the formation of a grey layer on the surface of the bulk magnets and that its thickness increased parabolically with oxidation time [3]. The result showed that rapid surface oxidation with formation of a thin powdered layer on uncoated REPM takes place within the first minute and this can be substantial as the average particle size decreases [4–6]. This surface oxide layer does not effectively inhibit further diffusion of oxygen. Later work on the oxide microstructure of bulk NdFeB magnets was carried out by Breton and Edgley et al. [7, 8] mainly using SEM, XRD, and conversion electron mössbauer spectroscopy (CEMS) to identify the oxidation products. It was indicated in the report that the main reaction was the dissociation of the $Nd_2Fe_{14}B$ phase into α-Fe nanocrystals, Fe_2B and small Nd-oxide particles as follows:

$$2Nd(s) + 3/2O_2(g) \rightarrow Nd_2O_3(s) \quad < 500\ K \tag{1}$$

$$Nd_2Fe_{14}B(s) + 3/2O_2 \rightarrow 14Fe(s) + B(s) + Nd_2O_3(s) \quad < 500\ K \tag{2}$$

$$2Fe(s) + 3/2O_2(g) \rightarrow Fe_2O_3(s) \quad < 750\ K \tag{3}$$

$$Fe_2O_3(s) + Nd_2O_3(s) \rightarrow 2FeNdO_3(s) \quad > 750\ K \tag{4}$$

Although the oxidation products were too fine, it was established later on using SEM and transmission electron microscopy (TEM) that the dominant grey zone formed on these alloys is, in fact, a zone of internal oxidation (IOZ) consisting principally of an α-Fe matrix containing Nd-oxide particles [3, 9–11]. Li et al. [3], Skulj et al. [12] and others [13–17] further studied the microstructures of the material to help understand the mechanism of oxidation. Based on their reports there was no degradation of the $Nd_2Fe_{14}B$ phase in the grains found using CEMS, and analysis by TEM confirmed that dissociation of the grain is unlikely and happens only when it reacts with oxygen [5].

The principal objective of this present study is to observe and evaluate the change in the microstructure of a commercial NdFeB alloy during oxidation in air at temperatures in the range 773–1273 K by using Scanning Electron Microscope (SEM) and X-ray diffraction (XRD). This will provide information that is useful for understanding the oxidation mechanism of waste NdFeB magnets at high temperature, and in return to improve the strategy and techniques in recovering the REE or recycling the REPM.

Experimental Methodology

Materials

Sintered N45-NdFeB magnets used in the current study were supplied by Alpha Magnetics Ltd. and had a nominal composition within the range given in Table 1 with platted nickel for coating. The clean magnet was thermally demagnetised at 573 K in a Nabertherm TR 60 oven furnace and samples with dimension of $10 \times 5 \times 5$ mm were produced by cutting the magnets using a Struers Secotom-15/-50 high performance cut-off machine. The nickel coating was removed manually after cutting using abrasive papers. The bulk composition of the demagnetised magnet sample, derived from inductively coupled plasma (ICP-AES) analysis is shown in Table 1. The typical cross sectional microstructure of the initial sample is shown in Fig. 1a, b. The back-scattered electron (BSE) micrograph of the unreacted sample shown in Fig. 1b indicates the two major phases typical of a Nd–Fe–B magnet system; the ϕ phase ($Nd_2Fe_{14}B$) matrix, indicated by (A) and Nd-rich phase at the grain boundary, indicated by (B). A small amount of η phase ($Nd_1Fe_4B_4$) is also present (indicated by C) as can be seen in Fig. 1a.

Approach and Parameters

For the high temperature experiments the surfaces of the samples were ground using a 1200-grade SiC paper and cleaned ultrasonically in acetone prior to oxidation. The samples were placed in a 4-cm diameter alumina crucible and oxidation was carried out in an air atmosphere in a Nabertherm LT 15/13/P330 muffle furnace. Four crucibles were placed in the hot zone of the furnace after the temperature of the furnace reach the set point and different samples were taken out after different set of hours. The samples were cooled in ambient air and then placed inside 1 ml LLG shell vials before conducting metallographic analysis. It is reasonable to assume that further oxidation do not occur under ambient conditions (e.g. at room temperature) due to kinetic limitation.

Table 1 Chemical composition of REPM sample (mass%)

	Nd	Pr/Dy	Fe	B
Typical composition commercial magnet	23–31	0–7	65–70	0.9–1.2
Bulk composition of sample used in the study	22.4	8.5	68	1

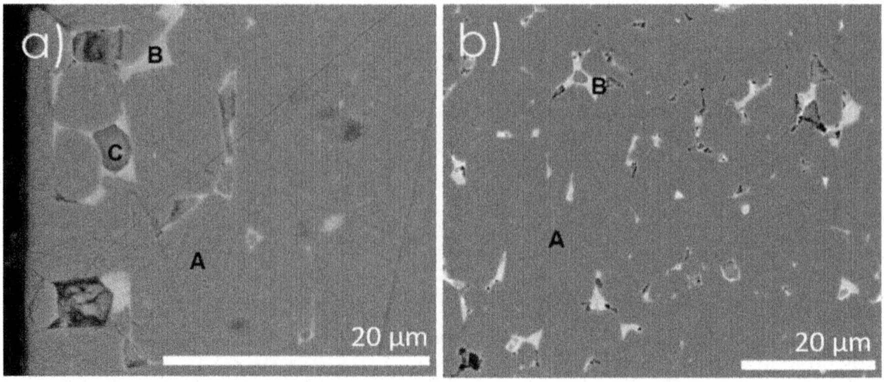

Fig. 1 Backscattered electron (BSE) images from the un-oxidised (Nd, Pr)2Fe14B starting material: **a** a region close to the sample surface, and **b** the central region of the sample. The dark region on **a** is the epoxy resin mounting medium

Post Experiment Sample Preparation

The samples pre and post-oxidation were analysed using different characterization techniques which included SEM, energy dispersive X-ray spectroscopy (EDX), XRD, and ICP. For the SEM and EDX analysis, a FEI Quanta 400F Environmental Scanning Electron Microscope (ESEM) was used. The samples preparation for the SEM analysis include mounting of the samples in an epoxy resin in 2.5-cm round blocks before being cured at 60 °C overnight, sectioned to expose a fresh surface, and then polished flat using successively finer diamond paste compounds down to a final polishing size of 1 μm. Immediately prior to analysis, each sample was coated with a ~5–10 nm thick carbon film to prevent charge build-up on the surface of the sample when probed by the electron beam. The SEM analysis was carried out using an accelerating voltage of 15 kV, a working distance of 10 mm, an emission current of 276 mA, and a vacuum of 2.8×10^{-6} Torr.

The samples were micronized in ethanol for 4 min g^{-1} and dried prior to XRD analysis. The XRD data were collected at 2θ angles from 5 to 140°, using a PANalytical MPD instrument fitted with a cobalt long-fine-focus X-ray tube operated at 40 kV and 40 mA. The incident beam path was defined using 0.04 rad Soller slits, a 20 mm mask, a 0.5° fixed divergence slit, and a 1° anti scatter slit. The diffracted beam incorporated a second set of Soller slits, a graphite monochromator to eliminate unwanted wavelengths and a 4.6 mm anti-scatter slit. An X'Celerator detector was used in scanning line (1D) mode with an active length of 2.122° 2θ. Approximate phase concentrations were calculated from the XRD data via the Rietveld method.

Results and Discussion

The XRD data for samples prior and after ~24 h oxidation at 973 and 1173 K are presented in Fig. 2. The approximate phase concentrations calculated from the XRD data using Rietveld method is presented in Table 2. The data for the unheated (un-oxidised) sample contain peaks indicative of $Nd_2Fe_{14}B$, as well as minor peaks (marked with * in Fig. 2) for which phase assignment could not be conclusively made. The XRD spectra for the 973 and 1173 K oxidised samples contain peaks for α-Fe, iron oxides and $NdFeO_3$, suggesting that the reaction from Eqs. (1) to (4) proceed and for the 1173 K sample also contains peaks indicative of $NdBO_3$. There was no evidence for the formation of pure Nd_2O_3 during oxidation but rather the formation of $FeNdO_3$ which suggests that Nd_2O_3 formed a solid solution with Fe_2O_3 in accordance with Eq. (4). However, the results show an increase of Fe_2O_3

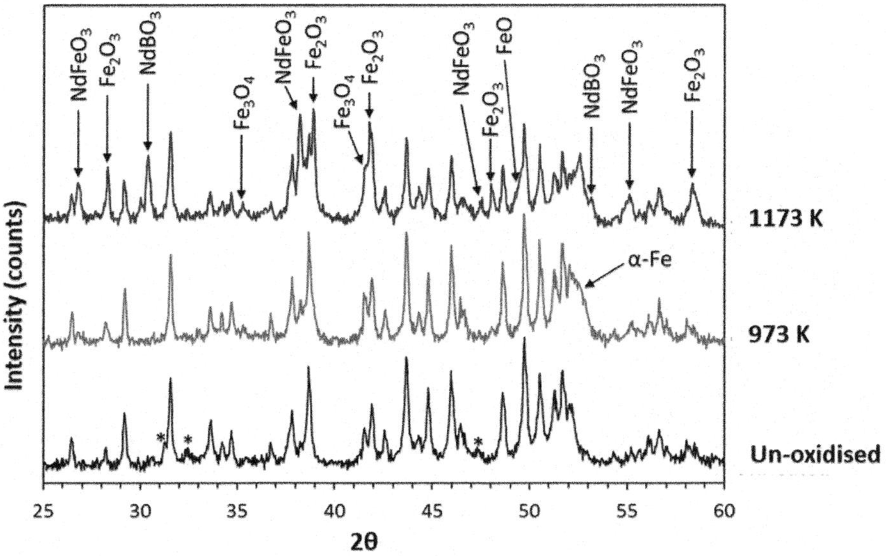

Fig. 2 XRD data collected for the unheated (un-oxidised) sample, and the samples oxidised at 973 and 1173 K. Datasets offset in the intensity axis for clarity. Significant peaks for the oxidisation phases are labelled in the 973 and 1173 K plots; in the un-oxidised plot all of the peaks are assigned to $Nd_2Fe_{14}B$ except the minor peaks annotated with *

Table 2 Summary of phase quantification results from the XRD data

Samples	Phase (wt%)						
	$Nd_2Fe_{14}B$	Fe_2O_3	Fe_3O_4	FeO	Fe	$NdFeO_3$	$NdBO_3$
Original	~100						
973 K	74	6	4		13	3	
1173 K	44	22	7	3	8	10	5

and Fe_3O_4 during oxidation, and the 1173 K sample also appeared to contain FeO that may indicate that the formation of $FeNdO_3$ (Eq. 4) may not be instantaneous or there were not enough Nd_2O_3 in the oxidised zone.

The typical morphology of the oxidation product at 773 K is shown in Fig. 3. The oxidation of Nd–Fe–B magnets at 773 K appeared to follow a typical gas–solid reaction. The oxidation reactions at this temperature resulted in the formation of an external scale (external oxidation zone—EOZ) and a grey dense internally oxidised zone (IOZ), which is an order of magnitude thicker than the surface oxide layers and represents the main form of degradation resulting from the oxidation reactions. The depth of external oxidation zone (EOZ) was found to be constant with increasing time and not effectively inhibit further diffusion as seen in Fig. 4. This observation is consistent with earlier investigations carried out by Li et al. [3], Skulj et al. [12] Edgley et al. [7, 8] and others [13–17].

The oxidation process at higher temperature above 973 K resulted in a more complex morphology as seen in Fig. 5, which shows a cross section of a sample oxidised for 60 h at 1173 K from its edge to the centre. On the surface of the sample, a darker layer can be seen. This dark surface layer or the EOZ was more

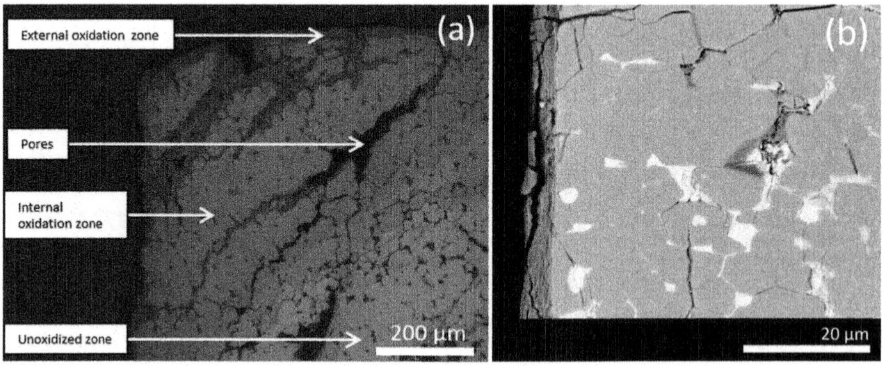

Fig. 3 Light microscope (**a**) and SEM backscattered electron image (**b**) showing typical cross-section microstructure of the oxidised layers of the NdFeB magnets heated at 773 K for 72 h

Fig. 4 Cross-section macrostructure showing the effect of oxidation time at 773 K, **a** t = 23 h, **b** t = 45 h, and **c** t = 72 h

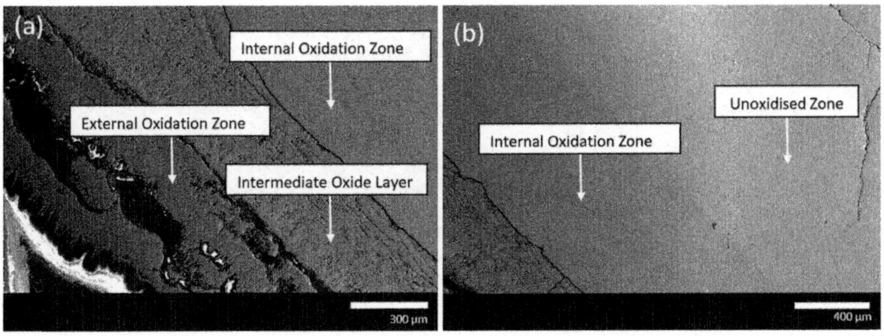

Fig. 5 Typical cross-section of the oxidised layers of the NdFeB magnets heated at 1173 K for 60 h showing **a** region close to the surface and **b** the central region of the sample

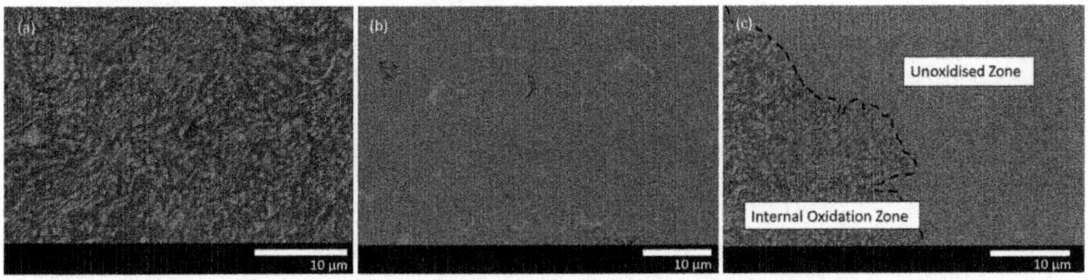

Fig. 6 SEM BSE images of microstructures of samples heated at 1173 K for 60 h **a** IOZ, **b** unreacted zone, and **c** transition between IOZ and unreacted zone (*dash line* for guide)

prominent compared to those observed at 773 K. This layer has grown outward considerably with time, unlike the EOZ observed in the oxidation at 773 K, and an expansion has taken place within this layer. Moving away from the edge the next layer is the second oxide layer (intermediate layer), which was not seen in the oxidation at 773 K, then followed by the grey IOZ. Unlike the microstructure at 773 K, the IOZ was completely different than the unreacted zone as seen in Fig. 6. The Nd-rich phase in the IOZ which was seen in the sample heated at 773 K (Fig. 3) appears to be reacted and dissolved into the IOZ in the sample heated at 1173 K. It should also be noted from the SEM topography that whilst the IOZ is fully dense, the EOZ and intermediate layer are porous with significant sign of presumably cracking due to density change upon cooling. An investigation using a high temperature microscope (result not shown here) supported the notion that this friability represents a mechanical damage induced during cooling.

Previous investigations at lower temperature indicate that there is no significant diffusion of Nd during oxidation [3, 4, 7, 8, 12]. Based on the investigation of the IOZ at temperature range of 600–773 K assuming that there is no access restriction of oxygen by the oxidised surface Li et al. [3] deduced that the permeability of oxygen in the oxidized zone is much larger than that of Nd. As such there will be minimum movement of Nd during the oxidation process unless microstructural

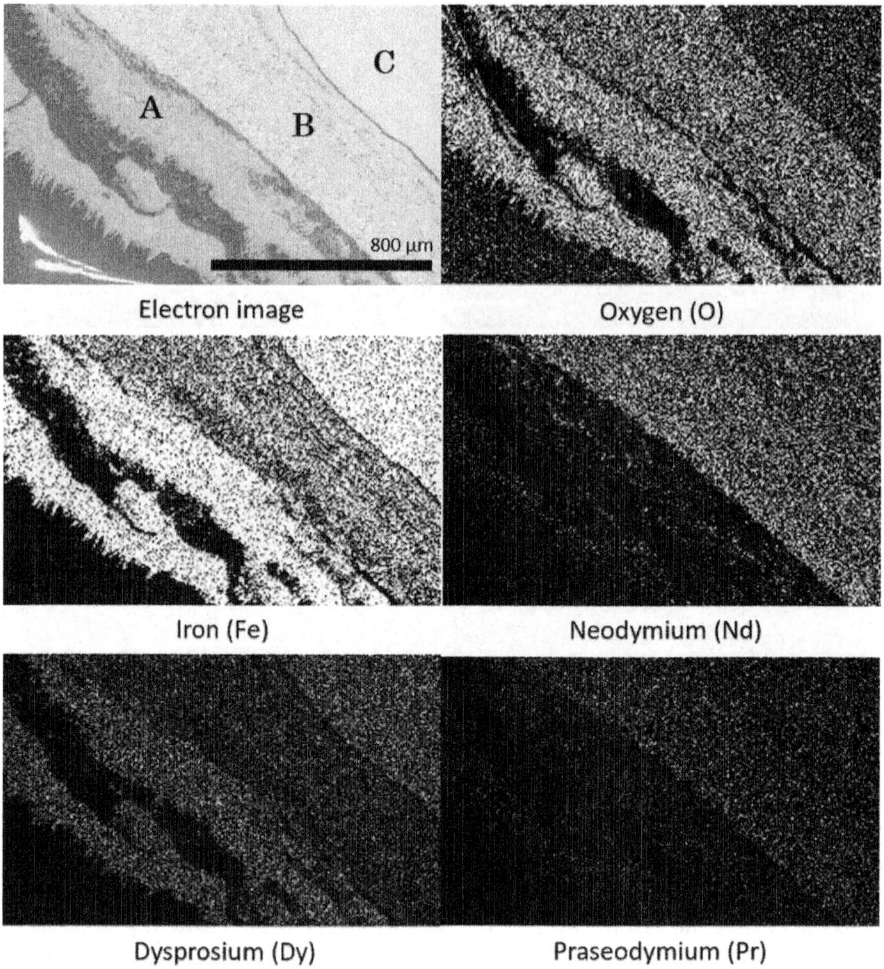

Fig. 7 EDS elemental mapping of the EOZ (**a**), intermediate layer (**b**) and IOZ (**c**) of sample heated at 1173 K, 60 h, brighter image represent high concentration

change occur. It is then expected that the effective Nd concentration remains essentially constant both within the oxidation zone and the un-oxidised substrate. This suggests that only oxygen or iron diffusing in or out of the core substance. It is believed that at temperature below 973 K the oxygen transport occurs by short-circuit diffusion and the most likely diffusion paths are the high-angle α-Fe grain boundaries. The α-Fe columnar grain structure coarsens as the oxidation front traverses an original alloy grain. This coarsening has been associated with the loss of relatively low-angle grain boundaries which carry less of the inward oxygen flux [12]. Furthermore, the previous studies on the oxidation at temperature below

973 K indicate that the ratio of iron to neodymium remains approximately constant, and that there is no evidence of outward diffusion of iron [3, 4, 7, 8, 12].

The EDS elemental mapping of the oxidation product layer depicted in Fig. 7 shows no neodymium in the outermost layer (EOZ) with considerably high concentration of iron and oxygen, suggesting that the layer consist of iron oxides (Fe_2O_3 or Fe_3O_4) as also indicated by the XRD result in Fig. 2. Figure 7 also shows that the intermediate layer (B) is an iron depleted region where the concentration of iron is lower compared to the EOZ (A) and IOZ (C). It is then evident with the growth of the EOZ then that there is outward diffusion of Fe from the intermediate layer to the EOZ. The oxygen concentration profiles across EOZ (A) and intermediate layer (B) are relatively constant with small reduction at the interface. The concentration reduces dramatically between the intermediate layer (B) interface and the IOZ (C). The interfaces are sharp and there is no gradual composition transition. Across the IOZ, moving further into the sample, there is a gradual reduction in oxygen concentration. It is also need to be noted that the distribution of Dy follows that of Fe whilst the distribution of Pr follows that of Nd.

Conclusions

Oxidations of NdFeB magnet have been carried out in an air atmosphere over the temperature range 773–1273 K. A different microstructural evolution was evident when comparing the samples oxidised at 773 and 1173 K. The microstructure of magnet oxidised at 773 K consists of dark outer layer (EOZ) and a grey internal oxidation zone that grow inwards with time. The oxidation process at higher temperature above 973 K resulted in a more complex microstructure with EOZ growing outwards and intermediate layer and IOZ growing inwards. Evidence of Fe depletion region shows that there is Fe diffusion to the outer scale (EOZ) which was not seen in the magnet oxidised at 773 K. The mechanism of oxidation may include both the diffusion of oxygen and iron into and out of the substrate, with further details on the matrix and diffusion path required. Further systematic study will need to be carried out to clarify the detailed oxidation mechanism of the NdFeB magnet.

Acknowledgements The authors are grateful to the CSIRO Minerals Resources for the financial support (top up scholarship for Mr Muhamad Firdaus) for this work. The authors would like to also thanks Dr. Mark Pownceby for the SEM analysis at CSIRO.

References

1. M. Firdaus et al., Review of high-temperature recovery of rare earth (Nd/Dy) from magnet waste, J. Sustain. Metall. (2016)
2. K. Binnemans et al., Recycling of rare earths: a critical review. J. Clean. Prod. **51**, 1–22 (2013)

3. Y. Li et al., The oxidation of NdFeB magnets. Oxid. Met. **59**(1–2), 167–182 (2003)
4. J. Jacobson, A. Kim, Oxidation behavior of Nd-Fe-B magnets. J. Appl. Phys. **61**(8), 3763–3765 (1987)
5. J.M. Le Breton, J. Teillet, Oxidation of (Nd, Dy)FeB permanent magnets investigated by 57Fe Mossbauer spectroscopy. IEEE Trans. Magn. **26**(5), 2652–2654 (1990)
6. E.D. Dickens, A.M. Mazany, The corrosion and oxidation of Nd-Fe-B magnets. J. Appl. Phys. **67**(9), 4613–4615 (1990)
7. D.S. Edgley et al., Dissociation of $Nd_2Fe_{14}B$ during high temperature oxidation. J. Magn. Magn. Mater. **128**, L1–L7 (1993)
8. D.S. Edgley et al., Characterisation of high temperature oxidation of $Nd_2Fe_{14}B$ magnets. J. Magn. Magn. Mater. **173**, 29–42 (1997)
9. K. Miura, M. Itoh, K.-I. Machida, Extraction and recovery characteristics of Fe element from Nd–Fe–B sintered magnet powder scrap by carbonylation. J. Alloy. Compd. **466**, 228–232 (2008)
10. T. Saito et al., The extraction of Nd from waste Nd–Fe–B alloys by the glass slag method. J. Alloy. Compd. **353**, 189–193 (2003)
11. M. Nakamoto et al., Extraction of rare earth elements as oxides from a neodymium magnetic sludge. Metall. Mater. Trans. B **43**(3), 468–476 (2012)
12. I. Skulj, H.E. Evans, I.R. Harris, Oxidation of NdFeB-type magnets modified with additions of Co, Dy, Zr and V. J. Mater. Sci. **43**(4), 1324–1333 (2008)
13. T.G. Woodcock et al., Understanding the microstructure and coercivity of high performance NdFeB-based magnets. Scripta Mater. **67**(6), 536–541 (2012)
14. H. Sepehri-Amin et al., Grain boundary and interface chemistry of an Nd–Fe–B-based sintered magnet. Acta Mater. **60**(3), 819–830 (2012)
15. L. Deshan et al., Grain boundary phase formation and magnetic properties of NdFeB/Nd multilayered films, Jpn. J. Appl. Phys. **48**(3) (2009)
16. E. Belin-Ferré, Basics of thermodynamics and phase transitions in complex intermetallics. Book Series on Complex Metallic Alloys. (World Scientific, Singapore, 2008)
17. W.Q. Liu et al., Oxidation kinetics of Nd-Fe-B permanent magnets prepared by spark plasma sintering. Corrosion **66**(5), 0550041–0550045 (2010)

Part II
Rare Earth Elements II and Platinum Group Metals

Electrochemical Behavior of Neodymium in Molten Chloride Salts

L. Diaz, P. Chamelot, M. Gibilaro, L. Massot and J. Serp

Abstract The discovery of NdFeB permanent magnets helps to perform technologies with reduced weight and gives them access to miniaturization. The industrial process of production of the Nd-Fe alloy by electrolysis used to manufacture of NdFeB magnets consists in reducing Nd^{3+} ions dissolved in a LiF-NdF_3-Nd_2O_3 salt on iron cathode at 1050 °C. This route was chosen since the electroplating of neodymium at low temperature molten chloride leads to low recovery yields. This is usually attributed to the comproportionation reaction between the electrodeposited metal and its chloride salt ($NdCl_3$) leading to the formation of $NdCl_2$. In this work, the neodymium electrochemical behavior is reviewed in order to understand its reduction mechanism. Meal addition to LiCl-KCl-$NdCl_3$ melt is usually used to simulated comproportionation reaction in solution. Neodymium deposition yields are studied. Stability of Nd^{2+} in two different solvent (LiCl-KCl and LiCl) is also discussed through electrodeposition tests on the gram scale.

Keywords Molten chloride · Rare earth · Neodymium · Reduction · Comproportionation reaction

Introduction

Rare earth elements (REE) take an important place in a economical point of view and are considered today as critical raw materials with high supply risk. They are used in various consumer materials and products, military equipment of high technology, wind turbines (permanent magnets), cars (Ni-MH batteries), computers

L. Diaz (✉) · J. Serp
Nuclear Energy Division, RadioChemistry & Processes Department,
CEA Marcoule, 30207 Bagnols-sur-Cèze, France
e-mail: laure.diaz@cea.fr

P. Chamelot · M. Gibilaro · L. Massot
Laboratoire de Génie Chimique UMR 5503 CNRS-INP-UPS,
Université de Toulouse III, 118 route de Narbonne, 31062 Toulouse, France

© The Minerals, Metals & Materials Society 2017
H. Kim et al. (eds.), *Rare Metal Technology 2017*,
The Minerals, Metals & Materials Series,
DOI 10.1007/978-3-319-51085-9_8

(hard-disk drives), etc. The discovery in the 1980s of NdFeB permanent magnets helps to perform technologies with reduced weight and give them access to miniaturization. The rare earths metals used in NdFeB alloys are either produced by metallothermy or by electrolysis in molten salts.

The industrial process for Nd-Fe alloy production by electrolysis used in the manufacture of NdFeB magnets consists in reducing Nd^{3+} ions dissolved in a LiF-NdF_3-Nd_2O_3 salt bath on an iron reactive cathode at 1050 °C. The high temperature is selected to obtain a liquid alloy Nd-Fe (80-20%) at the iron cathode surface. This route is chosen since the electroplating of neodymium in low temperature molten chloride leads to low recovery yields [1–3]. This is usually attributed to the comproportionation reaction between the electrodeposited Nd(0) metal and its chloride salt ($NdCl_3$) leading to the formation of $NdCl_2$:

$$2Nd^{3+} + Nd^0 \leftrightarrow 3Nd^{2+} \tag{1}$$

This work investigates the neodymium electrochemical behavior on inert cathode in different molten chloride salts using transient electrochemical techniques. The comproportionation reaction is highlighted after electrowining tests on the gram scale or simulated by metal addition in molten salt containing $NdCl_3$.

Experimental

Electrochemical Cell

All the experiments were carried out in a controlled purified argon atmosphere glove-box containing less than 10 ppm of water and oxygen. Chloride salts: LiCl and KCl (Sigma Aldrich: 99.0%) are stored in a drying oven at 150 °C under air and atmospheric pressure during several days. $NdCl_3$ (Sigma Aldrich: 99.99%) is used as received and kept in the glove box under Ar atmosphere.

Electrochemical Measurements

Salt chloride mixture was fused in an alumina crucible placed in a quartz cell inside a furnace. The cell was kept under Ar (99.999%).

Electrochemical behavior studies were performed with cyclic voltammetry (CV) and square wave voltammetry (SWV). The quasi reference electrodes (RE) for the LiCl-KCl and LiCl systems were W (Ø: 1.0 mm) and Pt (Ø: 1.0 mm) rod respectively. A W wire (Ø: 0.5 mm) was used as working electrode (WE) and a Mo spiral as counter electrode (CE).

Electrodeposition Tests

Electrowining tests of neodymium were carried out in LiCl-KCl eutectic salt on an inert Mo cathode (Ø: 5 mm, h: 10–20 mm) (Fig. 1). The eutectic salt containing $NdCl_3$ was electrolyzed under constant current at 450 °C. The anodic reaction: oxidation of chloride was into chlorine gas on glassy carbon anode. In this type of cell, the reaction between chlorine gas and the metal deposited at the cathode is the primary source of Faraday yield drop, this assembly used is effective in limiting this problem by a high argon flush (10 L/h) [4].

Results and Discussions

Electrochemical Measurements

Cyclic voltammetry and square-wave voltammetry (SWV) were used to study neodymium reduction mechanism. Cyclic voltammetry curves in two molten chloride salts, i.e. LiCl-KCl (450 °C) and LiCl (650 °C) containing $NdCl_3$ at 1 wt% are presented in Fig. 2. In both case, the reduction of Nd(III) to neodymium metal on the inert cathode occurs in a two-step process. The first step (peaks A and A′) is attributed to Nd(III)/Nd(II) reduction and the second step (peaks B and B′) to Nd (II)/Nd(0) transition leading to metal deposition. The anodic peak observed on the reverse scan corresponding to the second step is specific to a anodic striping (metal dissolution).

The shape difference between A and A′ can be attributed to the chloro-acidity of LiCl and LiCl-KCl salts which plays an important role on the stability of

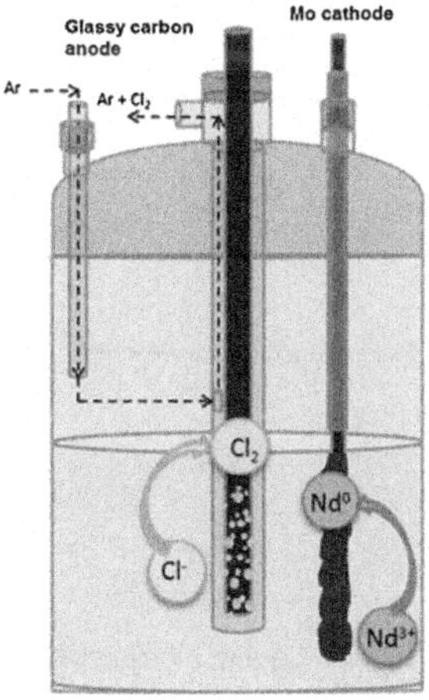

Fig. 1 Electrodeposition cell

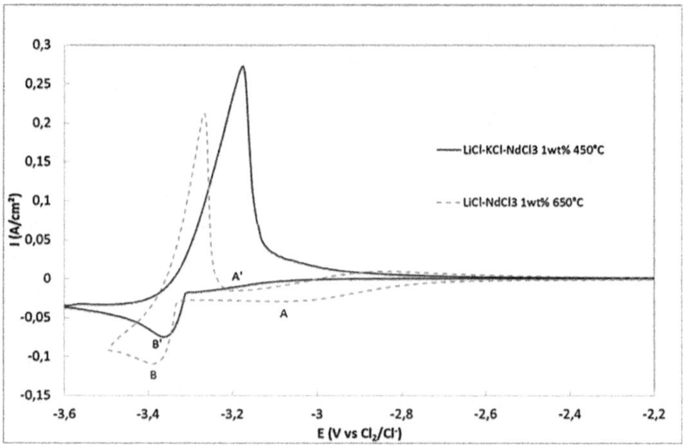

Fig. 2 Cyclic voltammograms of $NdCl_3$ (1 wt%) in LiCl-KCl and LiCl melts at v = 100 mV/s, working electrode: W

neodymium complexes and thus on its reduction potential [5]. According to Fig. 2, Nd(III) chloro-complex seems to be more stable in LiCl-KCl eutectics than in LiCl salt. In LiCl-KCl, Nd^{3+}/Nd^{2+} signal is not clearly defined since the potential difference between the Nd^{3+}/Nd^{2+} and Nd^{2+}/Nd^{0} signals is very low ($\Delta E < 100$ mV) whereas the difference in LiCl pure salt is larger than 300 mV.

This two-step neodymium reduction study is completed by square wave voltammetry analysis. Figure 3 shows square wave voltammograms of the

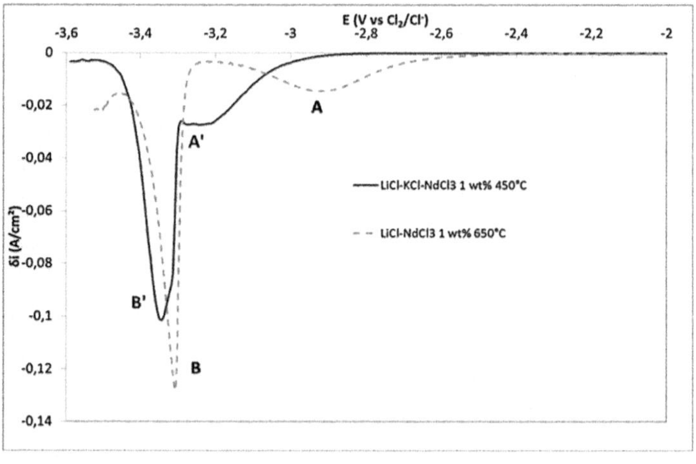

Fig. 3 Square ware voltammograms for the reduction of 1 wt% $NdCl_3$ at a W electrode in LiCl-KCl at 450 °C (step potential: 5 mV, amplitude: 20 mV, frequency: 25 Hz) and LiCl at 650 °C (step potential: 2 mV, amplitude: 20 mV, frequency: 25 Hz)

$LiCl$-KCl-$NdCl_3$ (1 wt%) at 450 °C and $LiCl$-$NdCl_3$ (1 wt%) at 650 °C systems on an inert W working electrode. Two reduction peaks are clearly observed corresponding to the two-step reduction of neodymium:

$$Nd^{3+} + 1e^- \rightarrow Nd^{2+} \quad (2)$$

$$Nd^{2+} + 2e^- \rightarrow Nd^0 \quad (3)$$

These results are consistent with the ones obtained from cyclic voltammograms. Each peak can be modeled with a Gaussian shape curve [6]. The mathematical analyses of the peaks allow calculating the number of exchanged electron by measuring half peak width ($W_{1/2}$) according to equation:

$$W_{1/2} = 3.52 \frac{RT}{nF} \quad (4)$$

The value of number of exchanged electron is close to 1 for the first peak (A and A′). If the curve-fitting of the first signal is not questionable, the results obtained on the second peak are neither consistent nor reproducible in particular for LiCl salt (Table 1). Moreover, the ratio between the differential current density of the two peaks should be equal to the ratio of exchanged electrons, i.e. 2 if Nd^{3+}/Nd^{2+} and Nd^{2+}/Nd^0 transitions are considered. The ratio found is rather 4 in LiCl-KCl and 8 in LiCl (Table 1). The comproportionation reaction between the electrodeposited Nd(0) metal and its chloride salt ($NdCl_3$) could have an influence on those calculations leading to incorrect number of electrons.

Lithium underpotential deposition takes place when the concentration of neodymium increases in the salt phase or at low scan rate in cyclic voltammetry. This characteristic alloy was also observed for several lanthanides elements such as lanthanum and cerium. This is identified on Fig. 4 at:

- Low Nd^{3+} concentration (1 wt%) at low scan rate (<10 mV/s) (Fig. 4a)
- High Nd^{3+} concentration (~3 wt%) at any scan rate.

Table 1 Variation of number of exchanged electron depending on frequency of SWV in $LiCl$-KCl-$NdCl_3$ (1 wt%) system

		Frequency (Hz)				
		9	16	25	36	64
LiCl-KCl at 450 °C	Electron number (1st step)	0.9	0.9	0.7	0.8	0.9
	Electron number (2nd step)	1.8	2.4	2.2	2	1.8
	Ratio δi(II/0)/δi(III/II)	4.4	4.1	3.7	3.4	2.8
LiCl at 650 °C	Electron number (1st step)	1	0.9	0.9	1	0.9
	Electron number (2nd step)	3.5	3.4	3.1	2.9	2.5
	Ratio δi(II/0)/δi(III/II)	10	8.9	8.8	8.8	6.6

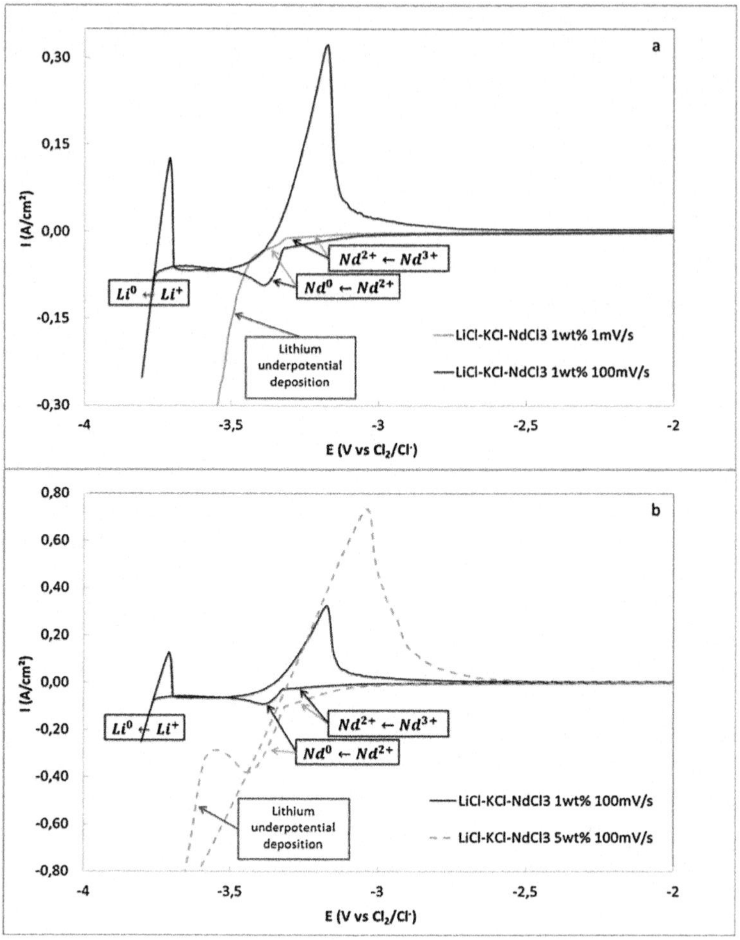

Fig. 4 Cyclic voltammograms in LiCl-KCl system at different scan rates (**a**) and concentrations of $NdCl_3$ (**b**); temperature: 450 °C, working electrode: W

Electrowining Test

Electrodeposition tests of neodymium metal is realized by electrowining under constant current density (cathodic current density: −0.250 to −0.450 A/cm^2) in molten LiCl-KCl-$NdCl_3$ (5 wt%) at 450 °C. Faradic yield is defined by the ratio of the mass of neodymium metal (determined by ICP-OES) to the theoretical mass calculated according to the Faraday's law with a current efficiency of 100%:

$$Faradic\,yield\,(\%) = \frac{mass\,of\,neodymium\,metal\,at\,the\,cathode}{theoretical\,mass\,(m = \frac{Q \times M}{z \times F})} \times 100 \quad (5)$$

where Q is the number of coulomb (C), M the molecular weight of neodymium (g mol^{-1}), z the number of electron and F the Faraday constant (C mol^{-1}).

The results show a low metal recovery with low faradic yields (lower than 50%). Novoselova and Smolenski [7] and Kvam et al. [8] attributed the faradic losses to the comproportionation reaction. In the tests carried out in this work, the metal is deposited as a fine divided powder unlike many rare earths metals (lanthanum and cerium) are deposited in the same experimental conditions (dendritic structure).

Stability of Nd(II) After Addition of Nd Metal

After several electrolysis, the open circuit potential (E_{OCP}) of the tungsten WE is shifted to the Nd(III)/Nd(II) redox potential (i.e. −3.3 V vs. Cl_2/Cl^-) whereas in LiCl-KCl-$NdCl_3$ without Nd metal, the E_{OCP} is rather −1.5 V versus Cl_2/Cl^-. The coexistence of Nd(III) and Nd(0) in the electrolyte leads to a spontaneous reaction:

$$2Nd^{3+} + Nd^0 \rightarrow 3Nd^{2+} \quad (6)$$

According to the Nernst equation, the presence of Nd(II) and Nd(III) in the electrolyte sets the E_{OCP} at the neodymium redox couple:

$$E_{OCP} = E^0_{Nd^{3+}/Nd^{2+}} + \frac{RT}{F} \ln \frac{a_{Nd^{3+}}}{a_{Nd^{2+}}} \quad (7)$$

where E° is the standard potential of Nd(III)/Nd(II) couple (V vs. Cl_2/Cl^-) and R the ideal gas constant (J K^{-1} mol^{-1}).

Moreover Nd(II) oxidation is clearly observed when the voltammogram is started in the anodic direction (grey curve on Fig. 5) with a positive residual current (I > 50 mA/cm^2) corresponding to the oxidation of Nd(II) to Nd(III). The Nd(III)/Nd(II) reduction signal tends to disappear on the cyclic voltammogram (black curve on Fig. 5). All these observations fit with the formation of Nd(II) when Nd(0) is in contact with Nd(III) as already observed by Yamana et al. [9] and Hayashi et al. [10].

This experiment has been repeated by adding a Nd metallic ingot (placed in a boron nitride basket) into LiCl-KCl-$NdCl_3$ solution. Similar changes of the open circuit potential is observed (Fig. 6), it went slowly (in 50 h probably because the metal surface was partially oxidized) from tungsten equilibrium potential in

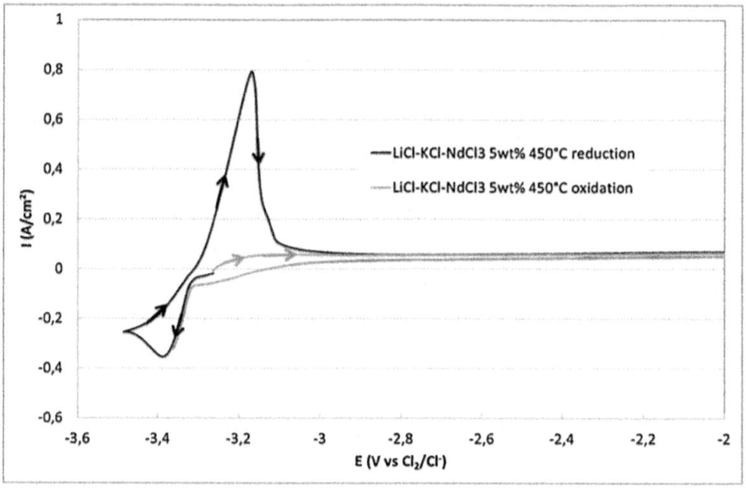

Fig. 5 Cyclic voltammograms of $NdCl_3$ (1 wt%) in LiCl-KCl at ν = 100 mV/s after several electrolysis, working electrode: W

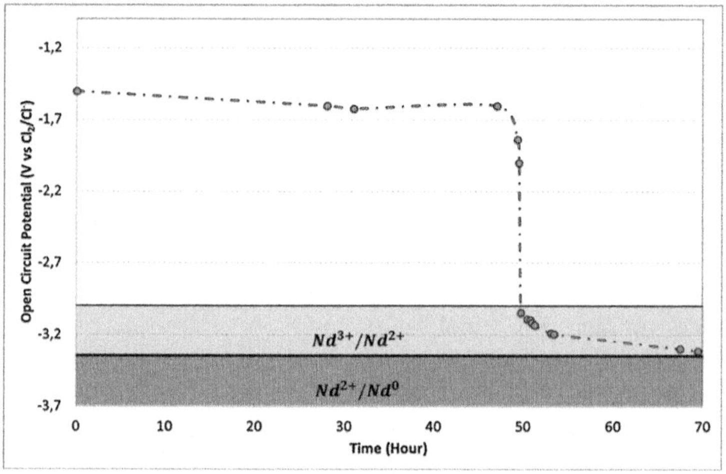

Fig. 6 Open circuit potential variation during presence of Nd metal in salt

LiCl-KCl-$NdCl_3$ (E_{OCP}: −1.5 V vs. Cl_2/Cl^-) to the Nd(III)/Nd(II) system (E_{OCP}: −3.3 V vs. Cl_2/Cl^-) which means that equilibrium of comproportionation reaction is shifted to the production of Nd^{2+} (Eq. 6).

E_{OCP} stays for several weeks at the Nd(III)/Nd(II) potential showing that Nd^{2+} is stable in reductive conditions (presence of an excess of Nd metal in the crucible). Surprisingly, Nd(III) signal doesn't completely disappear because an equilibrium is

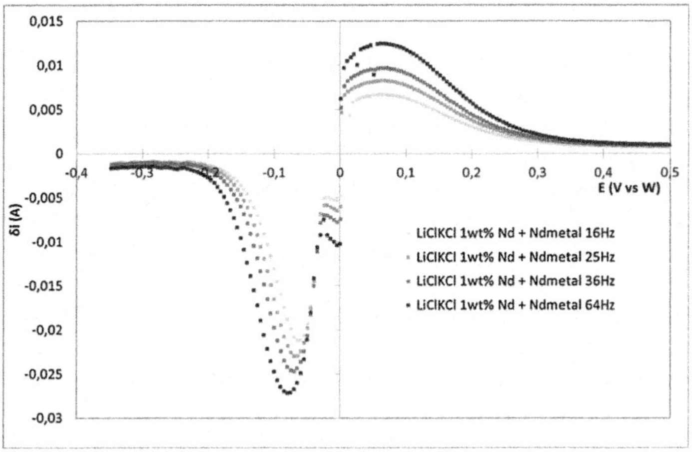

Fig. 7 Square ware voltammograms at different frequency for the reduction of LiCl-KCl-$NdCl_3$ (1 wt%) after metal addition at a W electrode

Table 2 Variation of number of exchanged electron depending on frequency of SWV in LiCl-KCl-$NdCl_3$ (1 wt%) system after metal addition, temperature: 450 °C

		Frequency (Hz)				
		9	16	25	36	64
LiCl-KCl at 450 °C	Electron number (oxidation)	0.8	0.9	0.9	0.9	0.8
	Electron number (reduction)	2.6	2.4	2.4	2	2
	Ratio δi(II/0)/δi(III/II)	3.5	3.2	2.8	2.6	2.2

established between the potential of the Nd(III)/Nd(II) and Nd(II)/Nd(0) couple and no more Nd(III) can be reduced.

Square wave voltammetry is carried out from E_{OCP} towards cathodic and anodic senses (Fig. 7). Oxidation of Nd(II) is clearly identified in the anodic one (exchange of 1 electron, Table 2):

$$Nd^{2+} \rightarrow Nd^{3+} + 1e^- \tag{8}$$

On the square wave voltammogram obtained from E_{OCP} to the cathodic direction, a first signal that could be attributed to the reduction of residual Nd(III) appears, followed by a reduction peak characteristic of metal deposition.

The calculation of number of electrons on the second reduction peak is close to a two electrons transfer corresponding to:

$$Nd^{2+} + 2e^- \rightarrow Nd^0 \tag{9}$$

Conclusion

The electrochemical behaviour of neodymium is carried out in LiCl-KCl and LiCl. Neodymium (III) reduction mechanism into metal occurs in two steps through the formation of divalent Nd(II) compound. The reduction potential of Nd(III) into Nd (II) is strongly dependent on the solvent composition (chloro-acidity). Lithium underpotential deposition is observed at high concentration of Nd(III) in LiCl-KCl melt. Nd metal electrowining tests shows that both recovery and faradic yields are low and metal is obtained as a fine powder contrary to other rare earth elements deposits that exhibit dendritic morphology. The low deposition efficiency is usually attributed to the chemical reaction between Nd(III) and Nd(0) leading to the formation of Nd(II). The formation of Nd(II) in LiCl-KCl is observed after several successive electrolysis of LiCl-KCl-$NdCl_3$ solution or by contacting Nd metal ingot with Nd(III) containing salt. The Nd(II) is stable in LiCl-KCl at 450 °C for several weeks as long as Nd metal is present in excess in the solution (reductive condition).

References

1. H. Vogel, B. Friedrich, *Development and Research Trends of the Neodymium Electrolysis—A Literature Review* (2015)
2. B.M.F. Chambers, *Electrolytic Production of Neodymium Metal From a Molten Chloride Electrolyte* (1991)
3. C. Pernel, J. Serp, M. Ougier, R. Malmbeck, E. Commission, Electrochemical investigations of La, Nd and Am in molten chloride salts in view of Am/Ln partitioning. Eur. Comm. Karlsruhe, 1–7 (2004)
4. M. Gibilaro, L. Massot, P. Chamelot, L. Cassayre, P. Taxil, Electrochemical extraction of europium from molten fluoride media. Electrochim. Acta **55**(1), 281–287 (2009)
5. K. Fukasawa, *Systematic Study on the Thermodynamic Stability of Lanthanides and Actinides in Molten Alkali and Alkaline Earth Chlorides* (2012)
6. L. Ramaley, M.S. Krause, Theory of square wave voltammetry. Anal. Chem. **41**(11), 1362–1365 (1969)
7. A. Novoselova, V. Smolenski, *Electrochemical behavior of neodymium compounds in molten chlorides*, Electrochim. Acta, vol 87, (Jan 2013), pp. 657–662
8. K.R. Kvam, D. Bratland, H.A. Øye, The solubility of neodymium in the systems NdCl3-LiCl and NdCl3-LiCl-KCl. J. Mol. Liq. **83**(1–3), 111–118 (1999)
9. H. Yamana, B.G. Park, O. Shirai, T. Fujii, A. Uehara, H. Moriyama, Electrochemically produced divalent neodymium in chloride melt. J. Alloys Compd. **408–412**, 66–70 (2006)
10. H. Hayashi, M. Akabori, T. Ogawa, K. Minato, Spectrophotometric study of Nd^{2+} ions in LiCl-KCl eutectic melt. Zeitschrift fur Naturforsch. Sect. A J. Phys. Sci. **59**(10), 705–710 (2004)

Novel Reactive Anode for Electrochemical Extraction of Rare Earth Metals from Rare Earth Oxides

Aida Abbasalizadeh, Seshadri Seetharaman, Prakash Venkatesan, Jilt Sietsma and Yongxiang Yang

Abstract Electrolytic production of metallic neodymium is carried out in fused fluoride salts containing neodymium oxide. Two major challenges pertaining to neodymium production are (a) low oxide solubility, (b) possibility of anodic fluorine gas evolution if the electrolysis rate exceeds feeding rate of neodymium oxide. In this study, a novel method is proposed in which iron fluoride (FeF_3) is used as a fluorinating agent to convert neodymium oxide into neodymium fluoride. Electron Probe Micro Analysis (EPMA) results of as-converted salt show a complete conversion of neodymium oxide into neodymium fluoride. In the electrolysis process, iron is used as a reactive anode with electrochemical dissolution of iron into the melt, thus preventing fluorine gas evolution at the anode. Therefore, the fluorinating agent is constantly regenerated in situ which enables the continuous conversion of neodymium oxide feed. The cathodic product is a Nd–Fe alloy which can be directly used as a master alloy for the production of NdFeB permanent magnets.

Keywords Rare earth · Electrochemical extraction · Reactive anode

Introduction

The dominant industrial method for the extraction of rare earth metals from their oxides is the molten salt electrolysis process [1–3]. In comparison to hydrometallurgical processes, lower energy consumption, higher efficiency and higher purity of the deposits are advantages in molten salt routes compared to hydrometallurgy processes [4–6]. Solubility of the rare earth oxide (REO) in the fluoride electrolyte is an important factor, since it is the dissolved oxide that is subjected to electrolysis.

A. Abbasalizadeh (✉) · P. Venkatesan · J. Sietsma · Y. Yang
Department of Materials Science and Engineering, Delft University of Technology (TU Delft), Mekelweg 2, 2628CD Delft, The Netherlands
e-mail: a.abbasalizadeh@tudelft.nl

S. Seetharaman
Royal Institute of Technology (KTH), Stockholm, Sweden

© The Minerals, Metals & Materials Society 2017
H. Kim et al. (eds.), *Rare Metal Technology 2017*,
The Minerals, Metals & Materials Series,
DOI 10.1007/978-3-319-51085-9_9

However, the solubility of rare earth oxide in molten alkali fluoride does not exceed 4 wt% [7–9]. Moreover, neodymium oxide forms likely neodymium oxyfluoride in molten fluoride salts. Oxyfluoride formation has been reported to affect the neodymium electrodeposition adversely [10]. Another challenge in neodymium production is fluorine gas evolution on the carbon anode when the electrolysis rate exceeds the rare earth oxide feeding and dissolution rate [11].

Herein we adopt a two-fold approach to address the solubility issue of neodymium oxides in the fluoride melts. Initially, REOs are treated with strong fluorinating agents like FeF_3 which converts REOs to REF_3. To overcome halogen evolution on the anode, a reactive anode was employed. Iron was anodically dissolved re-generating the fluorinating agent FeF_3 in situ in the electrochemical reactor. The rare earth fluoride thus formed can subsequently be processed through the electrolysis route in the same reactor to extract rare earth metal as the cathodic deposit. In this concept, the REO dissolution in the molten fluorides would become redundant due to the rapid and complete conversion of the oxide into the fluoride as REF_3.

Experimental

Lithium fluoride (98.5%-Alfa Aesar) was mixed with neodymium oxide (Rhodia) in a glove box. In the experiments for the conversion of neodymium oxide to neodymium fluoride, iron fluoride (97%-Alfa Aesar) was added to the mixture and the mixture was charged into a graphite crucible. For these experiments, the stoichiometry equivalent concentration of fluorinating agent was used in order to reach the complete conversion of Nd_2O_3 into NdF_3. A molybdenum rod (2 mm diameter) was used as cathode. The immersion depth of the Mo rod was measured after each experiment. An iron rod (98%-Salomon metals, 2 mm diameter) was used as anode. All experiments were performed at 950 °C under a purified argon atmosphere. The argon gas was purified by passing through KOH flakes in order to remove traces of CO_2 sulphur and also through silica gel and P_2O_5 to remove moisture. The gas was passed through a tube furnace containing Ti sponge which was held at 1123 K (850 °C) in order to remove the trace amount of O_2. The samples were removed from the furnace and immediately quenched in liquid nitrogen after the experiment.

The samples were analyzed by Electron Probe Micro Analysis (EPMA), in order to derive the phases that are formed during the experiments from the local compositions. WDS point analyses or mappings were performed at 15 kV–15 nA and for the quantification, standard samples were used.

Results and Discussion

EPMA mapping of the sample from the LiF-Nd_2O_3-FeF_3 salt bath that was quenched in liquid nitrogen after 3 h at 950 °C, is shown in Fig. 1. It can be seen that neodymium and oxygen do not occur in the same regions, whereas iron and oxygen

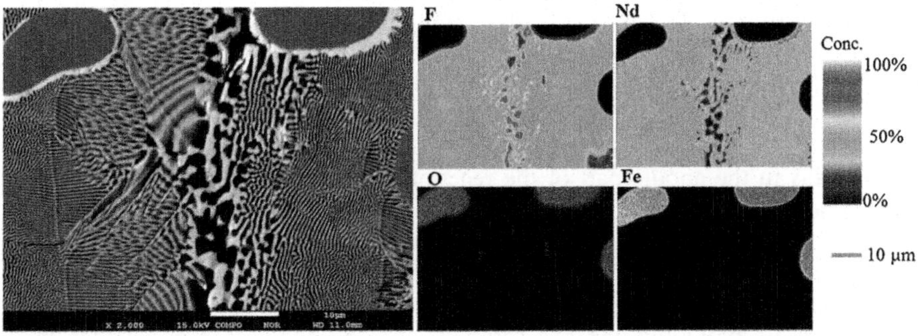

Fig. 1 EPMA mapping of LiF-Nd_2O_3-FeF_3 sample after fluorination treatment showing the presence of NdF_3 and FeO. Note that Li cannot be detected in these EPMA measurements

Table 1 EPMA quantification results of LiF-Nd_2O_3-FeF_3 sample (in at.%) after fluorination treatment showing the presence of NdF_3 and FeO_x

Element	NdF_3	FeO_x
F	72	–
O	–	54
Nd	26	–
Fe	–	45

are clearly distributed in the same area, confirming the iron oxide formation. Neodymium has a high concentration around the FeO boundaries. The light-grey lamellar phase is NdF_3 within the dark LiF, indicating LiF-NdF_3 eutectic microstructure. In the EPMA mapping, in which Li cannot be detected, complete conversion of neodymium oxide to neodymium fluoride is exhibited.

The EPMA quantitative results are presented in Table 1. The composition in the column NdF_3 was measured in the Nd-regions of Fig. 1 and the FeO_x composition in the Fe- and O-rich regions. In this table, the presence of NdF_3 (26 at.% Nd and 72 at.% F) and FeO_x (45 at.% Fe and 54 at.% O) are confirmed. The phase indicated as iron oxide can be a mixture of FeO and Fe_2O_3, since the concentrations can be related to magnetite phase (Fe_3O_4). Based on the calculated standard Gibbs energy (FactSage 7.0) for the reaction (1),

$$Nd_2O_3 + 2FeF_3(\text{salt}) \rightleftarrows 2NdF_3 + Fe_2O_3 \quad \Delta G^o(\text{at } 950^oC) = -222\,\text{kJ/mol} \quad (1)$$

conversion of neodymium oxide to neodymium fluoride using iron fluoride is feasible. According to this reaction, Fe_2O_3 formation is expected, but as EPMA results show, the oxide of iron which is formed can be wüstite or non-stoichiometric FeO_x. In the argon ambient and low oxygen pressure atmosphere, hematite (Fe_2O_3) will not be formed or depending upon the oxygen pressure prevailing, the formed Fe_2O_3 is decomposed to Fe_3O_4 and FeO. At high temperatures (more than ca. 560 °C), stable oxide is wüstite (FeO).

NdF_3, formed as the result of the reaction between Nd_2O_3 and FeF_3, will go through the electrolysis process. In this process, the cathodic reaction would be:

$$Nd^{3+} + 3e^- \rightleftarrows Nd \tag{2}$$

Whereas in the anodic part of the reaction, fluorine gas is generated if an inert anode such as graphite is used:

$$2F^-(salt) \rightleftarrows F_2(g) + 2e^- \tag{3}$$

In order to avoid fluorine gas formation on the anode, the novel idea in the present work is to use iron as reactive anode. In this process, iron in the anode is dissolved in the salt as the result of an anodic reaction which leads to in situ formation of iron fluoride. Besides avoiding the fluorine gas evolution, another advantage of this method is that iron fluoride is generated and used in the system, which leads to less material consumption, meaning that fluorinating agent is constantly generated and consumed in the system. In this case the anodic reaction will be iron dissolution:

$$Fe \rightleftarrows Fe^{3+}(g) + 3e^- \tag{4}$$

A series of experiments was carried out in order to examine the electrochemical dissolution of iron in the molten salt. In these experiments iron served as anode and Mo as cathode, which were immersed in LiF-Nd_2O_3 mixture. The experiment was performed in constant current mode at 1.5 A for 3 h at 950 °C. Based on Faraday's law, considering 3 exchanged electrons in the cathodic reduction of neodymium metal, mass of the dissolved iron can be calculated:

$$m = ItM/nF \tag{5}$$

where I is the electrolysis current in A, t is the time of the electrolysis in s, n is the number of exchanged electrons, F is the Faraday's constant (F = 96,500 °C), m is the mass of the element in gram and M is the atomic mass of the element.

The cathodic product on the molybdenum electrode was analysed by EPMA. The result is shown in Fig. 2.

From EPMA analysis result, it can be seen that a layer of iron is formed on the molybdenum cathode and neodymium is reduced onto this layer. Fe might be deposited from the iron fluoride, which is formed from dissolution of the iron anode, since iron fluoride has a less positive decomposition voltage compared to neodymium fluoride. Part of the iron fluoride which reacts with neodymium oxide based on reaction (1), forms iron oxide. The dissolved iron oxide can also participate in the electro-decomposition process, and co-deposition of Fe occurs at the cathode. Nd–Fe alloy production is desirable since it can be used as master alloy for magnet production. The challenge would be as to how to remove the undissolved iron oxide in order to minimize their side effect on cathodic deposition of Nd.

Fig. 2 EPMA mapping of the deposited product on the molybdenum cathode after 3 h electrolysis at 950 °C, using iron as anode

A new design of a two-chamber furnace system is one way for removing the undissolved oxide, which is presently under study. A possible alternative may be to take advantage of the density difference between the formed oxide and the fluoride bath in order to separate the oxides from the molten salt.

Conclusion

The problem of low solubility of the neodymium oxide in molten fluorides is solved to a large extent by converting neodymium oxide to neodymium fluoride in situ. Once neodymium fluoride is formed, it can subsequently be electrolyzed and extracted on the cathode. In this study, a novel method is studied in which iron is used as anode, promoting electrochemical dissolution of iron into the melt, thus preventing fluorine gas evolution at the anode. Therefore, the fluorinating agent is constantly regenerated in situ which enables the continuous conversion of neodymium oxide feed. The cathodic product is Nd–Fe alloy which can be directly used as a master alloy for the production of NdFeB magnets.

Acknowledgements This research has received funding from the European Community's Seventh Framework Programme ([FP7/2007-2013]) under grant agreement no. 607411 (MC-ITN EREAN: European Rare Earth Magnet Recycling Network, Project website: www.erean.eu). This publication reflects only the authors' view, exempting the Community from any liability.

References

1. S. Seetharaman, O. Grinder, US patent application no. 12/991128, ref. no.: 12057 (2010)
2. A. Abbasalizadeh et al., Highlights of the salt extraction process. J. Mater. **65**(11), 1552–1558 (2013)

3. A. Abbasalizadeh et al., Neodymium extraction using salt extraction process. Miner. Process. Extr. Metall. **124**(4), 191–198 (2015)
4. B. Mishra, D.L. Olson, Molten salt applications in materials processing. J. Phys. Chem. Solids **66**(2–4), 396–401 (2005)
5. W. Han et al., Electrodeposition of Mg-Li-Al-La alloys on inert cathode in molten LiCl-KCl eutectic salt. Metall. Mater. Trans. B **42**(6), 1367–1375 (2011)
6. Y. Castrillejo et al., Electrochemical behaviour of praseodymium (III) in molten chlorides. J. Electroanal. Chem. **575**(1), 61–74 (2005)
7. E. Morrice, T.A. Henrie, *Electrowinning high-purity neodymium, praseodymium, and didymium metals from their oxides* (Washington D.C., U.S. Department of the Interior, Bureau of Mines, 1967)
8. J.E. Murphy, D.K. Dysinger, M.F. Chambers, Electrowinning neodymium metal from chloride and oxide-fluoride electrolytes, in *Light Metals 1995, Warrendale, Pennsylvania*, 1995
9. B. Porter, E.A. Brown, *Determination of oxide solubility in molten fluorides* (Washington, D.C., U.S. Department of the Interior, Bureau of Mines, 1961), 3139
10. E. Stefanidaki, C. Hasiotis, C. Kontoyannis, Electrodeposition of neodymium from LiF-NdF3-Nd2O3 melts. Electrochim. Acta **46**(17), 2665–2670 (2001)
11. A. Kaneko, Y. Yamamoto, C. Okada, Electrochemistry of rare earth fluoride molten salts. J. Alloy. Compd. **193**, 44–46 (1993)

Electrochemical Formation of Nd Alloys Using Liquid Metal Electrodes in Molten LiCl–KCl Systems

Hirokazu Konishi, Hideki Ono, Eiichi Takeuchi, Toshiyuki Nohira and Tetsuo Oishi

Abstract We focused on the liquid metals as the alloy diaphragms. Liquid metals have higher diffusion rates of elements than those of solid metals during the electrolysis. For the first step, in this work Sn electrodes were used as liquid metal electrodes. The alloy samples were prepared by potentiostatic electrolysis using liquid Sn electrodes in molten LiCl–KCl added $DyCl_3$ (0.50 mol%) and $NdCl_3$ (0.50 mol%) at 723 K. As the first step, electrochemical formation of RE–Sn (RE = Dy, Nd) alloys were investigated in LiCl–KCl eutectic melts containing $DyCl_3$ (0.50 mol%) or $NdCl_3$ (0.50 mol%). The samples of Nd–Sn alloys were prepared by potentiostatic electrolysis at 0.60, 0.70, 0.80, 1.00 V (vs. Li^+/Li) for 0.50 h using liquid Sn electrodes in LiCl–KCl–$NdCl_3$ (0.50 mol%) melts. The formation of Nd–Sn alloys were confirmed by the XRF analysis of all samples. Moreover, potentiostatic electrolysis were conducted at 0.60 and 0.70 V for 0.50 h in LiCl–KCl–$DyCl_3$ (0.50 mol%)–$NdCl_3$ (0.50 mol%) melts. The formation of Dy–Nd–Sn alloys were confirmed in the bottom portion of liquid Sn electrodes. The atomic ratios of Nd/Dy in both alloy samples were 3.50.

Keywords Rare earth · Electrolysis · Molten salt · Liquid metal · Separation

H. Konishi (✉) · H. Ono · E. Takeuchi
Osaka University, 2-1 Yamadaoka, Suita, Osaka 565-0871, Japan
e-mail: konishi@mat.eng.osaka-u.ac.jp

T. Nohira
Kyoto University, Gokasho, Uji, Kyoto 611-0011, Japan

T. Oishi
National Institute of Advanced Industrial Science and Technology, 16-1 Onogawa, Tsukuba, Ibaraki 305-8569, Japan

© The Minerals, Metals & Materials Society 2017

H. Kim et al. (eds.), *Rare Metal Technology 2017*,
The Minerals, Metals & Materials Series,
DOI 10.1007/978-3-319-51085-9_10

Introduction

The use of rare earth (RE) -iron group (IG) alloys has increased significantly in a number of industrial fields over the past few decades. In particular, the demand for Dy-added Nd–Fe–B magnets is rapidly increasing because these magnets are indispensable for high-performance motors in electric vehicles (EVs) and hybrid electric vehicles (HEVs). These magnets need to possess sufficient thermal stability for use in such motors in high-temperature environments. The addition of Dy is necessary to improve the thermal stability of Nd–Fe–B magnets. However, there is the concern about a shortage of Dy resources in particular. Thus a worldwide need is being increasingly felt to augment the primary production of Nd and Dy by combining a suitable recycling method in order to reclaim these metals from their recyclable resources. In this context, it is worthwhile to mention that large Nd–Fe–B magnets are the only secondary resource materials as far as Dy is concerned. That is why, it is necessary to develop an inexpensive and environmentally friendly recovery/separation process for the recovery of Nd and Dy from a variety of scrap/waste magnets.

We already reported the separation and recovery process for RE metals from Nd magnet scraps using molten salt electrolysis [1–5] and an alloy diaphragm [6–11]. This process was first applied to chloride melts, and the separation of Dy and Nd were investigated using Ni and Cu cathodic electrodes in molten LiCl–KCl–$DyCl_3$–$NdCl_3$ systems [9–11]. Figure 1 shows the principle of process. A magnet scrap contained Dy and Nd is used as the anode. Dy alloy is used as an alloy diaphragm, which functions as a bipolar electrode. During the electrolysis, Dy and Nd are dissolved into the molten salt as Dy(III) and Nd(III). Dy(III) is selectively reduced to form Dy alloys on the alloy diaphragm according to their formation potentials and/or alloying rates. Subsequently, Dy atom chemically diffuses through the alloy diaphragm and is dissolved into the molten salt as Dy(III) in the cathode room. The permeated Dy(III) is finally deposited on the Mo or Fe cathode as RE metals. Nd remaining in the anode room can be collected by electrolysis using another cathode in the anode room. Almost all impurities remain in the anode room as residue or anode slime.

But in this process, a long electrolysis might be not able to be conducted due to cracks or splits of an alloy diaphragm. In order to solve this problem, we focused on a liquid metal as an alloy diaphragm. Figure 2 shows the principle of the process. Firstly, Dy and Nd are dissolved into the molten salt as Dy(III) and Nd(III) during the first electrolysis. Dy(III) is selectively reduced to form Dy alloys at the liquid metal electrode according to their formation potentials and/or alloying rates (Fig. 2a). Subsequently, Dy atom chemically diffuses through the liquid Sn electrode, or is transported by mechanical mixing in that electrode. Secondary, Dy atom is dissolved into the molten salt as Dy(III) in the other room during the second electrolysis. The dissolved Dy(III) is finally deposited on the Mo or Fe cathode as RE metals (Fig. 2b). This process using liquid metals has some advantages, i.e., high diffusion rates, no cracks or splits of liquid metals, possibilities of mechanical

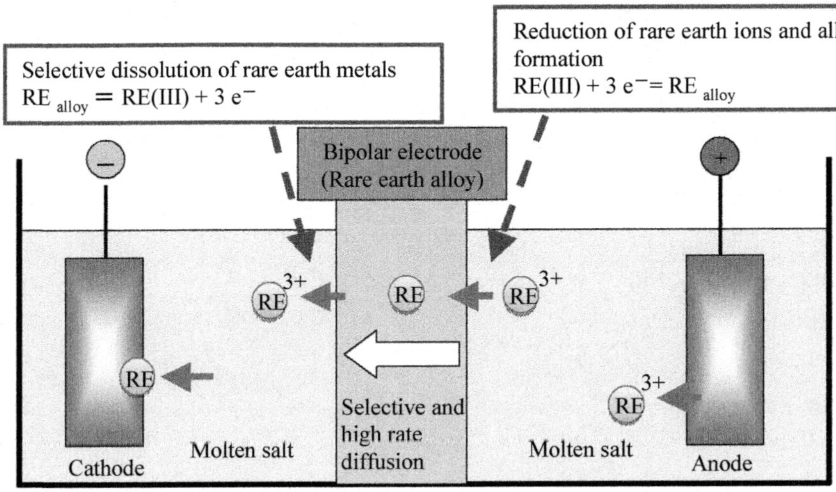

Fig. 1 Schematic drawing of the process for separation and recovery of rare earth metals

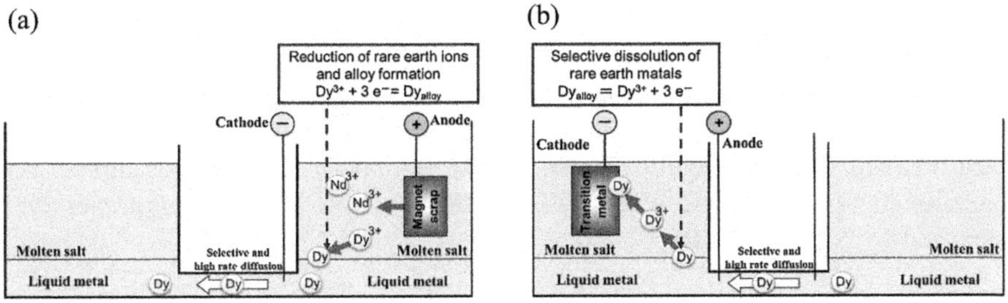

Fig. 2 Schematic drawing of the process for **a** separation and **b** recovery of rare earth metals by molten salt electrolysis and a liquid metal electrode

mixing. Therefore, we used the liquid metal electrodes in this process. As the first step, we investigated the electrochemical formation of Dy, Nd–Sn alloys using liquid Sn electrodes in LiCl–KCl eutectic melts containing $DyCl_3$ (0.50 mol%) or $NdCl_3$ (0.50 mol%) at 723 K.

Experimental

Figure 3 shows the experimental apparatus. All experiments were performed in LiCl–KCl eutectic melts under dry argon atmosphere at 723 K. The LiCl–KCl eutectic melts (LiCl:KCl = 58.5:41.5 mol%; Wako Pure Chemical Co., Ltd.) were placed in a high purity alumina crucible, and kept under a vacuum for more than

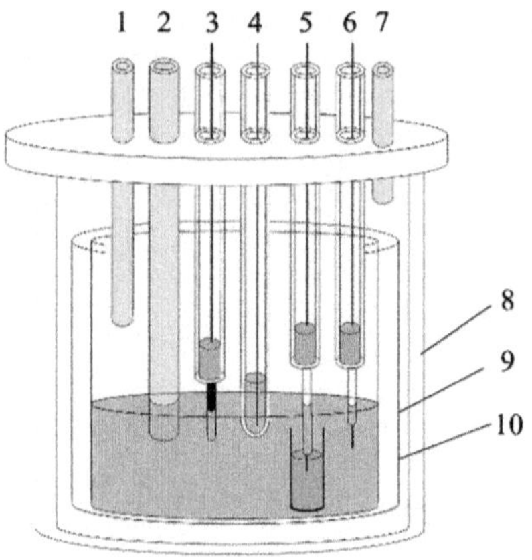

Fig. 3 Schematic drawing of the experimental apparatus. *1* Ar gas inlet, *2* thermocouple, *3* counter electrode (glassy carbon), *4* reference electrode (Ag^+/Ag), *5* working electrode (Sn), *6* Li^+/Li electrode (Mo), *7* Ar gas outlet, *8* pyrex holder, *9* alumina crucible, *10* LiCl–KCl–$RECl_3$ (RE = Dy, Nd)

24 h at 473 K to remove water. $DyCl_3$ or $NdCl_3$ (99.9%, Kojundo Chemical Laboratory Co., Ltd.) was added directly to these melts. The working electrodes were liquid Sn electrodes (5.5 g, Wako Co., Ltd.). The reference electrode was an Ag^+/Ag electrode consisting of a silver wire immersed in LiCl–KCl eutectic melts containing 1.0 mol% AgCl. All the potentials given hereafter were referred to Li^+/Li electrode potential on a Mo wire (5 mm × ϕ 1 mm, 99.95%, Nilaco Co., Ltd.). The counter electrode was a glassy carbon rod (50 mm × ϕ 5 mm, Tokai Carbon Co., Ltd.). Electrochemical behavior was analyzed by using an electrochemical measurement system (Hokuto Denko Corp., HZ-5000). The alloy samples were prepared by potentiostatic electrolysis. After the electrolysis, the samples were analyzed by SEM and XRF.

Results and Discussion

Electrochemical Behavior of RE(III) (RE = Dy, Nd) Using Liquid Sn Electrodes

As shown in Fig. 4, the phase diagrams of the Dy–Sn and Nd–Sn systems show the presence of intermetallic compounds (Dy_5Sn_3, Dy_5Sn_4, $Dy_{11}Sn_{10}$, DySn, Dy_4Sn_5, $DySn_2$) and (Nd_5Sn_3, Nd_5Sn_4, $Nd_{11}Sn_{10}$, NdSn, Nd_3Sn_5, $NdSn_2$, Nd_3Sn_7, Nd_2Sn_5 and $NdSn_3$) at 723 K [12]. From these phase diagrams, $DySn_2$ and $NdSn_3$ might be formed by potentiostatic electrolysis at most positive potential since that activity of Dy and Nd in these alloys are lowest, respectively. Taking into account the possibilities of the formation of these intermetallic compounds, cyclic voltammetry was conducted in LiCl–KCl–$RECl_3$ (0.50 mol%) melts at 723 K.

Before the investigation of RE–Sn alloys formation, the electrochemical behavior of Li^+ was investigated using a liquid Sn electrode as a working electrode in a molten LiCl–KCl system at 723 K. Figure 5 shows the obtained cyclic

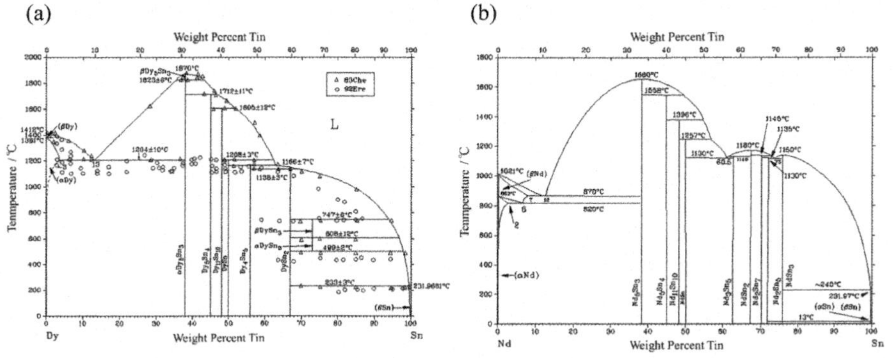

Fig. 4 Phase diagrams of the **a** Dy–Sn system and **b** Nd–Sn system (12)

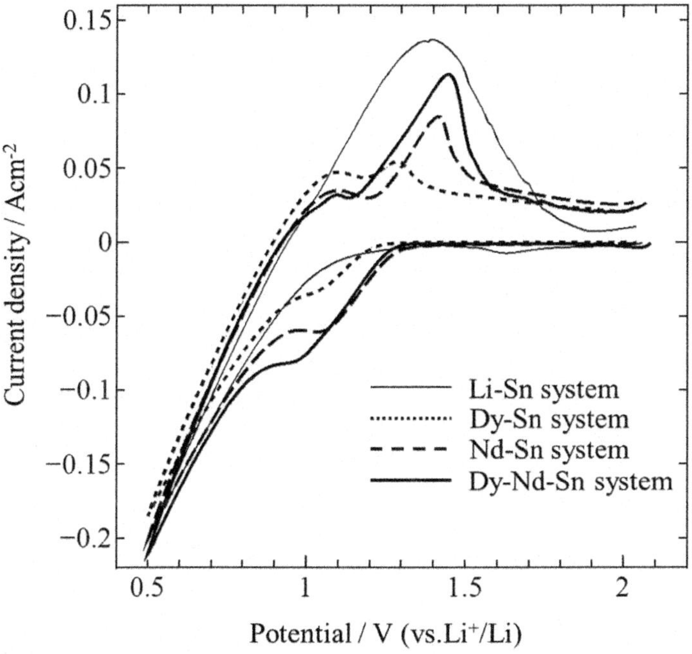

Fig. 5 Cyclic voltammograms for liquid Sn electrodes in LiCl–KCl–$DyCl_3$ (0.50 mol%)–$NdCl_3$ (0.50 mol%) melts at 723 K. Scanning rate: 0.05 V s^{-1}

voltammogram. The scanning rate was set at 0.05 V s^{-1}. The solid curve (Li–Sn system) represents the voltammogram. During the negative scanning from 2.10 V (vs. Li^+/Li), a large cathodic current was observed from 1.10 V. Since Li can form alloy with Sn, the cathodic current corresponds to the formation Li–Sn alloys.

Based on the above results, the electrochemical behaviors of Dy(III) and Nd (III) were investigated using liquid Sn electrodes in LiCl–KCl eutectic melts containing $DyCl_3$ (0.50 mol%) or $NdCl_3$ (0.50 mol%) at 723 K. In the negative scanning for LiCl–KCl–$DyCl_3$ melts, small and large cathodic currents were observed from 1.15 V and from 0.90 V, respectively. These cathodic currents might correspond to that formation of Dy–Sn alloys. On the other hand, in the negative scanning for LiCl–KCl–$NdCl_3$ melts, a cathodic current was observed from 1.30 V. This cathodic current might correspond to the formation Nd–Sn alloys. Furthermore, in the negative scanning for LiCl–KCl–$DyCl_3$–$NdCl_3$ melts, a cathodic current was observed from 1.30 V, and then a large current was observed from 0.90 V. This result indicted the synthesized voltammgrams recorded in LiCl–KCl–$DyCl_3$ and LiCl–KCl–$NdCl_3$ melts. From the above results, the formation of Dy–Sn and Nd–Sn alloys were suggested at more negative potential than 1.15 and 1.30 V in these melts, respectively.

Electrochemical Formation of RE–Sn (RE = Dy, Nd) Alloys Using Liquid Sn Electrodes

Based on the cyclic voltammograms shown in Fig. 5, in order to form Nd–Sn alloys, alloy samples were prepared by potentiostatic electrolysis at 0.60, 0.70, 0.80 and 1.00 V for 0.50 h using liquids Sn cathodes in LiCl–KCl–$NdCl_3$ (0.50 mol%) melts at 723 K. Figures 6, 7, 8 and 9 show results obtained by XRF analysis. The blue areas indicate the existence of Nd in all samples. The Nd–Sn alloys are observed at the bottom potion of all samples. These results suggested that Nd–Sn alloys were formed at the surface of liquid Sn electrode and then the alloys settled down since a density of Nd–Sn alloy is higher than that of Sn.

Table 1 shows results of atomic ratios of Sn and Nd in liquid Sn samples analyzed by XRF. Atomic ratio of Nd in the alloy samples decreases from 0.70 to 1.00 V. In contrast, atomic ratio of Nd in the alloy samples obtained at 0.60 V is lower than that of Nd obtained at 0.70 V. This result might be caused by the formation of Li–Sn alloys. As shown in the voltammogram of Li–Sn system in

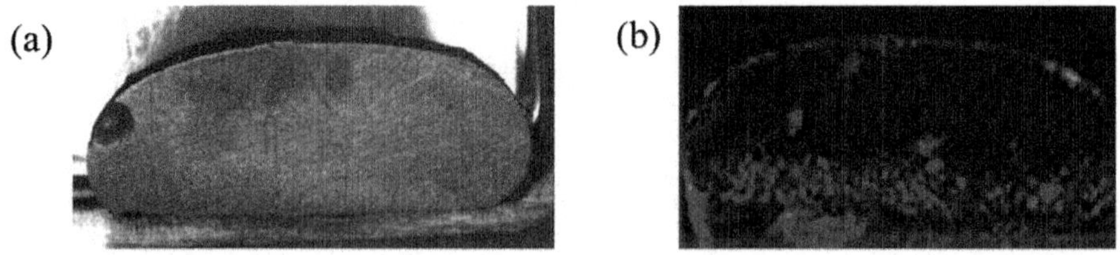

Fig. 6 **a** A cross-sectional real image and **b** an elemental mapping of Nd in a liquid Sn sample obtained by potentiostatic electrolysis at 0.60 V for 0.50 h in LiCl–KCl–$NdCl_3$ (0.50 mol%) melts at 723 K

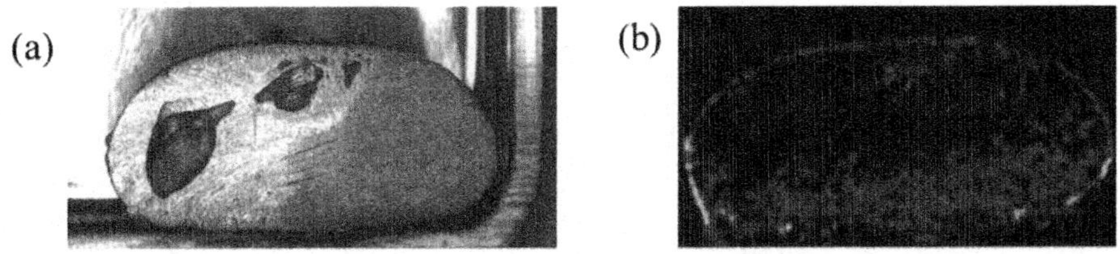

Fig. 7 **a** A cross-sectional real image and **b** an elemental mapping of Nd in a liquid Sn sample obtained by potentiostatic electrolysis at 0.70 V for 0.50 h in LiCl–KCl–$NdCl_3$ (0.50 mol%) melts at 723 K

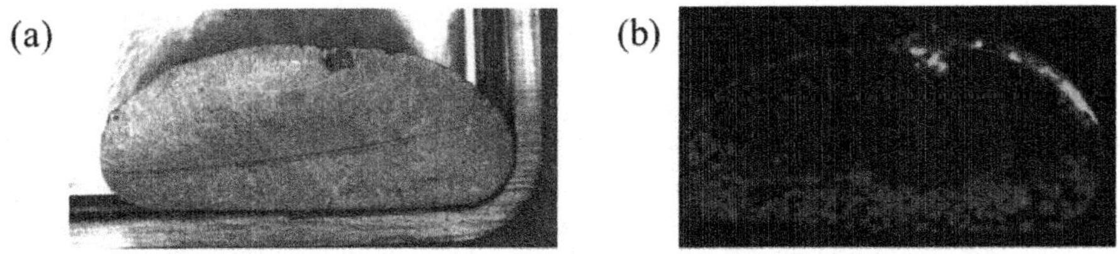

Fig. 8 **a** A cross-sectional real image and **b** an elemental mapping of Nd in a liquid Sn sample obtained by potentiostatic electrolysis at 0.80 V for 0.50 h in LiCl–KCl–$NdCl_3$ (0.50 mol%) melts at 723 K

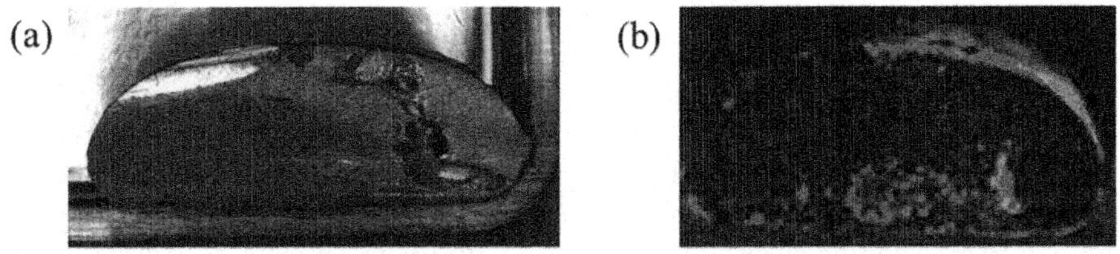

Fig. 9 **a** A cross-sectional real image and **b** an elemental mapping of Nd in a liquid Sn sample obtained by potentiostatic electrolysis at 1.00 V for 0.50 h in LiCl–KCl–$NdCl_3$ (0.50 mol%) melts at 723 K

Table 1 Atomic ratios of Sn and Nd in liquid Sn samples obtained by potentiostatic electrolysis at 0.60, 0.70, 0.80, 1.00 V for 0.50 h in LiCl–KCl–$NdCl_3$ (0.50 mol%) melts at 723 K

Electrolysis potential value (V) (vs. Li^+/Li)	Electrolysis time (h)	Sn (atom%)	Nd (atom%)
0.60	0.50	83.9	2.5
0.70	0.50	83.6	4.5
0.80	0.50	74.0	3.6
1.00	0.50	76.8	2.0

Table 2 Atomic ratios of Sn, Dy, Nd in liquid Sn samples obtained by potentiostatic electrolysis at 0.60 and 0.70 V for 0.50 h in LiCl–KCl–$DyCl_3$ (0.50 mol%)–$NdCl_3$ (0.50 mol%) melts at 723 K

Electrolysis potential value (V) (vs. Li^+/Li)	Electrolysis time (h)	Sn (atom%)	Dy (atom%)	Nd (atom%)
0.60	0.50	79.3	0.2	0.7
0.70	0.50	96.8	0.2	0.7

Fig. 5, a large cathodic current of Li–Sn alloys is confirmed at 0.60 V. When Li–Sn alloys are formed, the current efficiency of Nd–Sn alloys decreases. But the atomic ratio of Li couldn't be analyzed by XRF. And the main other element in theses samples was Si due to polishing these by sandpapers. The atomic ratio of Si was about 10–20%.

Furthermore, based on the results of cyclic voltammograms, alloy samples were prepared by potentiostatic electrolysis at 0.60, 0.70, 1.00 V for 0.50 h using liquid Sn electrodes in LiCl–KCl–$DyCl_3$ (0.50 mol%)–$NdCl_3$ (0.50 mol%) melts at 723 K. The formation of Nd–Sn alloys were confirmed by the XRF analysis of all samples. These results correspond to the results of Nd–Sn alloys in LiCl–KCl–$NdCl_3$ (0.50 mol%) melts. On the other hand, Dy–Nd–Sn alloys were formed at 0.60 and 0.70 V, and was not formed at 1.00 V. This indicated that a small cathodic current from 1.15 V in the negative scanning for LiCl–KCl–$DyCl_3$ melts (Fig. 5) didn't correspond to the formation of Dy–Sn alloys.

Table 2 shows results of atomic ratio of Sn, Dy, Nd in liquid Sn samples analyzed by XRF. The atomic ratios of Nd/Dy in the alloy samples were found to be 3.50 at 0.60 and 0.70 V. These results suggested that the formation potential of Nd–Sn alloys was more positive than that potential of Dy–Sn alloys. Considering these, the separation of Nd from Dy might be achieved by setting the optimum potential, i.e, over 1.00 V. But these results indicated that the separation of Dy and Nd highly by cathodic electrolysis (Fig. 2a), with liquid Sn metal was difficult. We will have to find other liquid metal by this process. On the other hand, as shown in Fig. 2b, selective Dy dissolution from Dy–Nd–Sn alloys might be possible by anodic electrolysis.

Conclusions

The electrochemical formation of RE–Sn (RE = Dy, Nd) alloys were investigated in LiCl–KCl eutectic melts containing $DyCl_3$ (0.50 mol%) and $NdCl_3$ (0.50 mol%). The results could be summarized as:

1. The formation of Nd–Sn alloys could be confirmed in all samples prepared by potentiostatic electrolysis at 0.60, 0.70, 0.80, 1.00 V for 0.50 h in LiCl–KCl–$NdCl_3$ melts.

2. The formation of Dy–Nd–Sn could be confirmed in the samples prepared by potentiostatic electrolysis at 0.60 and 0.70 V for 0.50 h in LiCl–KCl–$DyCl_3$–$NdCl_3$ melts. The atomic ratios of Nd/Dy in the alloy samples were found to be 3.50 at 0.60 and 0.70 V

Acknowledgements This work was supported by Grant-in-Aid for Scientific Research (B) from Japan Society for the Promotion of Science (JSPS).

References

1. H. Konishi et al., *J. Electrochem. Soc.*, vol. 148 (Pennington, NJ: Electrochemical Soc Inc, 2001), C506–511
2. H. Konishi et al., *Electrochem. Solid-State Lett.*, vol. 5 (Pennington, NJ: Electrochemical Soc Inc, 2002), B37–39
3. H. Konishi et al., *Electrochim. Acta*, vol. 47 (Kidlington, Oxford OX5 1GB: Pergamon-Elsevier Science Ltd, 2002), pp. 3533–3539
4. H. Konishi et al., *Electrochim. Acta*, vol. 48 (Kidlington, Oxford OX5 1GB: Pergamon-Elsevier Science Ltd, 2003), pp. 563–568
5. H. Konishi et al., *Electrochim. Acta*, vol. 48 (Kidlington, Oxford OX5 1GB: Pergamon-Elsevier Science Ltd, 2003), pp. 1403–1408
6. T. Oishi et al., *Kagaku Kogaku Ronbunshu*, vol. 36 (Soc Chemical Eng Japan, Bunkyo-ku, Tokyo, 2010), pp. 299–303
7. S. Kobayashi et al., *J. Electrochem. Soc.*, vol. 158 (Pennington, NJ: Electrochemical Soc Inc, 2011), E142–146
8. S. Kobayashi et al., *J. Electrochem. Soc.*, vol. 159, (Pennington, NJ: Electrochemical Soc Inc, 2012), E193–197
9. H. Konishi et al., *MOLTEN SALTS*, vol. 54 (Molten Salt Committee of The Electrochemical Society in Japan, Suita, Osaka, 2011), pp. 21–28
10. H. Konishi et al., *ECS transactions*, vol. 50 (Electrochemical Soc Inc, Pennington, NJ, 2012), pp. 463–472
11. H. Konishi et al., *ECS transactions*, vol. 53 (Electrochemical Soc Inc, Pennington, NJ, 2013), pp. 37–46
12. T.B. Massalski (ed.), *Binary alloy phase diagrams*, vols. 2, 3 (Materials Park, OH: ASM International, 1990), 1562, 2811

Challenges in the Electrolytic Refining of Silver—Influencing the Co-deposition Through Parameter Control

Ann-Kathrin Maurell-Lopez, Bernd Friedrich and Wolfgang Koch

Abstract Due to the high standards for silver in electronic uses, it is essential to control the behavior of the main impurities such as copper or palladium during the electrolytic refining. In regard to palladium, it is indispensable to control its complete transition into the anode slime. Established silver refining processes are currently operated with parameters that favor the silver cathodic deposition. Usually this requires a high precious metal stock and leads to low space-time yields. Parameter research studies show the possibility of changing these parameters to achieve a more efficient process without decreasing the silver grades. This research leads to the use of less pure electrolyte systems and more contaminated anodes. This also enables the use of increased current densities and accordingly increased production capacities. The research depicts the influence of the anode alloy, electrolyte composition, as well process parameters like current density and pH on the electrorefining. The applied electrode potential and therefore the applied current density shows great influence on the dissolution behavior of more noble metals than Ag at the anode. At the cathode, the applied current density influences the co-deposition of less noble metals, such as Cu. This phenomenon is based on the Nernst equation. The pH of the electrolyte system and thus the acid concentration influences the solubility behavior of the impurities. This leads to an influence on their concentration in the electrolyte and their co-deposition on the cathode.

Keywords Silver electrorefining · Palladium and copper deposition · Parameter influence

A.-K. Maurell-Lopez (✉) · B. Friedrich
IME Process Metallurgy and Metal Recycling,
RWTH Aachen University, 52056 Aachen, Germany
e-mail: amaurell@metallurgie.rwth-aachen.de

W. Koch
Agosi Allgemeine Gold-und Silberscheideanstalt AG,
Kanzlerstr. 17, 75175 Pforzheim, Germany

© The Minerals, Metals & Materials Society 2017

H. Kim et al. (eds.), *Rare Metal Technology 2017*,
The Minerals, Metals & Materials Series,
DOI 10.1007/978-3-319-51085-9_11

Introduction

From an economic point of view, it becomes more and more relevant to use less pure electrolyte systems and insert more contaminated anodes into the refining process. This is based on the otherwise high necessary silver stock. Additionally, economic reasons cause an increase in current densities and accordingly an increase in production capacities and a decrease in refining duration. Therefore, the impurity behavior needs to be controlled and influenced during the electrolytic refining process.

Basic electrolysis parameter through which the process can be influenced and controlled are the current density, hence the applied potential on the electrode, the acid concentration in the electrolyte, as well as the anode and electrolyte composition.

The paragenesis of silver in copper ores as well as the use of sterling silver as main jewelry alloy leads to high Cu contents in the crude silver anodes. Moreover, the use of silver palladium alloys in electrics and electronics leads to increasing Pd contents in the crude anodes. This explains the main impact these two elements have on the electrorefining and specifies the vital importance of considering their behavior and impact in the electrorefining process. Furthermore, the electrochemical similarity of Ag and Pd can lead to further economical disadvantages. The Pd content in the electrolyte needs to be monitored precisely in order to not be deposited on the cathode and therefore be handled as a loss. This research therefore identifies possible parameter ranges which allow a high grade silver production with less economic impacts.

Background

Electrolysis is a process which can be handled on different complexity levels. There are many parameters that show a great influence on the product. Furthermore, the parameters even have different and/or increased impact due to interdependencies on each other [1].

Independent of the silver origin, it is indispensable to refine silver electrochemically. This is necessary to produce high grade silver with an impurity content less than 0.01 wt%. These grades are needed especially for industrial use purposes. Furthermore, the silver electrolysis enables to separate the contained precious metals such as Au or PGM's. In the electrorefining, the crude silver anodes are dissolved in a nitric acid, silver nitrate solution and the fine silver is deposited on stainless steel cathodes. A typical cell arrangement, as shown in Fig. 1, is the Möbius cell.[1] Operating parameter such as $AgNO_3$ concentration, amperage or

[1]The dominant cell type in Europe is the Möbius cell, whereas in northern America and for anodes with high Au contents, the Balbach-Thum cells are preferred. Due to the cell construction, Möbius electrolysis only can handle up to 6-wt% Au, while anodes containing 20-wt% Au can be inserted in Balbach-Thum cells [2, 3].

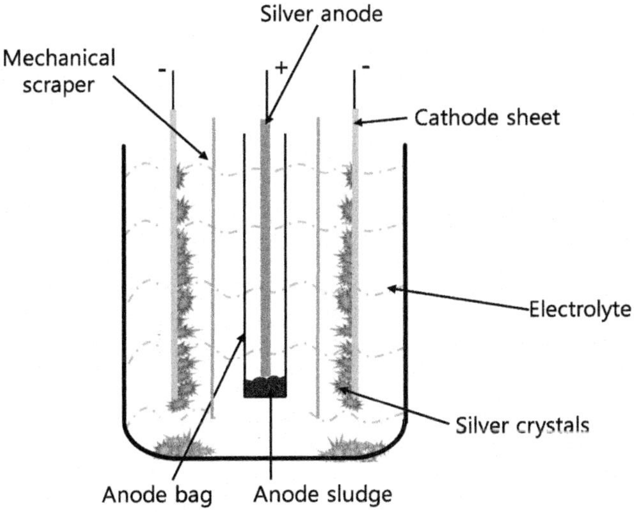

Fig. 1 Möbius silver electrolysis cell [7]

electrolyte temperature are set in a way, that ideally the less noble metals (e.g. Cu, Pb) will stay dissolved in the electrolyte. The more noble metals (e.g. Au, Pd, Pt) will not dissolve and pass into the anode slime. The anode slime is collected in anode bags around the dissolving anode. Under optimal conditions, only fine silver (purity >99.9%) will deposit on the cathode. An advantageous condition for the electrolytic deposition of Ag is the large potential difference between Ag and Cu (0.5 V). Thus, Cu should theoretically not co-deposit on the cathode. The containing Au and PGM's, however, are either insoluble in the electrolyte or have a too large standard potential and hence remain in the anode slime. Pd though behaves in a slightly different way. The higher the content of Ag in the electrolyte and the lower the current density, the purer is the deposited fine silver [2, 4–10].

Copper Co-deposition

Looking at the standard potential of Cu ($Cu/Cu^{2+} = 0.345$ V and $Cu/Cu^{+} = 0.522$ V), it is apparent that during electrorefining of Ag, Cu will mostly dissolve into and enrich in the electrolyte. Partial occurrence of Cu in the anode slime can be detected. This is more or less due to oxidic Cu present in the crude silver anodes. Theoretically, Cu should not be co-deposited on the Ag cathode, even when applying high current densities. Nevertheless, during operation, cathodic Cu co-deposition is always observed. This leads to the necessity of monitoring and controlling the Cu content in the electrolyte. Studies clearly depict a direct correlation of the Cu content in the electrolyte to the Cu co-deposition as shown in

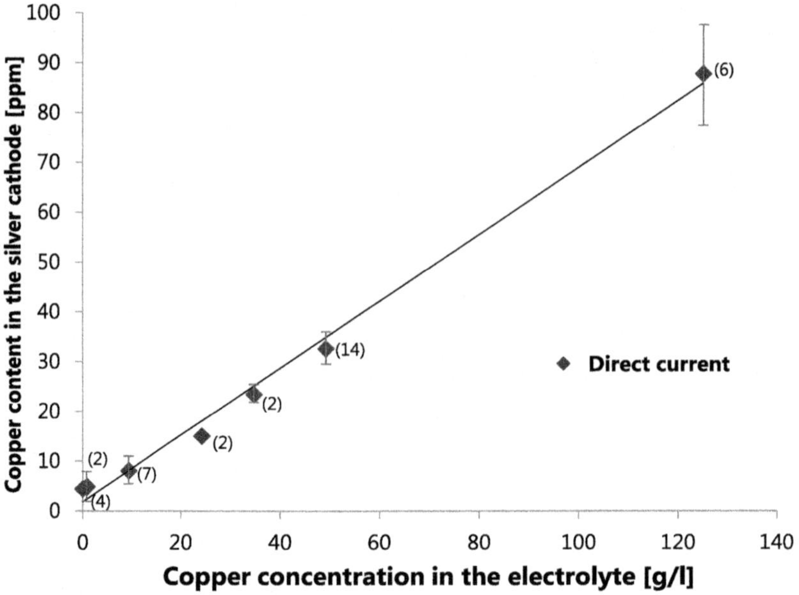

Fig. 2 Linear correlation of cathodic Cu co-deposition on the Cu concentration in the electrolyte (basic parameter: $c_{Ag} = 100$ g/l, $c_{HNO3} = 1$ g/l, $i_{cathode} = 550 - 600$ A/m^2, $T_{Electrolyte} = 25 \pm 1$ °C; set in *brackets*: quantity of trials to calculate mean value) [15]

Fig. 2. However, it is also apparent that the electrolytic Cu concentration can rise up to 50–100 g/l until it results in a significant Cu deposition (>50 ppm) on the cathode. This co-deposition on the cathode can either be electrochemically or due to an inclusion of Cu salt in the cathode silver [2, 11–17].

This shows the need to determine critical limits for Cu in the electrolyte as well as in the anodes due to anodic Cu dissolution. Ag anodes with a high Cu content also lead to high Cu impurity in the cathodes through the enrichment of Cu in the electrolyte (as discussed above). Moreover, in high Cu containing anodes (Cu > 30%, cf. [18]) cementation of Ag from the electrolyte can occur. This cemented Cu collects in the anode slime and therefore leads to Ag losses. Nevertheless, the critical Cu limits basically depend on the required purity of the cathodes. Three basic mechanism can lead to Cu impurity in the cathode based on high Cu contents in the electrolyte:

- Electrochemical co-deposition,
- Inclusion of Cu containing electrolyte in the cathode crystals,
- Hydrolysis reaction[2] of Cu salts and their subsequent precipitation as hydroxide during washing process of the fine silver crystals.

[2]Hydrolysis can occur during the washing process of fine silver, and therefore be accordingly eliminated by washing with acidified water followed by washing with a neutral hot water [15].

These mechanisms can be applied to all contained less noble elements in the electrolyte. Further studies depict a negligible influence of other electrolyte parameter such as cathodic current density variation (250–3000 A/m^2) or the electrolyte temperature on the Cu content in the cathode [7, 8, 11, 12, 15, 16, 18].

Palladium Co-deposition

Pd, however, shows a certain electrochemical similarity to Ag (standard potential 0.810–0.830 V). This leads to a definite Pd dissolution into the electrolyte during the electrorefining. Due to the small difference in the standard potential to Ag, small Pd concentrations lead to Pd co-deposition on the cathode. Ag rich electrolytes (min. 60 g/l Ag) as well as low acid concentrations can counteracted to this effect, as Pd (as well as Pt) preferably dissolves in electrolytes with low pH values. When looking at the dependency of the Pd concentration in the electrolyte on the content in the crude anode silver, a direct relation is visible. Plus, the Pd content in the cathode, directly connects to the Pd in the electrolyte. Both phenomena are displayed in Fig. 3. Research illustrates that even high Pd contents in the anodes (0.05–0.27%) do not necessarily lead to high Pd co-deposition. Even with this high Pd concentration, fine silver with <3 ppm Pd and <5 ppm Cu can be achieved. This is based on the use of slightly contaminated electrolytes (<20 mg/l Pd) and high pH values. Studies of Cu-Ni-PGM alloys show, that in these alloys the PGMs occupy atomic places. Therefore, parts of the alloy are more noble than others. This leads to a higher dissolution potential of these parts, than the rest of the crystal. When polarizing anodically, this part accordingly is not dissolved and remains as

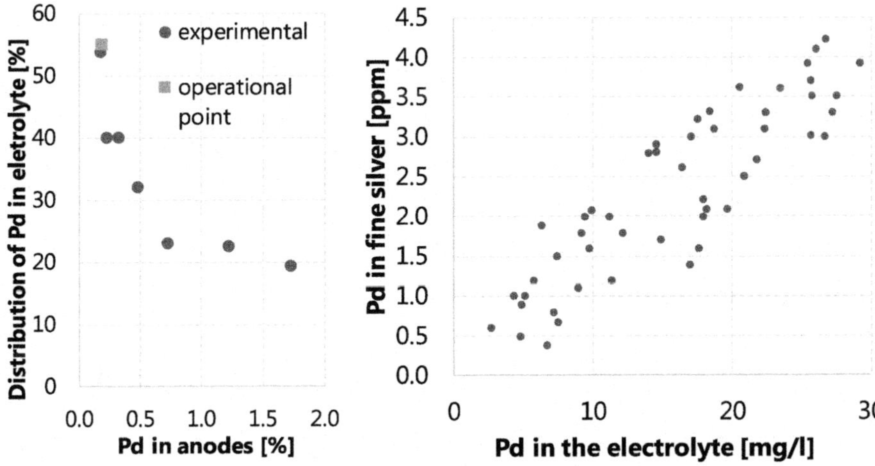

Fig. 3 Dependence of the Pd transition into the electrolyte on the Pd content in the anodes (*left*) and Ag quality depending on the Pd concentration in the electrolyte (*right*); both [16]

amorphous and very fine component in the anode slimes (very fine atomic agglomerations without crystalline structure) [11, 12, 15, 19–22].

Process Consideration and Research Strategy

The ultimate goal of the process improvement is to reduce the necessary metal stock while ensuring the required product purity. In silver refining electrolysis, Pd is a challenge which shall not be underestimated as the standard potential of Pd (E_0 = 0.987 V) is very close to Ag (E_0 = 0.799 V). Cu accumulates in the electrolyte and is also a challenge due to a co-deposition at the cathode when exceeding certain values that are too high in the electrolyte. Consequently, the electrolyte is continuously checked on the Cu content. When reaching certain limits, the electrolyte is exchanged. During the electrolysis process, the Ag concentration decreases while Cu increases. This offers two alternatives for handling the electrolyte; an entire removal of the electrolyte as soon as the Cu content in the cathode exceeds the specifications or a continuous exchange of part of the electrolyte. The latter results in a more uniform operational management and is therefore the preferable process handling. Cu in the electrolyte has indeed disadvantageous effects on the cathode product purity, but the presence of Cu ions increases the conductivity of the electrolyte. Hence, a certain presence of Cu ions during the process is desirable. This compensates for the loss of conductivity due to the decrease of the silver content (based on the created misbalance when anodically dissolving Cu^{2+}-ions while depositing Ag^{+}-ions on the cathode). The nitric acid content in the electrolyte should remain constant and the HNO_3 content of the electrolyte should not be too high (approximate value pH of 1.5–2.5). Otherwise this can lead to increased dissolution of Pd [2, 3, 11, 12].

The following Table 1 displays an overview on published processing data of industrial silver electrolysis. Based on this Table 1, the investigated parameter in this research were defined. The main focus lay on investigating high Cu (up to 100 g/l) and Pd (up to 1.4 g/l) contents in the electrolyte (and for Pd also in the anode) in order to still ensure a high grade silver production with moderate current densities and pH values whilst decreasing the Ag concentration in the electrolyte.

Experimental Validation

All experiments were conducted in an electrolytic refining laboratory cell, which is displayed in Fig. 4. The basic parameters are presented in Table 2. During the experiments, the acidic concentration, the anodic current density and the Ag-, Cu- and Pd-concentration in the electrolyte as well as the Pd-concentration in the anode

Table 1 Overview on published processing data for industrial silver electrolysis

Reference		[7]	[12]	[23]	[10]	[24]	[13]	[14]	[3]	[2]	[15]	[16]	[25]		[19]
i (A/m^2)	Anode	400	400–500	1000	<1000					400–500		360	500	430	1435[a]
	Cathode		460–500			300						335			645[a]
Ag concentration electrolyte (g/l)		50	>60		100–150	150	60–160		80–100	50	100	60	40	130	150–200
Cu concentration electrolyte (g/l)			Max. 50–80			Max. 45	60	Max. 60	5–10		15–50	3–5	35	10	
Pd electrolyte												20 mg/l			
Final electrolyte (g/l)		Ag: 50 Cu: 100							Ag: 30–40 Cu: 70						
Anode mass (kg)		15	3.9–4.2	45		21.77 (48 lbs)									12
Anodes	Ag (%)	98			95–99	86–92					>99%				99.30
	Cu	1%				0.5–1									0.4–0.6
	Pd					0.16–0.18%									68 ± 19 ppm
	Au (%)	0.50				8–9									0.04–0.07
Temperature electrolyte (°C)		45			<55	32						45–55			35–50
pH/c_{HNO3}						1–1.5	1.5–2.5		1–20 g/l	10 g/l	1 g/l	5 g/l	3.8	2 g/l	2.5 g/l
Anode rest (%)		5–10				15			25						
Current efficiency (%)		95			98–99										

[a]Calculated value from amperage and number anodes under the assumption that 60% of the surface are active

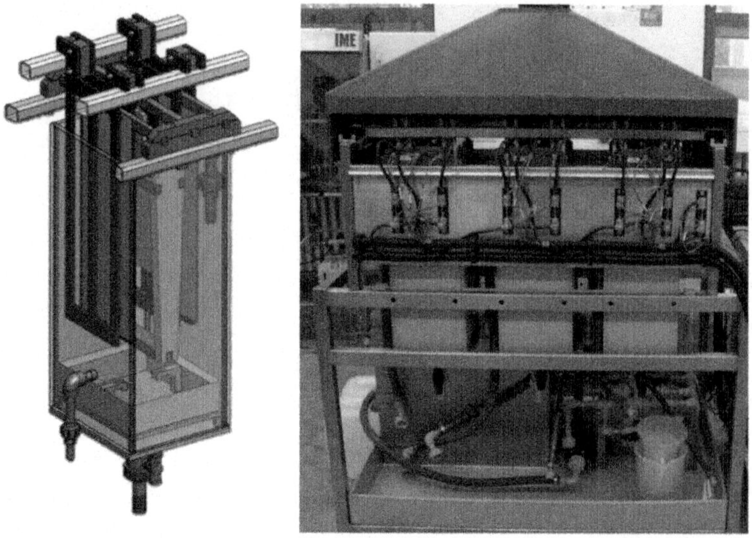

Fig. 4 Laboratory Ag electrorefining cells (3 cells in total) at IME Aachen

Table 2 Basic parameter

Parameter	Value	Electrolyte composition (varied values)		Anodic composition	
Anodes/cathodes per cell	1/2	Ag	65–120 g/l	Ag	96.5%
$V_{electrolyte}$ per cell	~27 l	Cu	45–98.6 g/l	Cu	3%
Electric charge per cell and trial	~580 Ah	Pd	0–1 g/l	Pd	0.05% (varied 0.01–0.1%)
Cathodic current density	335–535 A/m^2	HNO_3	0.6–10 g/l	Au	0.01%
Other parameter					
Anodic current density	500–800 A/m^2	Electrode distance anode-cathode		70 mm	
End of trials approx. i_A	660–1060 A/m^2	Active anode surface		25.7 × 10 × 2 cm	
Cathode active surface	15.6 cm × 30.5 cm	Electrolyte temp.		35 °C	
Cathode material	Stainless steel				

was varied. The main idea was, to increase the maximum Cu and Pd content in the electrolyte, as well as the maximum Pd content in the crude silver anodes. The main goal was to still produce high grade silver (>99.9%) even with the stated unfavorable conditions.

Results and Discussion

Taking into consideration the studied impurities Cu and Pd in Ag anodes, ideally Cu should dissolve and enrich in the electrolyte whereas Pd should be insoluble and pass completely into the anode slime.

Figure 5 presents the Pd concentration in the anode slime depending on different electrolyte compositions (with unvaried Pd impurity in the used anodes). The Pd content in the anode slime shows a significant dependence on its concentration in the electrolyte. Whereas the Cu content in the anode slime also shows a dependency on the Pd concentration in the electrolyte. With increasing Pd concentration in the electrolyte, the Cu content in the anode sludge decreases. Consequently, one can assume that palladium is anodically dissolved parallel to the Ag during electrolysis. Subsequently it reacts through cementation with the metallic Cu (or even Ag) present in the anode sludge according to the equation below. The structure of the anode sludge supports this assumption, since Cu only appeared as metal in the anode sludge.

$$Pd^{2+}_{electrolyte} + Cu^{0}_{anodesludge} \rightarrow Pd^{0}_{anodesludge} + Cu^{2+}_{electrolyte}$$

The subsequent Fig. 6 shows this dependence of the Pd and Cu content in the anode sludge on the Pd concentration in the electrolyte, without taking into account the concentrations of Cu and Ag in the electrolyte. In this diagram, the discussed influence becomes even more apparent.

With an increasing Ag concentration in the electrolyte, it basically can be observed, that the Cu co-deposition slightly decreases. An increasing Cu concentration in the electrolyte on the other hand leads to a slight increase in Cu co-deposition. If only the dependence of the Cu concentration in the produced fine

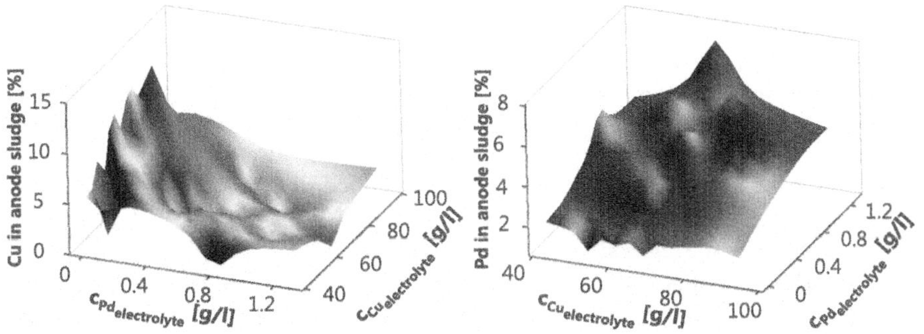

Fig. 5 Illustration of Cu (*left*) and Pd (*right*) content in the anode slime as a function of Cu and Pd electrolyte concentration (Ag influence on Cu and Pd anode sludge concentration was negligible at this point)

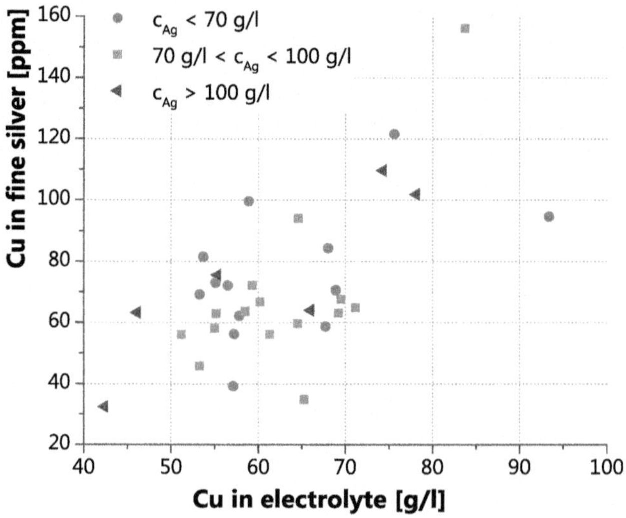

Fig. 6 Influence of the Cu concentration in the electrolyte on the Cu content in the fine Ag in electrolytes containing 60–100 g/l of Ag (parameter: c_{HNO3} = ~3 g/l, i_a = 700 A/m^2)

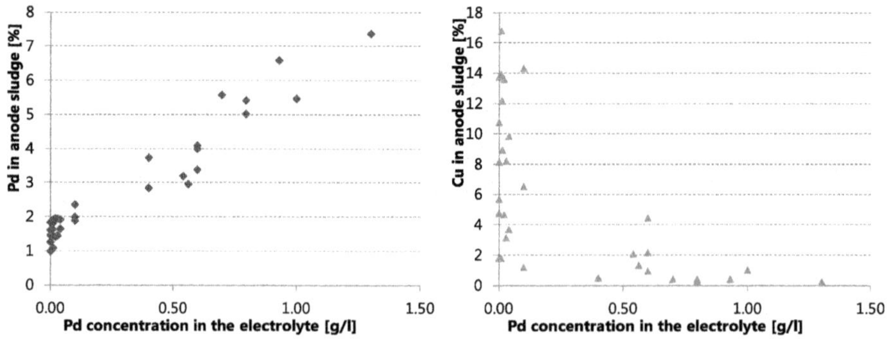

Fig. 7 Influence of Pd concentration in the electrolyte on the Pd (*left*) and Cu (*right*) content in the anode slime

silver on the Cu content in the electrolyte is applied, the Fig. 7 can be generated. With increasing Cu concentration in the electrolyte, a tendency of increasing Cu concentration in the cathode can be seen, regardless of the Ag electrolyte concentration. It is also apparent, that this is not a distinct linear dependency, rather a co-influence with the Ag electrolyte concentration. Whether the Cu concentration in the cathode is based on electrochemical deposition or rather based on electrolyte inclusions, cannot be clarified. Investigations of fine silver crystals via microprobe and SEM analysis were not able to identify whether Cu electrochemically alloyed in the silver matrix or was present as electrolyte inclusion. This is due to the detection

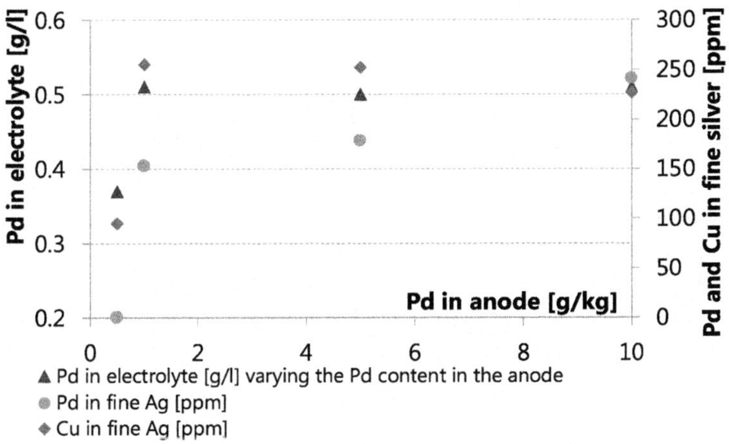

Fig. 8 Fine silver composition as well as palladium concentration in electrolyte depending on the Pd content in the anode (color figure online)

limits of the used analysis and the low Cu contents reached in the fine silver (<100 ppm).

The effect of the Pd content in the anode on the Pd and Cu concentration in the produced fine Ag as well as the electrolyte concentration is presented in Fig. 8. It is noticeable that elevated Pd contents in the anode lead to both increased Pd and Cu concentrations in the produced fine silver. This is based on the effect that an elevated palladium content in the anode should increase the anode potential. This causes a higher cathode potential and thus an increased deposition of Cu and Pd. Moreover, Pd contents >1 g/kg in the anode (green in Fig. 8) lead to higher Pd contents in the electrolyte (>0.5 g/l) and therefore also in the fine silver due to the dependency of the activity on the concentration of Pd in the electrolyte.

Figure 9 shows the generated results of the experiments dealing with the influence of the current density and the acid concentration on the Pd co-deposition. It can be seen that the acid concentration in the electrolyte is the main influence factor. Consequently, a higher acidity results in an increased Pd co-deposition on the cathode. This is based on the strong dependence of the Pd solubility in the electrolyte on the pH. The higher the acid content is, the greater concentrations of palladium in the electrolyte are reached. It can be concluded accordingly that the acidity is more or less an indirectly relevant parameter, whereas the palladium concentration in the electrolyte is crucial for Pd co-deposition. The anodic current density on the other hand tends to show a minor influence. Based on these results in order to not exceed critical Pd contents in the fine silver, the acid concentration generally should not extent 7 g/l.

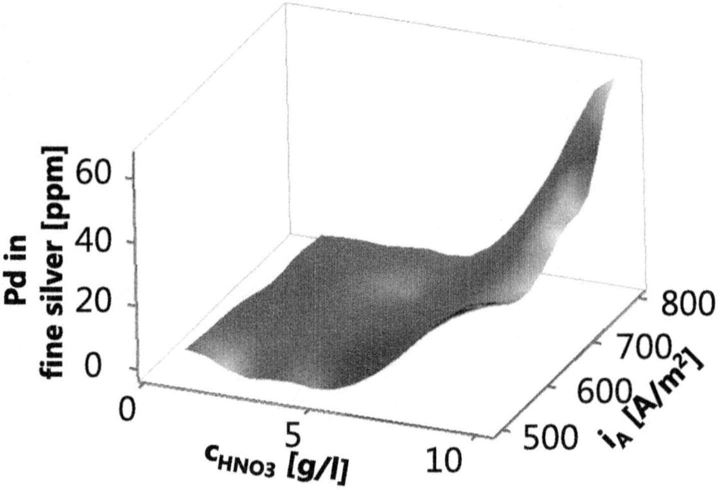

Fig. 9 Pd co-deposition dependent on the anodic current density and acid concentration

Conclusion

Concluding all test parameters, with respect to their influence on the copper content in produced the fine silver, the following Fig. 10 can be obtained. It is distinct that, although the copper concentration in the electrolyte is a non-negligible factor, an elevated palladium content in the anodes and a high acidity as well as high current densities also adversely affect the copper co-deposition. At least when the copper and silver concentrations in the electrolyte are almost at an equal level. This leads to the general conclusion, that in order to ensure low copper levels in the produced fine silver, the palladium content in the crude anodes should not be too high. Moderate current densities (i_a up to 600 A/m^2) and acid concentrations enable to operate electrolytic refining of silver with copper contaminated electrolyte without jeopardizing the silver quality.

Overall, the experimental results demonstrate, that the pH of the electrolyte system shows a great influence on the behavior of Pd. At least in a range of anodic current densities from 500 to 800 A/m^2 and a nitric acid concentration from 1 to 10 g/l. Consequently, it is indispensable to monitor the pH for an electrolyte refinery if the production rate and thus the current density shall be increased. However, monitoring and controlling the pH is a complex topic especially on a production site. One possibility is to add a manual acidic concentration titration in the electrolyte analysis. In order to continuously monitor the pH, an electrolyte resistant pH electrode can be included into the electrolyte flow streams.

As exemplified above, one of the main influence parameter on a palladium co-deposition is the palladium content in the crude anode. Concluding the varied parameter, Fig. 11 is prepared. This figure does not consider an elevated palladium

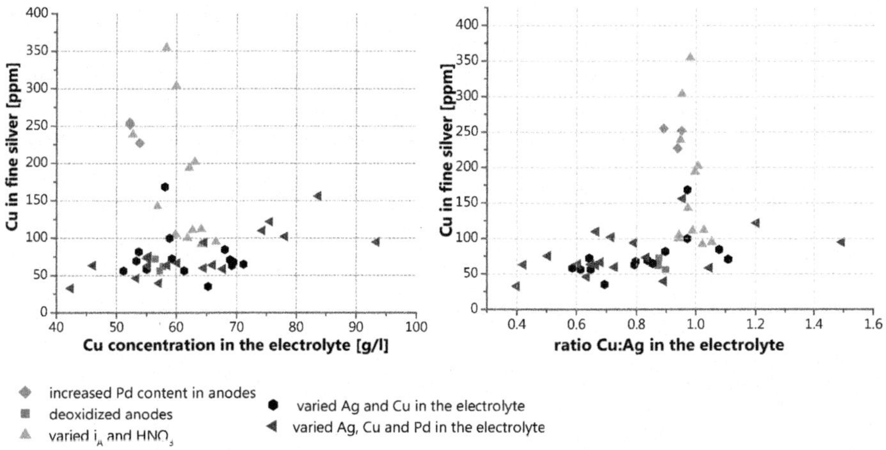

Fig. 10 Influence of the copper concentration in the electrolyte (*left*) and of the copper to silver ratio in the electrolyte (*right*) on the copper content in the fine silver as a function of the varied experimental parameters

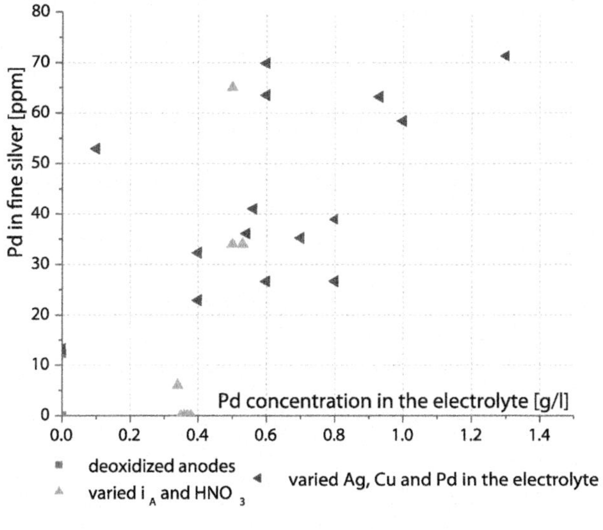

Fig. 11 Influence on the palladium content in the fine silver depending on the palladium content in the electrolyte as well as the acid concentration

anode concentration in order to identify other influencing parameter. Therefore, Fig. 11 prevails dependencies of the palladium content in the produced fine silver as a function of the palladium concentration in the electrolyte and therefore also the pH.

These results furthermore indicate that monitoring the palladium content in the anode and the electrolyte, as well as the acidity are essential for controlling the behavior of palladium. Ensuring low acid concentrations and thus higher pH during the refining, process lead to low palladium co-deposition. If, however, heavily palladium contaminated anode shall be inserted in the process, the use of separate anode and cathode chambers as a suggestion can be recommended. To avoid the contact of palladium-rich electrolyte and the cathode altogether, an intermediate palladium precipitation from the anolyte and a fine filtration is suggested. Afterwards this purified electrolyte can be supplied as a catholyte in the cathode chamber.

This research study is part of the authors dissertation thesis "Behavior of Copper and Palladium in the Electrolytic Refining of Recycling Silver", to reach the academic degree Dr.-Ing.

Acknowledgments The author would like to thank the Agosi, Allgemeine Gold- und Silberscheideanstalt, for enabling and supporting the entire collaborative research project.

References

1. M.J. Fischer, Beitrag zur Bildung und zum Verhalten des Anodenschlammes bei der Kupferraffinationselektrolyse, Dissertation, Technischen Universität Bergakademie Freiberg, Juli 1998
2. H. Renner et al., Silver, silver compounds, and silver alloys, in *Ullmann's Encyclopedia of Industrial Chemistry* (Wiley-VCH Verlag GmbH & Co. KGaA, Weinheim, Germany), ISBN 3-527-30673-0
3. J. Hoffmann, The process chemistry of the extraction and electrowinning of silver, p. 273–287, in *Electrometallurgy: Proceedings of 31st annual Hydrometallurgy Meeting*, 26–29 September 2001
4. A. Holleman, E. Wiberg, N. Wiberg, *Lehrbuch der anorganischen Chemie*, 101st edn. (de Gruyter, Berlin [u.a.]). ISBN 978-3-110-12641-9
5. Umicore, *Precious materials handbook*, 1st edn. (Vogel Business Media, Würzburg). ISBN 3-834-33259-3
6. A.G. Degussa, *Edelmetall-Taschenbuch*, 2nd edn. (Hüthig, Heidelberg). ISBN 978-3-778-52448-0
7. G. Cornelius, Die Raffination von Gold und Silber durch Elektrolyse, p. 215–226, in *Elektrolyse der Nichteisenmetalle: 11. Metallurgisches Seminar*, Schriftenreihe der Gesellschaft deutscher Metallhütten- und Bergleute, *GDMB*, vol. 37. (Clausthal-Zellerfeld)
8. J. Wanderer, Silber & gold—elektrolyse. Metall. J. **1**(4), 2–27 (1997)
9. J. Falbe, M. Regitz, H. Römpp, *Römpp Chemie Lexikon*, 10th edn. (G. Thieme, Stuttgart). ISBN 3-137-35009-3
10. M. Maliarik et al., High current density silver electrorefining process: technology, equipment, automation and Outotec's Silver Refinery Plants, in *Hydrometallurgy: Proceedings of the 7th International Symposium*, vol. 2, pp. 91–100
11. V. Tafel, *Lehrbuch der Metallhüttenkunde Band I*, 2nd edn. (S. Hirzel Verlagsbuchhandlung, Leipzig)
12. F. Heinz, *Erzeugung von NE-Metallen*, 1st edn. (Deutscher Verlag für Grundstoffindustrie, Leipzig)
13. W. Hunter, *Verfahren zum Verfeinern bzw. Reinigen von Silber*, Patent DT 25 42 020 A1, 1975
14. A. Prior, *Elektrolytisches Silberraffinationsverfahren*, Patent 0 214 116 A1, 1986
15. I. Haranczyk, J. Sedzimir, Copper and Lead in the Electrorefined Silver Their Recovery from the Waste Solution. Arch. Metall. **40**(3), 285–295 (1995)

16. T. Shibasaki et al., Recent improvements in silver electrolysis at Mitsubishi's Osaka Refinery, in *Precious Metals, TMS*, pp. 419–430
17. G. Eger, Fortschritte in der Scheidung edelmetallhaltiger Legierungen. Zeitschrift für angewandte Chemie **37**(11), 137–144 (1924)
18. A. Prior et al., *Verfahren einer kombinierten Silber-Kupfer-Elektrolyse unter Einsatz einer Flüssig-Flüssig-Extraktion*, pp. 243–251, in *Raffinationsverfahren in der Metallurgie* (Verlag Chemie, Weinheim). ISBN 3-527-26127-3
19. M. Jaskula, R. Kammel, Untersuchungen zur Verbesserung des Platinmetallausbringens bei der industriellen Silberraffination. Metall **51**(7–8), 393–400 (1997)
20. V. Borbat et al., Forms of occurence of platinum metals in electrolytic slimes, Zh. Prikl. Chim. (Journal of Applied Chemistry of the USSR, transl. from Russian), **55**(11), 2235–2238 (1982)
21. R. Kammel et al., Das Anodische Verhalten von Palladium in Wässrigen Lösungen. Metalloberfläche **30**(9), 414–418 (1976)
22. M.J.N. Pourbaix et al., Electrochemical Properties of the Platinum Metals. Platin. Met. Rev. **3**, 100–106 (1959)
23. B. Ludvigsson, S. Larsson, Anode slimes treatment: the boliden experiment, J. Metals, 41–44 (2003)
24. A. Leigh, Precious metals refining practice, in *2nd International Symposium on Hydrometallurgy*, pp. 95–110
25. O. Johnson, *Chapter 5—Refining Processes*, in *Silver: Economics, Metallurgy, and Use, Butts, A; Coxe, C. D.* (D. Van Nostrand Company, Inc., Princeon, New Jersey)

Vapor Treatment for Alloying and Magnetizing Platinum Group Metals

Yu-ki Taninouchi and Toru H. Okabe

Abstract Among platinum group metals (PGMs), Pt, Pd, and Rh are the essential constituents of automotive catalysts. Recovery of PGMs from catalyst scrap is important, but it is difficult because PGMs are chemically stable and are present as minor components in the scrap. In this study, the authors investigated the reaction between PGMs and $FeCl_2$ vapor in order to develop a novel method for the separation and concentration of PGMs directly from catalyst scrap. The wire samples of Pt, Pd, and Rh were reacted with $FeCl_2$ vapor in a steel vessel maintained at 1200 K. After the heat treatment, the surfaces of PGM wires were alloyed with Fe, and these samples became magnetized. The results obtained in this study suggest that a $FeCl_2$ vapor treatment followed by magnetic separation can be utilized as an effective technique for the separation of PGMs directly from catalyst scrap.

Keywords Platinum group metals · Recycling · Iron chloride · Alloying treatment · Vapor treatment

Introduction

Although platinum group metals (PGMs) are widely used in a variety of industrial applications such as catalysts, electrodes, and crucible materials, they are limited in supply and are expensive [1–3]. PGMs are among the rarest elements in the Earth's crust, and their mineral resources are highly localized in South Africa and Russia [1–3]. Furthermore, the PGM concentration in natural ores is as low as 5 ppm, even in the highest-grade ores [2, 4, 5]; hence, their mining and smelting generates large quantities of waste and consumes a huge amount of energy. The recovery of PGMs from end-of-life products or scraps is therefore important for ensuring a steady supply of PGMs and for reducing the environmental burden.

Y. Taninouchi (✉) · T.H. Okabe
Institute of Industrial Science, The University of Tokyo,
4-6-1 Komaba, Meguro-ku, Tokyo 153-8505, Japan
e-mail: taninou@iis.u-tokyo.ac.jp

© The Minerals, Metals & Materials Society 2017

H. Kim et al. (eds.), *Rare Metal Technology 2017*,
The Minerals, Metals & Materials Series,
DOI 10.1007/978-3-319-51085-9_12

Among various industrial applications, automotive catalysts account for a large portion of the world's consumption of Pt, Pd, and Rh [1]. Automotive catalysts are used for detoxifying the exhaust gas from automobiles and usually consist of a honeycomb-structured ceramic body and a porous catalyst layer supporting the fine particles of Pt, Pd, and Rh. The total PGM concentration in spent automotive catalyst is approximately 500–5000 ppm [6], which is nearly hundred to thousand times higher than that in their natural ores.

As shown in Fig. 1, the demand of PGMs for automotive catalysts has been steadily increasing because of the increasing automobile production and stricter environmental regulations [1, 7]. The amount of PGMs recovered from automotive catalyst scrap will increase markedly in the future.

Automotive catalyst scrap is the most important secondary resource for PGMs. Figure 2 shows the typical steps involved in the recovery of PGMs from catalyst scrap [2, 3, 6, 8–10]. In general, the catalysts collected from used automobiles are mechanically pulverized and are then subjected to pyrometallurgical or hydrometallurgical recycling processes.

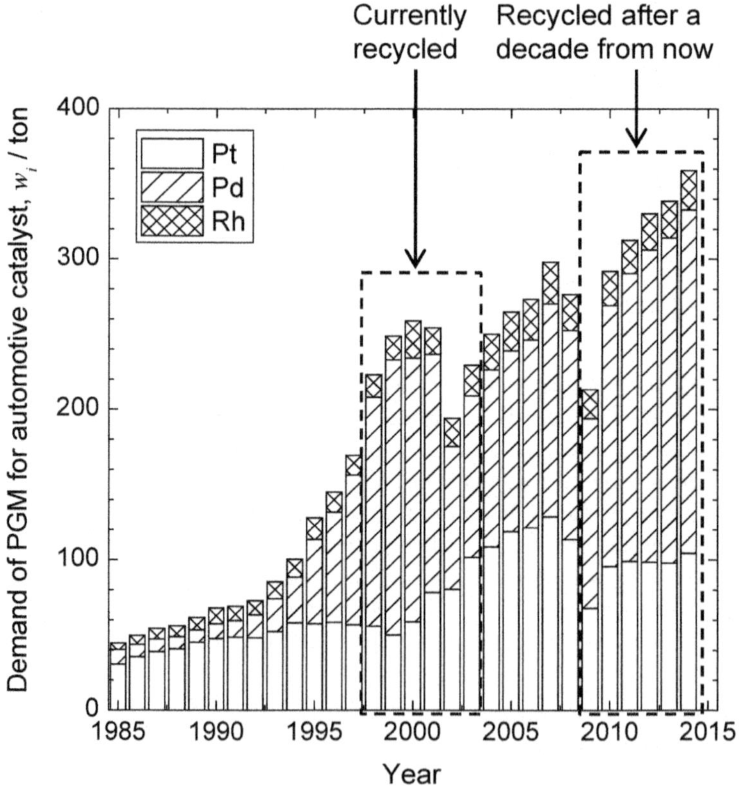

Fig. 1 Demand of Pt, Pd, and Rh for use in automotive catalysts [1, 7]

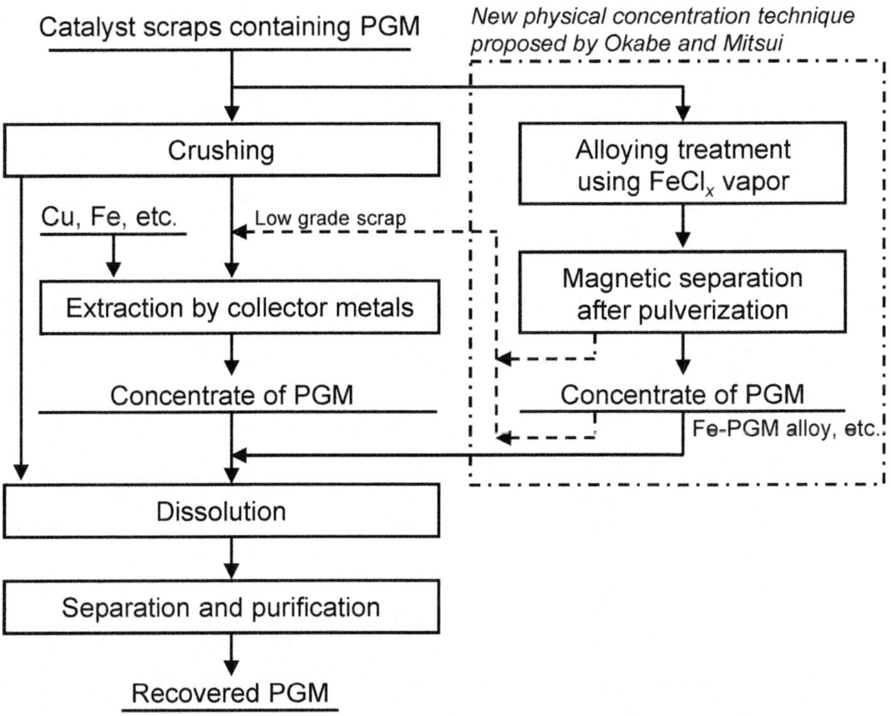

Fig. 2 Typical steps involved in the recovery of PGMs from catalyst scrap [2, 3, 6, 8–10] and a new technique for physical concentration of PGMs directly from catalyst scrap [18]

In pyrometallurgical recycling processes, the PGMs in the scrap are extracted and concentrated using a molten metal such as Cu or Fe as the collector, and the ceramic components in the catalyst scrap are removed as slag waste. Subsequently, the extracted PGMs in the form of metal alloys are dissolved in an aqueous solution, followed by their separation and purification using various techniques such as solvent extraction, precipitation, and ion exchange. The benefits of employing pyrometallurgical processes include their high PGM recovery rate and high throughput. However, these processes require large-scale equipment and consume large amounts of energy.

In hydrometallurgical recycling processes, the PGMs are dissolved directly from the spent catalysts using acids containing strong oxidants such as aqua regia. These processes can be carried out in a small-scale plant with less energy consumption. However, because PGMs are chemically stable and are present as minor components in scrap, direct-dissolution processes have drawbacks such as low PGM recovery rate (especially for Rh) and the generation of large volumes of toxic waste solutions and gases.

The total PGM concentration in spent catalysts is less than 1 mass%, and the remaining mass usually consists of ceramics such as cordierite and alumina, which are converted into waste during the pyrometallurgical and/or hydrometallurgical processes. Physical concentration of PGMs prior to metallurgical processes is therefore advantageous for improving the efficiency and treatment speed of the existing recycling processes [11].

Various techniques such as magnetic separation after pulverization [12, 13], flotation after pulverization [13, 14], selective grinding followed by size separation [15, 16], and selective grinding after a heating/quenching treatment [17] have been used for the physical concentration of PGMs from catalyst scrap. However, these techniques are still in the research stage or are not practical for use in industrial operations. Recently, Okabe and Mitsui proposed new physical concentration techniques which involve chemical pretreatment of PGMs [18]. In their patent, they claimed that PGMs in catalyst scrap can be alloyed with Fe by making them react with $FeCl_x$ ($x = 2, 3$) vapor and the alloyed PGMs can then be effectively concentrated by magnetic separation (see Fig. 2). This new technique is attractive because PGMs in the scrap can be concentrated in the form of Fe-PGM alloys, which are easily dissolved in acid. However, experimental results about this technique are quite limited, and they include some uncertainties. Previously, the alloying treatment was carried out using $FeCl_3$ (ferric chloride) as the vapor source [18, 19]. Powder samples of PGMs (Pt, Pd, and Rh) were heated in the presence of $FeCl_3$ vapor at 973–1273 K, and Fe-alloyed PGMs were obtained [18, 19]. Fe-alloyed Pt was also obtained when the mixture of Pt powder and Fe plate was heated in the presence of $FeCl_3$ vapor at 973–1173 K [19]. In this study, the reaction between PGMs (Pt, Pd, and Rh) and $FeCl_2$ (ferrous chloride) vapor was investigated in order to further examine the feasibility and effectiveness of the alloying treatment using $FeCl_x$ vapor.

Experimental

Figure 3 shows the experimental apparatus used in this study. The wire samples of Pt, Pd, and Rh (diameter (ϕ) = 0.5 mm; length (l) = $\sim$20 mm; Tanaka Kikinzoku Kogyo K.K.) were placed in a quartz crucible, which was then placed in a vessel made up of mild steel (see Fig. 3a). At the bottom of the steel vessel, a mixture of Fe (1.06 g; >95%; Wako Pure Chemical Industries, Ltd.) and $FeCl_2$ powders (6.05 g; 99.9%; Wako Pure Chemical Industries, Ltd.) was placed as the $FeCl_2$ vapor source. The PGM wire samples were not physically connected to the vessel or the mixture of Fe and $FeCl_2$ powders. The steel vessel containing the samples was covered by steel cap, and then positioned at the bottom of a vertical gas-tight quartz tube (see Fig. 3b). After the interior of the tube was evacuated, the quartz tube was then introduced into a vertical furnace at an elevated temperature to maintain the steel vessel at 1200 K for approximately 1 h. During the heat treatment, the top part of the quartz tube was cooled by a fan.

After the heat treatment, the quartz tube was removed from the furnace and cooled in air. The PGM wire samples were then recovered from the quartz crucible, and their cross-sections were analyzed by scanning electron microscopy (SEM) and energy-dispersive X-ray spectroscopy (EDS) using a JSM-6510LV (JEOL) system.

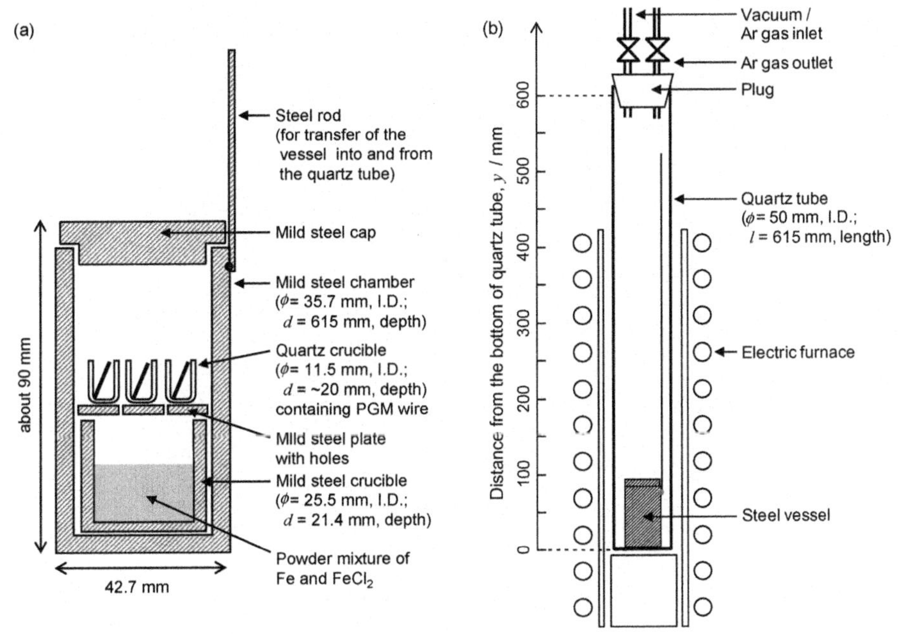

Fig. 3 Schematic illustration of the experimental apparatus. **a** Steel vessel. **b** Gas-tight quartz tube

Results and Discussion

After the heat treatment, $FeCl_2$ had disappeared from the bottom of the steel vessel and was deposited at the top part of the quartz tube. This is because the vapor pressure of $FeCl_2$ at 1200 K is as high as 0.3 atm [20] and the generated $FeCl_2$ vapor diffused upward according to the temperature gradient in the quartz tube.

Figure 4 shows the Rh wire obtained after the heat treatment. The heat treatment in the presence of $FeCl_2$ vapor caused the Pt, Pd, and Rh wires to become magnetized. Figures 5, 6, and 7 show the results of the SEM/EDS analysis of the cross-sections of the Pt, Pd, and Rh wires, respectively. The results indicated that the surface of the PGM wires (approximately 10 μm) was alloyed with Fe.

The results of this study suggest that upon reacting with $FeCl_2$ vapor at around 1200 K for approximately 1 h, Pt, Pd, and Rh are alloyed with Fe and are magnetized. Therefore, $FeCl_2$ vapor treatment followed by pulverization and magnetic separation (see Fig. 2) is a feasible and useful technique for the physical concentration of PGMs from catalyst scrap. The authors speculate that the alloying of the PGM samples with Fe proceeded with the disproportionation of $FeCl_2$ vapor:

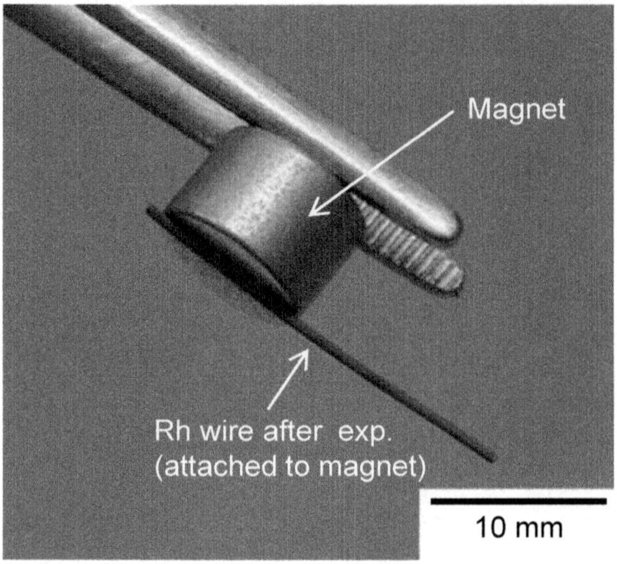

Fig. 4 Rh wire after reacting with $FeCl_2$ vapor at 1200 K (cf. Fig. 7c)

$$3\,FeCl_2(g) = Fe(s,\ in\,PGMs) + 2\,FeCl_3(g) \tag{1}$$

$$3\,FeCl_2(g) = Fe\ (s) + 2\,FeCl_3(g) \tag{2}$$

$$\Delta G^\circ_{(2)} = 126\ \text{kJ at 1200 K [20]}.$$

However, the detailed mechanism of the alloying reaction is still under investigation. As the disproportionation of $FeCl_2$ (reaction (1)) proceeds, $FeCl_3$ vapor is generated as a by-product. In parallel with the disproportionation of $FeCl_2$ vapor, the thermal decomposition of $FeCl_3$ vapor may proceed at the surface of PGMs.

$$FeCl_3(g) = Fe(s,\ in\,PGMs) + 3/2\,Cl_2(g) \tag{3}$$

$$FeCl_3(g) = Fe\ (s) + 3/2\,Cl_2(g) \tag{4}$$

$$\Delta G^\circ_{(4)} = 229\ \text{kJ at 1200 K [20]}.$$

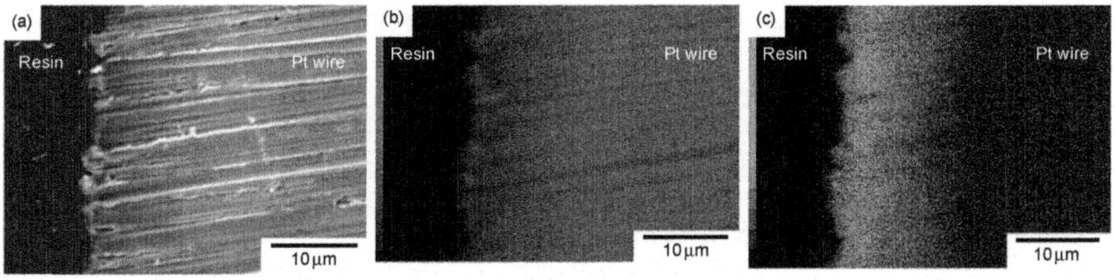

Fig. 5 **a** SEM image of the cross-section of Pt wire after the heat treatment, and the corresponding EDS images of **b** Pt and **c** Fe

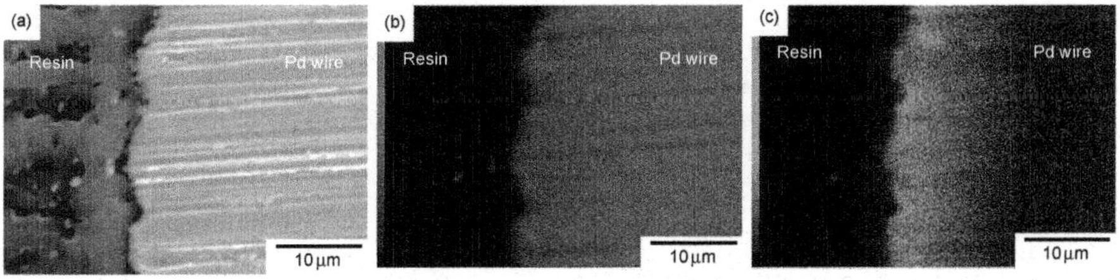

Fig. 6 **a** SEM image of the cross-section of Pd wire after the heat treatment, and the corresponding EDS images of **b** Pd and **c** Fe

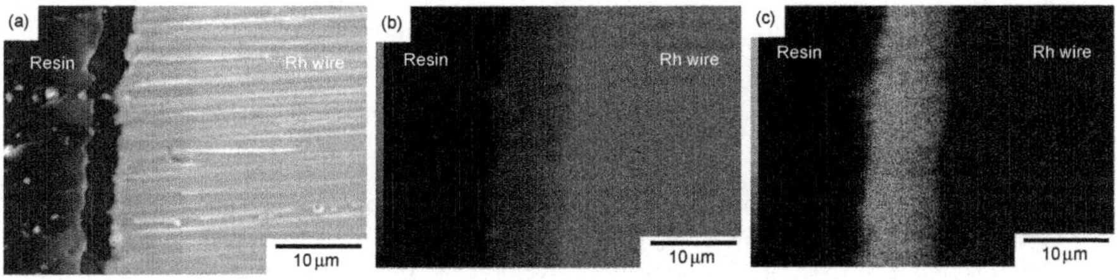

Fig. 7 **a** SEM image of the cross-section of Rh wire after the heat treatment, and the corresponding EDS images of **b** Rh and **c** Fe (cf. Fig. 4)

Summary

In order to develop a novel technique for the physical concentration of PGMs from catalyst scrap in a more efficient and environmentally friendly manner, the authors investigated the reaction between PGMs (Pt, Pd, and Rh) and $FeCl_2$ vapor. The wire samples of Pt, Pd, and Rh were reacted with $FeCl_2$ vapor in a steel vessel maintained at 1200 K. The heat treatment resulted in the Fe-alloying of the surface of wire samples. As a result, the samples became magnetized. The results obtained in this study are still in the fundamental stage and further studies are needed to elucidate the mechanism of the alloying reaction. However, it can be concluded that an alloying treatment using $FeCl_2$ vapor followed by pulverization and magnetic separation is a feasible technique for effectively extracting PGMs from catalyst scrap.

Acknowledgments The authors thank Dr. Akihiro Yoshimura and Mr. Ryohei Yagi (The University of Tokyo) for their valuable suggestions and comments. This research was financially supported by the Japan Society for the Promotion of Science (JSPS) through a Grant-in-Aid for Scientific Research (S) (KAKENHI Grant No. 26220910).

References

1. J. Matthey, *PGM marker report May 2016 summary of platinum supply & demand in 2015* (Johnson Matthey PLC, Royston, UK, 2016). http://www.platinum.matthey.com/services/market-research/pgm-market-reports
2. F.K. Crundwell et al., *Extractive metallurgy of nickel, cobalt and platinum-group metals* (Elsevier, Oxford, UK, 2011)
3. F. Habashi (ed.), *Handbook of extractive metallurgy* (VCH Verlagsgesellschaft mbH, Weinheim, Germany, 1997), pp. 1269–1326
4. A.J. Naldrett et al., Compositional variation within and between five sudbury ore deposits. Econ. Geol. **77**, 1519–1534 (1982)
5. R.T. Jones, An overview of Southern African PGM smelting, in *Proceedings of the International Symposium on Nickel and Cobalt* (2005), pp. 147–178
6. R.K. Mishra, A review of platinum group metals recovery from automobile catalytic converters, in *Proceedings of the 17th International Precious Metals Conference* (1993), pp. 449–474
7. Johnson Matthey Platinum Today. http://www.platinum.matthey.com/
8. J.E. Hoffmann, Recovering platinum-group metals from auto catalysts. J. Met. **40**(6), 40–44 (1988)
9. H. Dong et al., Recovery of platinum group metals from spent catalysts: a review. Int. J. Miner. Process. **145**, 108–113 (2015)
10. M.K. Jha et al., Hydrometallurgical recovery/recycling of platinum by the leaching of spent catalysts: a review. Hydrometallurgy **133**, 23–32 (2013)
11. T.H. Okabe, K. Nose, Y. Taninouchi, Development of environmentally sound recycling process of precious metals, in *Proceedings of MMIJ Spring Meeting* (2014) (in Japanese)
12. S. Owada et al., The existent state of platinum group metals in spent automobile catalysts and the possibility of the recovery by magnetic separation, in *Proceedings of the 2nd International Symposium on East Asian Resources Recycling Technology* (1993), pp. 69–77
13. S. Owada, Y. Tsubuku, H. Nakayama, Concentration of PGMs from automobile catalysts by magnetic separation or flotation, in *Proceedings of MMIJ Spring Meeting* (1994), pp. 282–283 (in Japanese)
14. S. Owada, K. Shinoda, Fundamental study on flotation of automobile catalysts for separating alumina coat layer from cordierite body, in *Proceedings of MMIJ Spring Meeting* (2006), pp. 69–70 (in Japanese)
15. W. Kim et al., Selective recovery of catalyst layer from supporting matrix of ceramic-honeycomb-type automobile catalyst. J. Hazard. Mater. **183**, 29–34 (2010)
16. G. Liu et al., Surface-grinding kinetics for the concentration of PGMs from spent automobile catalysts by attritor surface grinding. Mater. Trans. **55**(6), 978–985 (2014)
17. G. Liu, A. Tokumura, S. Owada, Separation of PGMs bearing alumina phase from cordierite in spent automobile catalyst by thermal shock. Resour. Process. **60**, 28–35 (2013)
18. T.H. Okabe, J Mitsui, Japan Patent, No. P2013-147702A (2013), P5946034 (2016) (in Japanese)

19. J. Mitsui, Development of new recycling techniques for platinum group metals by utilizing physical separation. Master thesis, The University of Tokyo (2012) (in Japanese)
20. I. Barin, *Thermochemical data of pure substances*, 3rd edn. (VCH Verlagsgesellschaft mbH, Weinheim, Germany, 1995)

Biotechnological Recovery of Platinum Group Metals from Leachates of Spent Automotive Catalysts

Norizoh Saitoh, Toshiyuki Nomura and Yasuhiro Konishi

Abstract This paper describes a new recycling method, based upon biotechnology, in order to extract platinum group metals (PGMs) from post-consumer products. The metal ion-reducing bacterium, *Shewanella algae*, was found to exhibit the ability to reduce and deposit the PGMs ions [platinum (IV), palladium (II) and rhodium (III)] into metal nanoparticles at room temperature and neutral pH within 60 min, using formate as the electron donor. We have collected fundamental data demonstrating that the ability of *S. algae* cells can be applied to the recovery of PGMs from dilute solutions. When targeting leachates of spent automotive catalysts, *S. algae* cells can successfully allow rapid reduction and deposition of PGM ions, proposing a new bio-recovery system of PGMs from spent automotive catalysts (three-way catalysts for minimizing automobile emissions). Our proposed biotechnology is linked to the development of low cost, eco-friendly recycling technology that enables the rapid recovery of PGMs utilizing microbial reactions at ambient temperature and ambient pressure.

Keywords Recycling · Platinum group metals · Automotive catalysts · Biomineralization

Introduction

It is widely accepted that about 30% of platinum group metals (PGMs) consumption comes from recycled end-of-use-scrap. Production of PGMs from scrap has the advantages of using less energy than metal production from ore and avoiding the production of mine waste products. We believe that the most important domestic source of PGMs can be found in increased recycling, but it will require more research and further development to recycle such metals.

N. Saitoh · T. Nomura · Y. Konishi (✉)
Department of Chemical Engineering, Osaka Prefecture University,
1-1 Gakuen-cho, Sakai, Osaka 599-8531, Japan
e-mail: yasuhiro@chemeng.osakafu-u.ac.jp

© The Minerals, Metals & Materials Society 2017
H. Kim et al. (eds.), *Rare Metal Technology 2017*,
The Minerals, Metals & Materials Series,
DOI 10.1007/978-3-319-51085-9_13

Although conventional chemical or thermal recycling techniques are often the most appropriate means of recovering precious and rare metals, biological methods provide an attractive and eco-friendly alternative strategy, in which microorganisms are used to separate and concentrate soluble metals from dilute solutions into microbial cells. Biological systems generally operate at room temperature and atmospheric pressure. Consequently, microbial recovery of metals would be an environmentally friendly method for recycling PGMs from post-consumer products.

We have recently developed new recovery technologies that are designed to reduce and deposit the precious metal ions of platinum (IV), palladium (II) and rhodium (III) as metallic nanoparticles, a process known as "biomineralization" [1–7]. This paper describes a new recycling method, based upon biotechnology, in order to recover PGMs from leachates of spent automotive catalysts utilizing microbial reactions at room temperature and atmospheric pressure.

Experimental Section

Ceramic Substrate Catalysts

Ceramic car catalyst substrate blocks were crushed, and the PGMs-coated substrates used in this study were ground to obtain the size range <30 μm. The crushed catalytic materials contain a large proportion of synthetic cordierite ($(Mg, Fe)_2Al_4Si_5O_{18}$) and only a small amount of PGMs (platinum, palladium and rhodium). The metal contents in the crushed catalytic materials were 0.2 wt% PGMs (platinum, palladium and rhodium), 20.9 wt% aluminum, 12.1 wt% silicon, 4.7 wt% magnesium, 4.8 wt% rare earths (cerium and lanthanum), and 1.9 wt% iron, respectively.

Microorganism and Growth Conditions

For the recovery of PGMs from aqueous solutions, we have focused on the metal ion-reducing bacteria, *Shewanella algae* (ATCC 51181) and *Shewanella oneidensis* (ATCC 700550). The *Shewanella* bacteria (Fig. 1) are a type of ferric iron-reducing microorganism and thus are able to reduce iron (III) ions to iron (II) ions by electron transfer. The electrons are obtained from the oxidation of organic acid salts such as lactate and formate. The *Shewanella* bacteria are safe non-pathogenic microorganisms that can multiply rapidly in the presence of general inexpensive nutrients. Therefore, it is not difficult to prepare the large quantities of *Shewanella* bacteria necessary for the microbial recovery of precious metals.

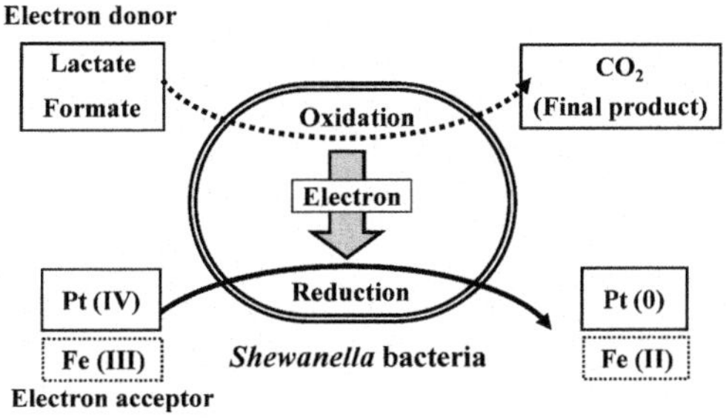

Fig. 1 A conceptual diagram for the microbial reduction of metal ions by *Shewanella* bacteria

S. algae was obtained from the American Type Culture Collection (ATCC 51181) and grown anaerobically in ATCC medium 2 containing sodium lactate and iron (III) citrate at 25 °C and pH 7.0 [1–4]. After 24 h of batch inoculation, *S. algae* cells were harvested by centrifugation, re-suspended in KH_2PO_4/NaOH buffer (pH 7.0), and pelleted again by centrifugation. This procedure was repeated twice, and the washed cells were subsequently re-suspended in KH_2PO_4/NaOH buffer (pH 7.0). The cell suspension was bubbled with N_2 for 10 min and immediately used for microbial recovery of PGMs.

An anaerobic glovebox was used to carry out microbial recovery experiments, as described previously [1–4]. For a typical recovery experiment at 25 °C, 5 cm^3 of *S. algae* cell suspension was added to 25 cm^3 of aqueous PGMs solution under N_2–CO_2 (80:20, v/v). The solutions were buffered at pH 6.0–7.0 with KH_2PO_4/NaOH, and the cell concentration was 5.0×10^{15} cells/m^3. The initial PGMs concentration was about 200 g/m^3, with 200 mol/m^3 sodium formate as the electron donor. To follow the time course of the microbial PGMs recovery, an aliquot of this mixture was periodically withdrawn and analyzed for PGMs. The concentration of PGMs in the liquid samples was determined by inductively coupled plasma atomic emission spectroscopy. The number of *S. algae* cells in the solution was counted in a Petroff-Hausser counting chamber with an optical microscope. The *S. algae* cells and biogenic particles were observed by transmission electron microscopy (TEM).

Results and Discussion

Chemical Leaching of Ceramic Substrate Catalysts

To extract PGMs from the crushed catalytic materials into aqueous solution, we proposed atmospheric acid leaching using a 50% aqua regia which is diluted with pure water. Using this method, above 95% of PGMs was leached out within 24 h at 60 °C and atmospheric pressure (Fig. 2). However, there was a marked difference in the leaching percentages between PGMs and other metals. When the chemical leaching with a 50% aqua regia was performed at 60 °C and atmospheric pressure,

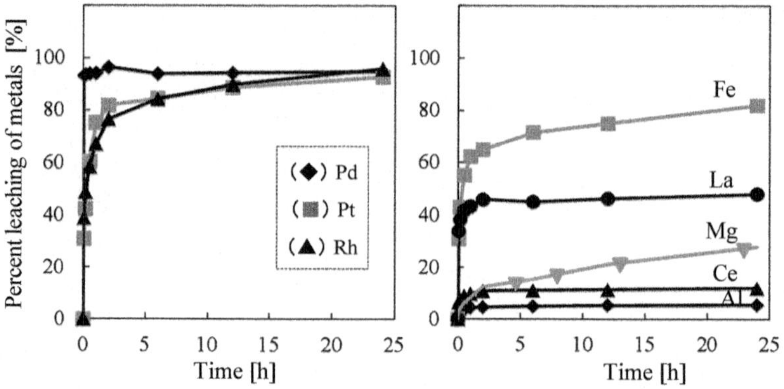

Fig. 2 Chemical leaching of ceramic substrate catalyst with a 50% aqua regia at 60 °C and atmospheric pressure

Fig. 3 Microbial recovery of PGMs (palladium, platinum and rhodium) from aqua regia leachate of spent automotive catalysts at 25 °C, pH 6, an initial formate concentration of 200 mol/m^3, and 5.0×10^{15} cells/m^3 *S. algae* cells

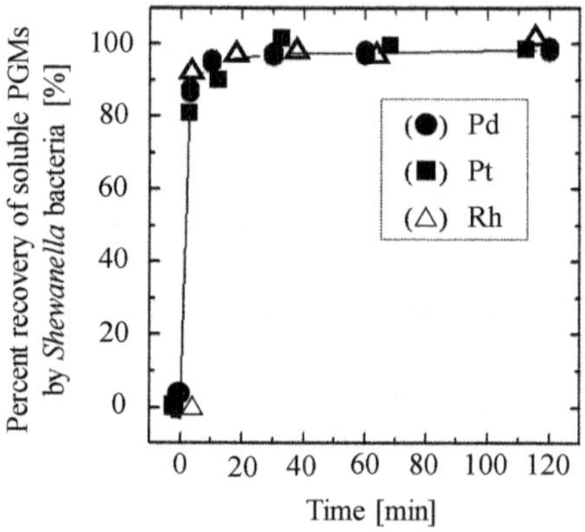

rare earths (cerium and lanthanum), magnesium and aluminum were leached less than 50% in 24 h, and 80% leaching of iron was attained in 24 h.

Microbial Reduction and Deposition of Soluble PGMs

The pH of aqua regia leachate was adjusted to 6.0–7.0 by the addition of aqueous sodium hydroxide solution. When processing the aqua regia leachate of catalytic materials, *Shewanella* bacteria were able to successfully perform above 95% bioreductive recovery of PGMs ions such as platinum (IV), palladium (II) and rhodium (III) (Fig. 3). To confirm the bioreductive deposition of PGMs, *S. algae* cells were observed using transmission electron microscopy (TEM) after 120 min of exposure to aqueous solutions of soluble PGMs. As shown in Fig. 4, *Shewanella* bacteria are rod-shaped cells, approximately 0.5 μm wide and 2 μm long. The TEM

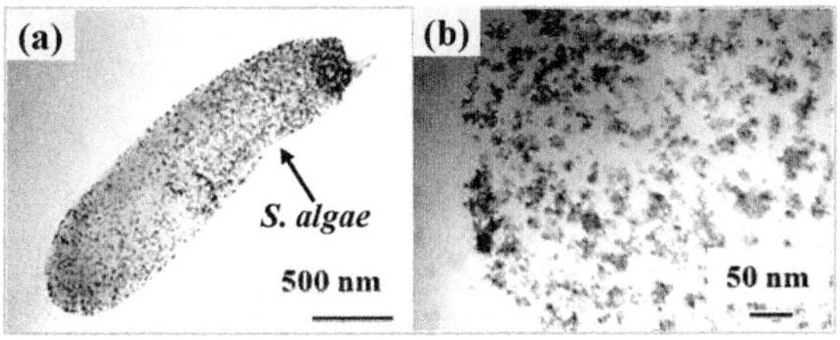

Fig. 4 Low- and high-magnification TEM images for biogenic PGMs nanoparticles deposited in *S. algae* cells at 25 °C and pH 6

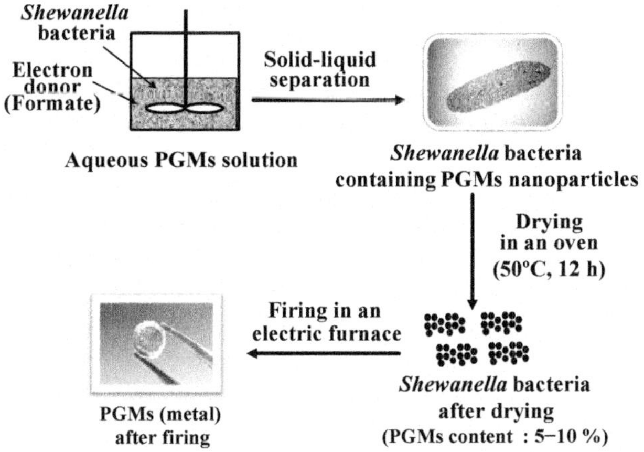

Fig. 5 Flow diagram of method for concentrating PGMs collected in bacterial cells

images show individual discrete PGMs nanoparticles about 10 nm in diameter in the bacterial cells. The ability to collect PGM ions suggests that this represents a new process for the microbial recovery of PGMs from spent automotive catalysts.

Recovery of PGMs from Microbial Cells

Figure 5 shows a flow diagram of the method for concentrating and recovering PGMs in *S. algae* cells. By drying wet, PGMs-containing cells at 50 °C for 12 h, the PGMs content in the dried cells was found to be 10%, which is 200 times greater than the 500 ppm concentration of PGMs in the starting solution. Furthermore, firing the dried cells in an electric furnace produced metallic PGMs, which is a mixture of platinum, palladium and rhodium.

Recycling Process of PGMs Using the Biomineralization Method

Figure 6 shows flow diagrams comparing the conventional chemical process and the new bioprocess for the recycling of precious metals. In contrast to the conventional process, the new bioprocess is integrated, unifying a multi-step method into a single-step procedure that separates and concentrates precious metals from a dilute solution, enabling the formation of metal nanoparticles. However, the addition of an electron donor and pH adjustment is required for the bioprocess to operate. Generally, microbial treatment has the disadvantage of being very slow but this microbial procedure has the capacity to perform above 95% recovery of soluble precious metals within 120 min at room temperature.

Compared to existing recycling methods, the new bio-recycling method has the following features: Enables a multi-step process of nanoparticles synthesis via separation and concentration of a precious metal in dilute solution to be accomplished in a single step; The bio-reduction and bio-precipitation is completed rapidly at room temperature (within 120 min for batch operation); An energy-saving recovery method with a low environmental load and functioning as a high value-added recycling process for the room temperature synthesis of metal nanoparticles. Introducing biotechnology into the process of recovering precious metals from urban mining is directed at building a foundation for the development of eco-friendly material recycling technologies.

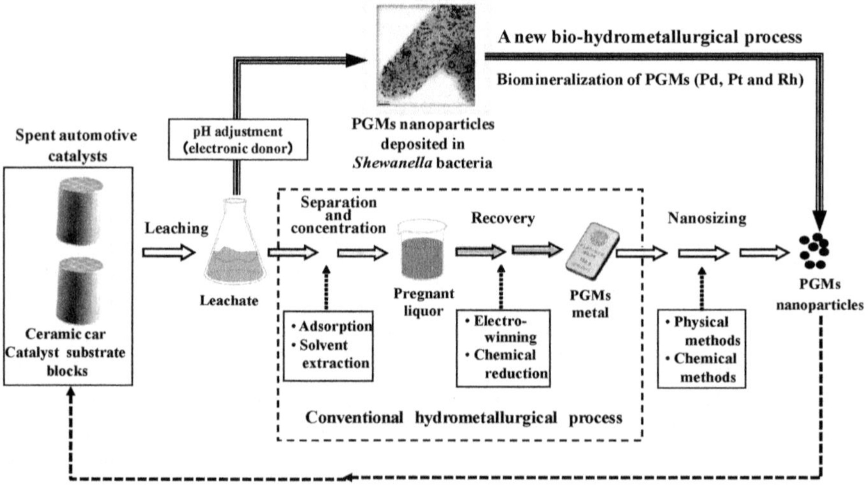

Fig. 6 Recycling flows of precious metals using the new and conventional technologies

Conclusions

We successfully developed a new recycling technology based upon eco-friendly biotechnology, in order to extract PGMs (palladium, platinum and rhodium) from spent automotive catalysts. Our proposed recycling bioprocess is linked to the development of low-cost and eco-friendly recycling processes that enable the low-energy and rapid recovery of the PGMs utilizing microbial reactions conducted at ambient temperature and at ambient pressure.

Acknowledgments This work was supported by JSPS KAKENHI Grant Number 20360411and 23360406.

References

1. Y. Konishi, T. Tsukiyama, K. Ohno, N. Saitoh, T. Nomura, S. Nagamine, Intracellular recovery of gold by microbial reduction of $AuCl_4^-$ ions using the anaerobic bacterium *Shewanella algae*. Hydrometallurgy **81**, 24–29 (2006)
2. Y. Konishi, T. Tsukiyama, N. Saitoh, T. Nomura, S. Nagamine, Y. Takahashi, T. Uruga, Direct determination of the oxidation state of gold deposits in the metal-reducing bacterium *Shewanella algae* using X-ray absorption near-edge structure spectroscopy (XANES). J. Biosci. Bioeng. **103**, 568–571 (2007)
3. Y. Konishi, T. Tsukiyama, T. Tachimi, N. Saitoh, T. Nomura, S. Nagamine, Microbial deposition of gold nanoparticles by the metal-reducing bacterium *Shewanella algae*. Electrochim. Acta. **53**, 186–192 (2007)
4. Y. Konishi, K. Ohno, N. Saitoh, T. Nomura, S. Nagamine, H. Hishida, Y. Takahashi, T. Uruga, Bioreductive deposition of platinum nanoparticles on the bacterium *Shewanella algae*. J. Biotechnol. **128**, 648–653 (2007)
5. T. Ogi, N. Saitoh, T. Nomura, Y. Konishi, Room-temperature synthesis of gold nanoparticles and nanoplates using *Shewanella algae* cell extract. J. Nanopart. Res. **12**, 2531–2539 (2010)
6. T. Ogi, R. Honda, K. Tamaoki, N. Saitoh, Y. Konishi, Direct room-temperature synthesis of a highly dispersed Pd nanoparticle catalyst and its electrical properties in a fuel cell. Powder Technol. **205**, 143–148 (2011)
7. K. Tamaoki, N. Saitoh, T. Nomura, Y. Konishi, Microbial recovery of rhodium from dilute solutions by the metal ion–reducing bacterium *Shewanella algae*. Hydrometallurgy **139**, 26–29 (2013)

Recovering Palladium from Chloridizing Leaching Solution of Spent Pd/Al_2O_3 Catalyst by Sulfide Precipitation

Qian Li, Qiang Zou, Bin Xu, Yongbin Yang, Xuefei Rao, Long Hu and Tao Jiang

Abstract The recovery of palladium from the chloridizing leaching solution of spent Pd/Al_2O_3 catalyst was studied mainly by the sulfide precipitation tests and chemical composition analysis, and the relevant mechanism was also investigated in this paper. Results showed that when NaOH was previously added to adjust the leaching solution to pH increasing from 1.21 to 4, the recovery of palladium was nearly 100% whilst the dosage of precipitant Na_2S decreased by 75% comparing with direct precipitating by Na_2S solution without previously adjusting the initial pH. However, at pH > 11.0, the excessive addition of NaOH resulted in the formation of some $Pd(OH)_2$ from $[PdCl_4]^{2-}$, then the recovery rate of palladium was 98.18%. When the palladium lixivium concentration is too low or the sodium sulfide concentration is overhigh or the adding rate of sodium sulfide solution is too fast, $[PdCl_4]^{2-}$ could be transformed into soluble $[PdS_2]^{2-}$ because of partial over-dosage of sodium sulfide. Under the appropriate control of the above three factors, nearly 100% of palladium could be recovery successfully.

Keywords Palladium · Spent catalyst · Extraction · Sulfide precipitation

Introduction

Palladium is one of the longest known and most studied of the six platinum group metals, a reflection of its abundance and consequent availability. The main uses of palladium are in the electronics and electrical industries, in circuitry and in dental alloys. It finds many catalytic applications in industry, as well as in diffusion cells for the synthesis of hydrogen, and in automobile catalysts [1, 2]. The catalytic ability of palladium is of considerable industrial importance, so the demand for palladium has been increasing in recent years [3]. However, because of the

Q. Li (✉) · Q. Zou · B. Xu · Y. Yang · X. Rao · L. Hu · T. Jiang
School of Minerals Processing & Bioengineering,
Central South University, Changsha 410083, Hunan, China
e-mail: csuliqian@126.com

© The Minerals, Metals & Materials Society 2017

H. Kim et al. (eds.), *Rare Metal Technology 2017*,
The Minerals, Metals & Materials Series,
DOI 10.1007/978-3-319-51085-9_14

continuous adsorption of reactive solvents on the palladium catalyst surface and the loss of metal or active component during use, the catalytic performance of palladium catalyst gradual failure [1, 4]. As a result, the waste catalysts containing palladium accumulate continuously, and not only huge waste of palladium resources but also possible environmental pollution will be caused. Hence, it is urgent to recycle palladium from the spent catalysts, which can lower the production cost and raise the utilization rate of palladium as well as avoid unnecessary environmental pollution [5, 6].

The spent Pd/Al_2O_3 catalyst with Al_2O_3 as carrier used in oil chemical industry was an important resource for recovering palladium. Currently, the palladium in the spent Pd/Al_2O_3 catalyst is usually pretreated by oxidative roasting—acid leaching into leachate, often a chloridizing leaching solution. Various methods mainly including precipitation, solvent extraction, displacement, ion exchange and hydrolysis, have been used to separate palladium from the leachate. Among them, the precipitation method has been used most extensively in industrial production [3, 7]. Wada once pointed out that palladium could be completely precipitated by H_2S under cold state and in faintly acidic conditions [8]. Later, Beamish designed the technological process for separating palladium using H_2S [9]. However, the usage of H_2S cannot be controlled easily and severe co-precipitation of palladium with aluminum ion is accompanied. Hence, the separation effect is unsatisfactory. It was researched by Taimni et al. [10] that a large amount of sodium sulfide solution could be used to convert palladium complex into its $[PdS_2]^{2-}$; then acetic acid could be used to decompose the $[PdS_2]^{2-}$ of palladium into insoluble precipitate of palladium sulfide, achieving an effective separation of palladium from the leachate.

In this paper, sulfide precipitation with Na_2S was adopted to recover Pd from the chloridizing leaching solution of spent Pd/Al_2O_3 catalyst. The various precipitation behaviors of palladium with concentration of Pd and Na_2S as well as the addition method, pH, temperature and agitation speed were studied systematically, and the relevant mechanism was also discussed.

Experimental

Physical and Chemical Properties and Components of Lixivium

The chloridizing leaching solution of palladium was provided by Kunming institute of precious metals. The palladium in the solution occurs in the form of H_2PdCl_4. Contents of various elements in the lixivium are shown in Table 1. In H_2PdCl_4 solution, palladium content was 44 mg L^{-1} but many other impurities also existed. The existence of impurities would influence the precipitation effect and dosage of sodium sulfide for palladium to some extent. Particularly, Al was the major impurity element in lixivium and its content reached 3754 mg L^{-1}.

Table 1 Content of elements in palladium leaching solution (mg L^{-1})

Element	Na	Mg	Al	Si	P	S	K	Ca
Content	124	15	3754	13.15	121	93	117	134
Element	Cr	Mn	Fe	Ni	As	Sn	Pb	Pd
Content	10	18	128	3.18	1.41	0.065	1.85	44

Reagent and Detection Method

(1) Sodium sulfide solution: Analytically pure (AR) reagent (molecular formula $Na_2S{\cdot}9H_2O$, molecular weight 240.19, colorless and transparent crystal) was used for preparation.
(2) pH measurement: Rex PHSJ-5 laboratory pH meter was used to measure pH value of the solution. E-201-D pH electrode assembly and T-820 °C temperature electrode were adopted. The measurement range of pH value is −2.000 to 18.000 and the fundamental error is ±0.002.
(3) Palladium content measurement: ICP-AES analysis was conducted to measure the components of lixivium. Palladium content in the filter liquor and precipitate was measured via atomic absorption spectrometry (AAS). The lower limit of analysis is good.

Experimental Principle

HS^- can be gained through hydrolysis of S^{2-}, and HS^- possesses quite strong electron-donating ability. $5P_z$ of complex ion $[PdCl_4]^{2-}$ having a square planar structure is a vacant orbital. Owing to sulphophile affinity of Pd atom, HS^- will directly contact Pd atom, and a pair of electrons in S atom will transfer to Pd. As a result, the binding force of Pd on four Cl^- ligands is weakened. Hence H^+ and Cl^- will be decomposed, and PdS can be formed.

$$S^{2-} + H_2O \leftrightarrow HS^- + OH^- \tag{1}$$

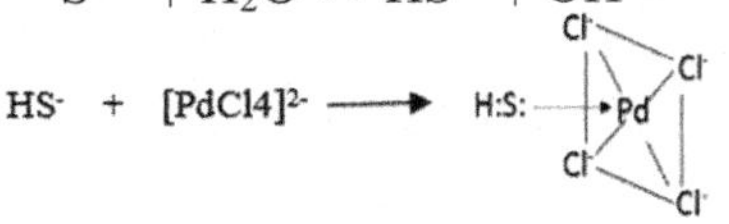

(2)

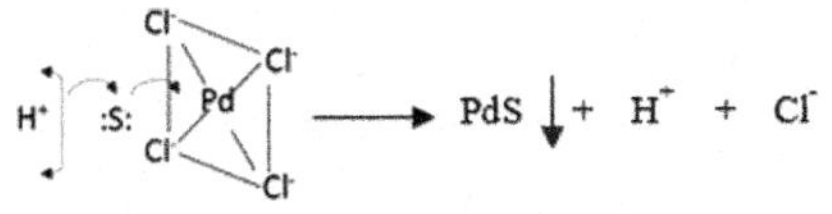

(3)

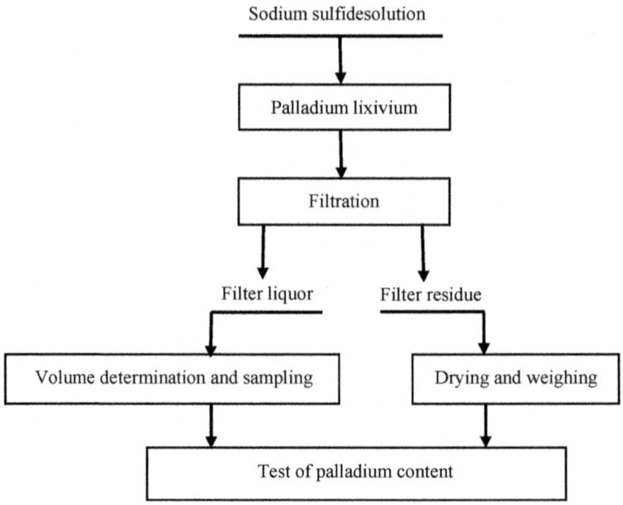

Fig. 1 Flow chart of test

Experimental Process

(1) 25 ml of palladium lixivium was taken out and put into a smaVll beaker of 100 ml;
(2) A rotor of 15 mm was used for stirring in the electromagnetic stirrer;
(3) An infusion tube which could be used to adjust the addition rate of Na_2S solution was utilized to fix the addition rate in the range of 0.1–0.5 ml s^{-1}, and the concentration of Na_2S solution 0.05–1.0 mol L^{-1};
(4) pH electrode was put into the beaker, and real-time measurement was conducted for pH;
(5) Filtration was conducted after the reaction, and the filter liquor was analyzed.

The flow chart of test is shown in Fig. 1.

Results and Discussion

Unless specified otherwise, the basic test conditions were dropwise addition rate 0.1 ml s^{-1}, agitation rate 1000 rpm, temperature 25 °C, initial pH 1.21, terminated pH 4.5, sodium sulfide concentration 0.5 mol L^{-1}, $[PdCl_4]^{2-}$concentration 44 mg L^{-1}.

Table 2 Leachate concentration on Pd recovery

Group	1	2	3	4
Leachate concentration of pd (mg/L)	44.0	22.0	14.7	11.0
The amount of sodium sulfide (ml)	8.0	4.0	2.7	2.0
Filtrate concentration of Pd (mg/L)*	0	0.1	0.1	0.2
Pd recovery (%)	100	98.18	98.18	96.36

"*" filtrate with 200 ml volumetric flask

Table 3 Effect of sodium sulfide concentration on Pd recovery

Group	1	2	3	4
Na_2S concentration (mol/L)	0.05	0.25	0.5	1.0
Termination pH	4.5	4.5	4.5	4.5
Na_2S amount (ml)	81.0	16.5	8.0	4.0
Pd recovery (%)	96.36	97.09	100	73.63

Effect of Lixivium Concentration on Palladium Precipitation

Changing concentration of Pd in leachate respectively group 1, 2, 3, 4 during the test to obtain the following data.

According to Table 2, with the decrease of palladium lixivium concentration, the dosage of sodium sulfide solution reduced proportionally. The recovery rate of palladium declined slightly with the decrease of lixivium concentration. As for the reasons, when lixivium concentration is excessively low, partial over-dosage of sodium sulfide will be caused. Hence, some of $[PdCl_4]^{2-}$ will produce thiofate of palladium, influencing precipitation of PdS. The reaction equation is indicated as follows:

$$[PdCl_4]^{2-} + 4HS^- \rightarrow \left[Pd(HS)_4\right]^{2-} + 4Cl^- \quad (4)$$

$$\left[Pd(HS)_4\right]^{2-} + 2OH^- \rightarrow [PdS_2]^{2-} + 2HS^- + 2H_2O. \quad (5)$$

Effect of Sodium Sulfide Concentration on Palladium Precipitation

Under the basic test conditions, changing Na_2S concentrations of the test to obtain the following data:

According to Table 3, with the rise of sodium sulfide concentration, the dosage of sodium sulfide diminished proportionally. When the sodium sulfide concentration (0.05 mol L^{-1}) was too low, the volume dosage of sodium sulfide was high

Table 4 Effect of terminated pH on Pd recovery

Group	1	2	3	4
Termination pH	3.8	4.0	4.5	5.0
Pd recovery (%)	98.18	100	100	100

and the reaction speed was relatively slow. When the sodium sulfide concentration (1 mol L^{-1}) was overhigh, $[PdCl_4]^{2-}$ would generate soluble $[PdS_2]^{2-}$, and thus the palladium precipitation would be incomplete.

Effect of Terminated pH During Reaction on Palladium Precipitation

The initial pH of lixivium was 1.21. When the stirring speed was 1000 rpm and the temperature was 25 °C, 0.5 mol L^{-1} sodium sulfide solution was added at the speed of 0.1 ml s^{-1}. The addition of sodium sulfide was stopped when the pH values reached 3.8, 4.0, 4.5 and 5, respectively. The measurement results are presented in Table 4.

According to Table 4, when solution pH was 4, palladium could be precipitated completely. When sodium sulfide was added continuously till pH became 5, PdS produced is still insoluble. When pH was 4, owing to the disturbance of aluminum ion, a large amount of colloid was produced, and the solution became brownish black gel. Consequently, the filtering separation was difficult. When pH was greater than 5, the formed colloid was destroyed, and obvious precipitate of black fine grains could be seen by the naked eye. If left to stand for 3 min the suspension would become obviously layered, and the separation effect was satisfactory.

Effect of Initial pH of Lixivium on Palladium Precipitation

As for the sample of Group 1, 0.5 mol L^{-1} HCl solution was used to adjust the initial pH decreasing from 1.21 to 0.5. As for the sample of Group 2, 0.5 mol L^{-1} NaOH solution was used to adjust pH increasing from 1.21 to 4.0, and the solution became milk white and cloudy due to the formation of insoluble aluminum hydroxide. As for the sample of Group 3, 0.5 mol L^{-1} NaOH solution was used to convert aluminum hydroxide into meta-aluminic acid; then the solution became clear again with a small amount of precipitate appearing in the bottom and pH shooting up to 11.0. As for the sample of Group 4, 0.5 mol L^{-1} NaOH solution was used to adjust pH to 12.0, and the solution property was the same as that of Group 3. The measurement results and content of elements in the filtrate are presented in Tables 5 and 6.

Table 5 Effect of initial pH on Pd recovery

Group	1	2	3	4
Initial pH	0.5	4.0	11.0	12.0
Termination pH	4.5	4.5	12.60	12.60
Na_2S amount (ml)	9.8	2.0	3.2	1.2
Pd recovery (%)	100	100	98.18	98.18

Table 6 Content of elements in the filtrate of group 2 and group 3 (mg L^{-1})

Element	Na	Mg	Al	Si	P	S	K	Ca
Group 2 content	1207	0.98	387	3.51	13.29	218	18	19
Group 3 content	3410	0	402	3.55	7.46	198	19	0
Element	Cr	Mn	Fe	Ni	As	Sn	Pb	Pd
Group 2 content	1	1.97	11	0.11	0.23	0.063	0.26	0
Group 3 content	0	0	0	0	0.25	0.053	0.26	0.1

According to Table 5, the palladium precipitation would not be affected under general acidic conditions. Moreover, palladium precipitation would also not be affected when NaOH solution was used to replace Na_2S solution to adjust pH to 4.0. Meanwhile, H_2S generation could be reduced, and the dosage of Na_2S solution would be decreased by 75% comparing with direct precipitating by Na_2S solution without previously adjusting the initial pH. This is primarily because that NaOH solution will precipitate impurity ions like Al^{3+}, Fe^{3+} and Ca^{2+}, etc. and meanwhile neutralize a large amount of H^+.

Under alkaline conditions, many aluminum ions existed in the form of meta-aluminic acid radical, and the dosage of Na_2S could be reduced by 85%. Meanwhile, no H_2S would be generated. However, the addition of a large amount of NaOH will result in the formation of some $Pd(OH)_2$ from $[PdCl_4]^{2-}$.

Effect of Adding Rate of Sodium Sulfide Solution on Palladium Precipitation

As for the samples of Group 1, 2 and 3, Na_2S solution was added into palladium lixivium at the rates of 0.1 0.2 and 0.5 ml s^{-1}, respectively. When the rate was higher, more bubbles could be produced. As for the sample of Group 4, palladium lixivium was added into 0.5 mol L^{-1} Na_2S solution, and the reaction was fast. When one drop was added, the entire system became brown and cloudy. The solution presented a colloidal and cloudy state after the reaction (Table 7).

Table 7 Effect of addition rate of Na_2S solution on Pd recovery

Group	1	2	3	4
Dropwise addition rate (ml s^{-1})	0.1	0.2	0.5	gigantic
Pd recovery (%)	100	98.18	89.09	27.27

Table 8 Effect of temperature on Pd recovery

Group	1	2	3	4
Temperature (°C)	25	50	80	100
Pd recovery (%)	100	100	100	100

Table 9 Effect of stirring speed on Pd recovery

Group	1	2	3	4
Stirring speed (rpm)	Without stirring	800	1000	1200
Pd recovery (%)	89.09	98.18	100	100

Under the conditions of maintaining lixivium concentration, Na_2S solution concentration and stirring speed unchanged, the recovery rate of palladium decreased slightly with the increase of adding rate of sodium sulfide solution. When lixivium was directly added into the sodium sulfide solution, the recovery rate of palladium was only 27.27%. This further shows that excessive sodium sulfide results in the generation of soluble thiofates of palladium. Hence, $[PdCl_4]^{2-}$ could not be completely precipitated in the form of PdS (Table 8).

Effect of Temperature on Palladium Precipitation

Under the temperature of 25 °C, palladium could precipitate completely. The rise of temperature had little influence on precipitation.

Effect of Stirring Speed on Palladium Precipitation

The stirring speed was related to the concentration of palladium lixivium and sodium sulfide as well as the adding rate of sodium sulfide solution, only 89.09% of palladium recovery percent without stirring, this is related to the generation of thiofate of palladium, so it was better to control the stirring speed within 1000 rpm (Table 9).

Conclusions

(1) The main controlling factors influencing precipitation of palladium from its lixivium as palladium sulfide is pH. When terminated pH during reaction is in the range of 4–5, it can precipitate completely. However, when pH was less than 4.5, the filtration is difficult due to the generation of colloid. Therefore, the optimal terminated pH should be about 4.5. After complete precipitating in the form of PdS, the generated precipitate is still insoluble even if excessive sodium sulfide is added.
(2) Under alkaline conditions of pH > 11, most palladium complex ions can precipitate in the form of PdS with little loss of palladium as $Pd(OH)_2$, and no H_2S will be produced. A large amount of NaOH was consumed previously, but the sequent dosage of sodium sulfide can be reduced by 85%.
(3) When the palladium lixivium concentration is too low or the sodium sulfide concentration is overhigh or the adding rate of sodium sulfide solution is too fast, partial over-dosage of sodium sulfide will be caused easily, but $[PdCl_4]^{2-}$ will be transformed to soluble $[PdS_2]^{2-}$ inevitably. In this situation, the precipitation of palladium is affected severely.
(4) In order to eliminate the influence of aluminum ion and reduce the generation of H_2S, NaOH can be used in advance to adjust the pH of palladium lixivium to 4.0. Hence, palladium can precipitate completely, and the dosage of Na_2S will be reduced by 75%.

Acknowledgments This work was financially supported by National Natural Science Foundation of China (Grant No. 51574284), National Science Foundation for Distinguished Young Scholars of China (Grant No. 51504293), Science and Technology Program of Yunnan (No. 2013IB020).

References

1. I.J. Ferrer et al., An investigation on palladium sulphide (PdS) thin films as a photovoltaic material. Thin Solid Films **515**(15), 5783–5786 (2007)
2. D.S.A. Cotton, *Chemistry of Precious Metals* (Chapman and Hall, London, UK, 1997)
3. W. Xu et al., Tailoring supported palladium sulfide catalysts through H2-assisted sulfidation with H_2S. J. Mater. Chem. A. **1**(41), 12811–12817 (2013)
4. A.A. Zirka, A.V. Mashkina, Liquid-phase hydrogenation of thiophenes on palladium sulfide catalysts. Kinet. Catal. **41**(3), 388–393 (2000)
5. A. Birri et al., Allyl palladium dithiocarbamates and related dithiolate complexes as precursors to palladium sulfides. J. Organomet. Chem. **692**(12), 2448–2455 (2007)
6. C. Fontàs, V. Salvadó, M. Hidalgo, Selective enrichment of palladium from spent automotive catalysts by using a liquid membrane system. J. Membr. Sci. **223**(1–2), 39–48 (2003)
7. R.S. Marinho, J.C. Afonso, Recovery of platinum from spent catalysts by liquid–liquid extraction in chloride medium. J. Hazard. Mater. **179**(1–3), 488–494 (2010)

8. J. Chen, Mechanism of sodium sulfide reacting with tetrachloropalladate and its application —I: quantitative precipitation of palladium by sodium sulfide. Precious Metals **1**(1), 7–15 (1980)
9. F.E. Beamish, A critical review of gravimetric methods for determination of the noble metals—II. Talanta **13**(6), 773–801 (1966)
10. I.K. Taimni, S.N. Tandon, Thermolysis of sulphides precipitated by sodium sulphide: sulphides of selenium, tellurium, gold, platinum, ruthenium, lead, bismuth, cadmium, silver, palladium and indium. Anal. Chim. Acta **22**(1), 553–557 (1960)

Mechanism of Intensifying Cyanide Leaching of Gold from a Calcine by the Pretreatment of Acid or Alkali Washing

Yan Zhang, Qian Li, Xiaoliang Liu, Yongbin Yang, Bin Xu, Tao Jiang and Hongwei Li

Abstract The direct cyanide leaching rate of gold from a gold calcine was only 71.33%, due to 19.51% of gold being encapsulated in the iron oxides. To enhance the cyanide leaching of gold, the calcine was pretreated by acid and alkali washing and the relevant intensifying mechanism was also researched. The results showed that the cyanide leaching rate of gold after acid or alkali washing was increased to 82.24 and 87.07%, respectively. It was found that acid washing could remove iron oxides, but the dissolution of iron oxide particles was incomplete that part of gold was still encapsulated. However, alkali washing could effectively remove silicates and obtain rough and porous calcine particles. It was found that the blockage of pores in iron oxides by silicates was the primary reason for the low gold leaching rate of calcine and alkali washing was more favorable for exposing gold than acid washing.

Keywords Gold · Calcine · Cyanide · Acid or alkali washing · Intensifying mechanism

Introduction

A gold mine whose proven reserves exceed 40 t is located in Yunnan province of China. A gold concentrate was obtained by floatation, where most of gold (about 87%) occurs within associated sulfide minerals like pyrite and arsenopyrite and about 10% of gold exists in oxides. According to our previous research [1, 2], it was found (see Table 1) that gold inclusions in the gold concentrate could be exposed to a great extent via two-stage oxidative roasting, but over 20% of gold was still

Y. Zhang · Q. Li (✉) · X. Liu (✉) · Y. Yang · B. Xu · T. Jiang · H. Li
School of Minerals Processing and Bioengineering, Central South University, Changsha 410083, Hunan, People's Republic of China
e-mail: csuliqian@126.com

X. Liu
e-mail: sklxl1987@163.com

© The Minerals, Metals & Materials Society 2017

H. Kim et al. (eds.), *Rare Metal Technology 2017*,
The Minerals, Metals & Materials Series,
DOI 10.1007/978-3-319-51085-9_15

Table 1 Occurrence of Au in the calcine

Phase of gold occurrence	Free and associated gold	Sulfides	Oxides	Silicates	Total
Content (g/t)	17.60	0.24	4.39	0.27	22.50
Distribution (%)	78.22	1.07	19.51	1.20	100.00

locked in oxides and silicates. Secondary encapsulation of gold is unavoidable during the oxidative roasting of refractory sulfuric gold ores [3–6]. Consequently, the gold leaching rate by cyanidation is unsatisfactory, which makes it necessary to pretreat the calcine prior to leaching by cyanidation.

Commonly, pretreatment of acid washing is used to expose gold inclusions in oxides (mainly iron oxides) [7]. Specifically, as expressed by Eqs. (1) and (2), dilute sulfuric acid or hydrochloric acid is usually used to render iron oxides soluble to some degree into the aqueous phase, accompanying with the exposure of gold. The acid washed residue can be utilized to extract gold and other valuable metals can also be recovered from the washing solution [8].

$$Fe_2O_3 + 3H_2SO_4 = Fe_2(SO_4)_3 + 3H_2O \tag{1}$$

$$Fe_2O_3 + 6HCl = 2FeCl_3 + 3H_2O \tag{2}$$

Moreover, it was found that the pores of oxides forming during the roasting of associated sulfide minerals could be blocked by the complex silicates often containing Al, Ca and Mg, etc. [2]. Considering that silicates are hardly destroyed by common acids such as sulfuric acid and hydrochloric acid, etc., except hydrofluoric acid which is much more expensive, pretreatment of acid washing is not a viable option to open the locked gold in silicates [9, 10]. However, it was indicated that silicates could be effectively dissolved by strong base like sodium hydroxide [11].

In view of the above, pretreatment of the calcine by acid or alkali washing before cyanidation was conducted to enhance the gold leaching rate. Additionally, the relevant intensifying mechanism was elucidated in this paper by the analyses of chemical composition, particle size distribution, SEM-EDS and XRD of pretreated residues.

Experimental

Materials and Reagents

The calcine was obtained from two-stage oxidation roasting of a gold concentrate which contained As, S and C and was supplied from a plant of Yunnan, China. The particle size of the calcine was $d_{95} > -75$ μm, and its chemical composition is indicated in Table 2. The content of Au was 22.50 g/t and the content of C, As and

Table 2 Chemical composition of the calcine (%)

Constituent	Au[a]	SiO_2	CaO	MgO	As	Fe	S	Al_2O_3	C
Content	22.50	34.61	4.58	1.96	0.42	25.44	1.83	7.87	0.13

[a]Unit g/t

S was 0.13, 0.42 and 1.83%, respectively. Also, the total content of alkaline gangue including SiO_2, CaO, MgO and Al_2O_3 was as high as 49.02%.

XRD spectrogram of the calcine is shown in Fig. 1a, and it can be seen that the calcine mainly consisted of quartz, hematite and calcium sulfate. The occurrence of Au in the calcine is also presented in Table 2. The free and associated gold accounted for 78.22%. However, 19.51% of gold was still encapsulated in the oxides. Only 1.07 and 1.20% of gold was locked in sulfides and silicates, respectively. It is demonstrated that the sintering phenomenon is inevitable during oxidative roasting, which leads to the secondary encapsulation of gold by iron oxides. The gold inclusions in gangue minerals can be exposed hardly during roasting; on the other hand, the high content of the gangue minerals, i.e., silicates, possibly cause the blockage of particle porosity, thus hindering the exposure of gold from the calcine. The XRD spectrograms of pretreated residues by acid or alkali washing, i.e., Fig. 1b, c, will be discussed in Section "Chemical Compositions of Gold Ores".

The reagents used in this study, such as sodium cyanide, sodium hydroxide and sulfuric acid, were all in analytically pure grade. De-ionized water was used throughout all experiments.

Experiment Methods

The pretreatment tests of acid washing and alkali washing were both conducted in 1-L jacketed glass reactors each equipped with overhead stirrer, condenser and thermometer, and the reactors were connected with thermostatic water baths to

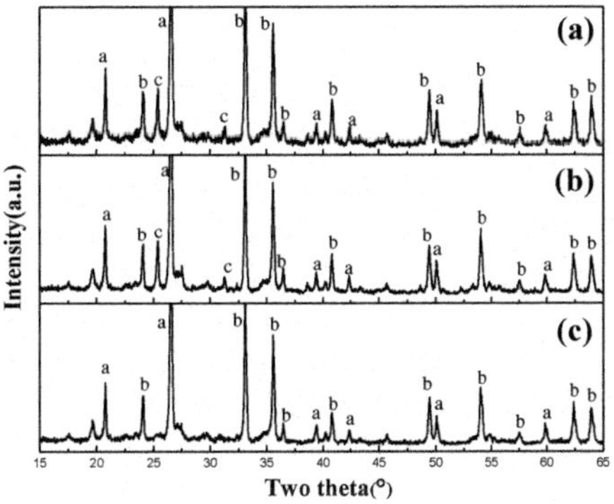

Fig. 1 XRD spectrogram of **a** the calcine, **b** the residue after acid washing with 15% H_2SO_4 and **c** the residue after alkali washing with 20 g/L NaOH (**a**—quartz; **b**—hematite; **c**—calcium sulfate)

control reaction temperature. The reaction conditions were liquid-solid ratio 3:1, temperature 80 °C, stirring speed 300 rpm, H_2SO_4 0–20% (mass fraction)/NaOH 0–30 g/L and leaching time 1 h. Gold cyanide leaching tests were implemented in 1-L jacketed glass reactors whose structure were analogous to the ones used in pretreatment tests, and under the conditions of liquid-solid ratio 2.5:1, temperature 25(±0.5) °C, stirring speed 600 rpm, NaCN 0.4% (mass fraction), leaching time 36 h and pH 11 adjusted by the careful addition of 1.0 M NaOH solution. When the pretreatment or leaching test was completed, the pulp was filtered by a vacuum filter and the obtained filter residue was washed adequately, among which the pretreated residues were first dried in a vacuum oven and then were used for gold cyanide leaching tests whilst the leached residues were also dried for the subsequent assay.

Analytical Methods

S and C contents in the gold concentrate and calcine were determined using a high frequency IR carbon and sulfur analyzer (HW2000B, Wuxi Yingzhicheng). The other elements were all analyzed using acid digestion and an atomic absorption spectrometer (AA-6800, Shimadzu). Mineralogical compositions of the caline and pretreated residues were obtained by an X-ray diffractometer (D/Max 2500, Rigaku). The size distributions and specific surface areas of pretreated residues were detected by the laser particle size analyzer (Mastersize2000, Malvern). Morphological studies on the pretreated residues were carried out on SEM coupled with EDS (JSM-6360LV).

Results and Discussion

Acid or Alkali Pretreatment-Cyanide Gold Leaching

As we can see from Fig. 2, the gold leaching rate of calcine by direct cyanidation was just 71.33%. So we used H_2SO_4 and NaOH separately to treat the calcine, and the effects of the concentration of H_2SO_4 and NaOH on gold leaching with cyanide are shown in Fig. 2a, b. It is clear that the gold leaching rates both increased as the concentration of H_2SO_4 (within 15%) and NaOH (within 20 g/L) increased. But when the addition of H_2SO_4 and NaOH exceeded 15% and 20 g/L respectively, no further increase of gold leaching rate occurred and it basically kept steady. The optimum extraction of gold from the calcine pretreated by H_2SO_4 and NaOH could amount to 82.24 and 87.07%, respectively.

Although the pretreatments of acid (H_2SO_4) washing and alkali (NaOH) washing both intensified significantly the gold leaching, it was more effective for exposing gold encapsulated in the caline via the alkali washing.

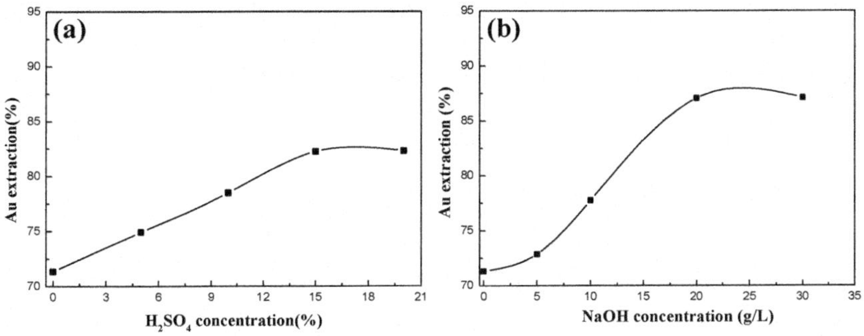

Fig. 2 Variations of Au extraction with the concentration of **a** H_2SO_4 (%, mass fraction) and **b** NaOH (g/L)

Intensifying Mechanism of Acid or Alkali Pretreatment

Under the optimum conditions of acid or alkali pretreatment, the intensifying mechanism of gold leaching from the washed residues was investigated from the following three respects.

Chemical Compositions of Gold Ores

The chemical compositions of calcine and pretreated residues are presented in Table 3.

As seen from the XRD spectrogram of Fig. 1b, the main phase of acid washed residue was nearly the same as that of caline. However, Table 3 indicated that the content of Fe in the acid washed residue declined remarkably from 25.44% of the calcine to 14.24%. Additionally, the content of As and C also decreased to a lower level.

The XRD spectrogram of the alkali washed residue is shown in Fig. 1c. Comparing with Fig. 1a, the phase of calcium sulfate ($CaSO_4$) disappeared from the

Table 3 Chemical compositions of the calcine and pretreated residues (%)

Gold ore type	Au[a]	Fe	SiO_2	Al_2O_3	CaO	MgO	As	S	C
Calcine	22.50	25.44	34.61	7.87	4.58	1.96	0.42	1.83	0.13
Acid washed residue	24.43	14.24	42.40	8.21	4.56	2.01	0.15	1.73	0.05
Alkali washed residue	23.29	26.44	30.93	6.87	2.30	1.28	0.30	0.30	0.10

[a]Unit g/t

calcine after alkali washing. Meanwhile, the content of CaO and S in the alkali washed residue also reduced significantly as shown in Table 3. This is primarily because that $CaSO_4$ can translate into $Ca(OH)_2$ which is slightly soluble in strong base solutions and may act as pH regulator during the cyanidation according to the following reactions as Eqs. (3) and (4) [12]:

$$CaSO_4 + 2OH^- = Ca(OH)_2 + SO_4^{2-} \tag{3}$$

$$Ca(OH)_2 = Ca^{2+} + 2OH^- \tag{4}$$

Particularly, the content of SiO_2, Al_2O_3 and MgO all decreased from the calcine in varying degrees after the alkali washing, and the relevant reactions are shown as the following Eqs. (5) and (7).

$$SiO_2 + 2OH^- = SiO_3^{2-} + H_2O \tag{5}$$

$$Al_2O_3 + 2OH^- = 2AlO_2^- + H_2O \tag{6}$$

$$xCaO \cdot yAl_2O_3 \cdot zSiO_2 + 2(y+z)OH^- = 2yAlO_2^- + zSiO_3^{2-} + xCa(OH)_2 + (y+z-x)H_2O \tag{7}$$

As a result, the acid washing is more favorable to remove Fe while the alkali washing is more effective for removing the gangue minerals, i.e., SiO_2, CaO, Al_2O_3 and MgO, etc.

Size Distributions and Specific Surface Area of Gold Ores

After the optimum acid washing or alkali washing, the size distributions and specific surface area of pretreated residues as well as calcine are shown in Table 4. In comparison with the calcine, the particle sizes of acid washed residue became slightly finer and there was only a slight degree of increasing of the specific surface area, which is primarily due to the dissolution of iron oxides. As for the alkali washed residue, the size distributions changed little but the specific surface area grew significantly from 6900 cm^2/g of the calcine to 7739 cm^2/g. So, it is

Table 4 Size distributions and specific surface areas of the calcine and pretreated residues (%)

Gold ore type	Size fraction/(μm) and distribution					Specific surface area/(cm^2/g)
	+75	−75 to +45	−45 to +37	−37 to +25	−25	
Calcine	4.15	16.52	9.05	13.82	56.46	6755
Acid washed residue	3.11	15.19	8.81	13.94	58.95	6900
Alkali washed residue	4.50	16.93	9.19	13.97	55.41	7739

demonstrated that the mineral particles became rough and porous after the alkali washing, which is theoretically more beneficial for exposing the enclosed gold.

SEM-EDS Analysis

The particle morphology after the pretreatment of acid or alkali washing was also studied by SEM-EDS, and the results are shown in Fig. 3 and Table 5. As indicated in Fig. 3a, it is clearly shown that the particle surface of acid washed residue was relatively compact and was adhered by some tiny particles. Additionally, the EDS spectra of the acid washed residue, i.e., Fig. 3c, d, and the quantitative EDS analysis (see Table 5) indicate that the content of Fe_2O_3 in the detection micro-area B of acid washed residue was very low (6.86%) and its composition was mainly SiO_2 and Al_2O_3, etc., which is consistent with the results of Section "Chemical Compositions of Gold Ores". However, the detection micro-area A of the small particle adhering on the surface is shown that the content of Fe_2O_3 was very high (66.44%). Consequently, acid washing could remove part of iron oxides from the calcine, but the effect on damaging the surface structure of calcine particles was quite limited. This primarily lies in the restricted kinetics of the dissolution of iron oxides by acid.

In contrast, as shown in Fig. 3b, the particle surface of calcine tended to be rough and porous after the alkali washing, which agrees well with the results of

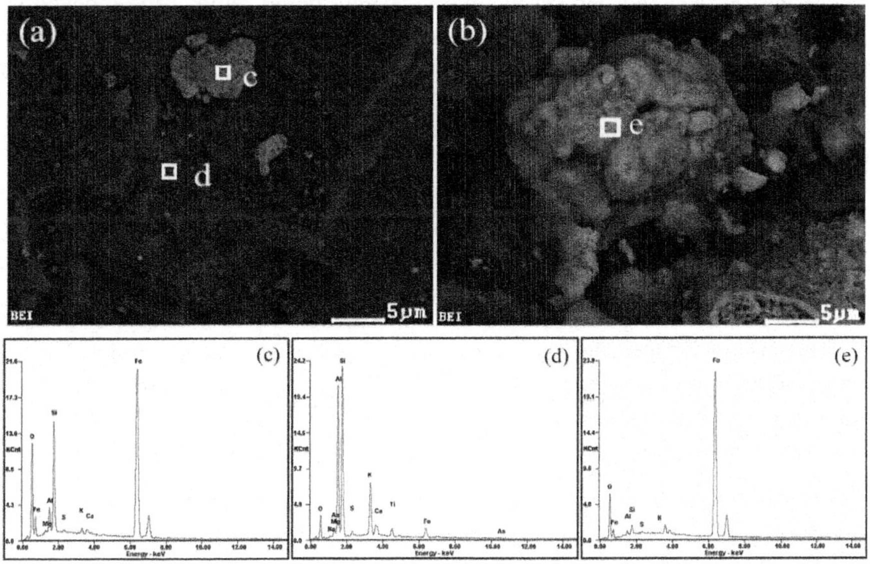

Fig. 3 SEM images and EDS spectra of the pretreated residues. SEM image (**a**) and EDS spectra (**c**), (**d**) of the acid washed residue; SEM image (**b**) and EDS spectrum (**e**) of the alkali washed residue

Table 5 Results of quantitative EDS analysis of the pretreated residues (%)

Pretreated residue	Detection micro-area	Fe_2O_3	SiO_2	CaO	MgO	Al_2O_3	K_2O
Acid washed residue	c	66.44	19.29	2.90	2.73	6.72	1.92
	d	6.86	43.58	7.08	2.79	28.99	10.70
Alkali washed residue	e	93.10	5.16	0.01	0.01	1.15	0.39

Section "Size Distributions and Specific Surface Area of Gold Ores". The EDS analysis including Fig. 3e and Table 5 also suggested that the content of Fe_2O_3 in the detection micro-area C of alkali washed residue reached up to 93.10%, but the main composition of gangue, i.e., SiO_2, Al_2O_3, CaO and MgO all decreased to much lower levels. Hence, alkali washing is favorable for not only removing the gangue of Calcine but more importantly generating a rough and porous structure, which can substantially improve the leaching kinetics of gold. Actually, according to the results from Section "Acid or Alkali Pretreatment-Cyanide Gold Leaching", it was more effective to extract gold via the pretreatment of alkali washing than the pretreatment of acid washing.

Based on the above analysis, we could infer the possible intensifying mechanism of gold leaching from the calcine by acid or alkali washing. During oxidative roasting of the gold concentrate, secondary encapsulation of gold in the calcine is inevitable and most of gold inclusions occur in the iron oxides whose porous structure is possibly blocked by the silicates containing Al_2O_3, CaO and MgO, etc. It is restricted from the kinetics through directly dissolving iron oxides by acid to expose the locked gold. However, alkali washing could remove the blocked silicates so effectively that more gold inclusions were exposed. Hence, leaching gold from the calcine intensified by alkali washing to open the original rough and porous structure of iron oxides is much more favorable than by acid washing to directly create new structure damages in iron oxides.

Conclusions

(1) The gold leaching rate from a calcine (Au 22.50 g/t) by direct cyanidation only was 71.33%. Most gold inclusions occurred in the iron oxides and high content of gangue, i.e., silicates (49.02%) containing Al, Ca and Mg, etc. was present in the calcine. After acid (H_2SO_4 15%) or alkali (NaOH 20 g/L) washing, the gold extraction from the calcine could be improved from 71.33 to 82.24 and 87.07%, respectively.

(2) Acid washing could remove the iron oxides of calcine to directly open the enclosed gold, but it was restricted by the dissolution kinetics of acid dissolving iron oxides. Consequently, the surface structure of acid washed residue was relatively compact and part of gold was always encapsulated. In contrast, alkali washing could not only remove the silicates from the calcine

but more importantly achieve a relatively rough and porous structure, which is more favorable for exposing the encapsulated gold.

(3) The possible intensifying mechanism of leaching gold from the calcine via the pretreatment of acid or alkali washing was that the presence of complex silicates possibly resulted in the blockage of porous structure of the calcine during the roasting process. In comparison with acid washing the calcine to directly expose the gold inclusions in the iron oxides, alkali washing was more advantageous.

Acknowledgements This work was financially supported by National Natural Science Foundation of China (Grant No. 51574284), National Science Foundation for Distinguished Young Scholars of China (Grant No. 51504293), Science and Technology Program of Yunnan (No. 2013IB020).

References

1. L. Hou et al, Volatilization behavior and mechanisms of arsenic, sulfur and carbon in the refractory gold concentrate, in *TMS, 6th International Symposium on High-Temperature Metallurgical Processing* (2015), pp. 163–170
2. J. Hu, Research on the process of roasting pretreatment-cyanide extracting au from sulfide-bearing refractory gold concentrate containing arsenic and high-carbon. Master thesis, Central South University (2015), pp. 10–20
3. L. Wang et al., Research on encapsulation of roasted refractory gold concentrate. Chin. J. Rare Metal. **29**, 424–428 (2005)
4. W. Kou, G. Chen, Research and practice on improving cyanidation recovery of the refractory gold bearing sulfide concentrate from a two-stage roasting process. Gold **33**, 47–49 (2012)
5. C. Komnitsas, F.D. Pooley, Mineralogical characteristics and treatment of refractory gold ores. Miner. Eng. **2**, 449–457 (1989)
6. K.G. Thomas, A.P. Cole, Roasting developments-especially oxygenated roasting. Dev. Miner. Process. **15**, 403–432 (2005)
7. Xu Bin et al., Fluidized roasting-stage leaching of a silver and gold bearing polymetallic sulfide concentrate. Hydrometallurgy **147–148**, 79–82 (2014)
8. Yongbin Yang et al., Pretreatment by sulfuric acid-curing of calcine roasting for gold ores with high sulfur and arsenic contents. Chin. J. Nonferrous Met. **24**, 2380–2386 (2014)
9. Hu Yaoqin et al., Mineralogical study of roasting of gold concentrate and analysis of difficulty in leaching gold of residue. Nonferrous Met. **51**, 52–56 (1999)
10. Y. Li et al, Roasting technology of refractory and complex gold ore using circulating fluid bed. Nonferrous Met. (Extr. Metall.) (2011), 31–33
11. H. Li, W. Wang, X. Zhang, Method of gold leaching by pretreating gold bearing tailing or cyanidation residue to remove silicon. China patent. No. CN201410412434.2 (2014)
12. J. Wang, B. Li, The stability and solubility of $CaSO_4 \cdot 2H_2O$ and $Ca(OH)_2$ in high concentrate KOH–NaOH–H_2O system. Chem. Eng. **S2**, 413 (2009)

Part III
Base and Rare Metals

Disclosure of the Kinetic Relations of Semidirect Cemented Carbide Leaching in Acid Media

Gregor Kücher, Stefan Luidold, Christoph Czettl and Christian Storf

Abstract The purpose of this investigation is to provide a basis for the semidirect recycling of hard metals. Accordingly, the present investigation focuses on revealing the fundamental chemical kinetic aspects during cemented carbide leaching in hydrochloric acid including an oxidant. In contrast to literature, the experiments are performed with pieces of cemented carbide specimen. This displays a crucial condition since the disintegration of hard metal scrap causes trouble and the tests should provide feasible information. Appropriate preliminary sample preparations combined with an adequate arrangement in the reaction vessel eliminate shrinking surface effects on the leaching kinetics. Moreover, an implementation of a design of experiments allows a statistical evaluation and regression analysis of the obtained results. During the experiments the influence of several parameters such as the temperature, the acid- as well as the oxidant concentration on the leaching behavior is examined. Finally a proposed empirical kinetic law describes the lixiviation performance of cobalt out of the hard metal substrate.

Keywords Cemented carbide · Hard metal · Semidirect recycling · Hydrometallurgy · Leaching · Reaction rate

G. Kücher (✉) · S. Luidold
CD Laboratory for Extractive Metallurgy of Technological Metals,
Chair of Nonferrous Metallurgy, Montanuniversitaet Leoben, Franz-Josef-Str. 18,
8700 Leoben, Austria
e-mail: gregor.kuecher@unileoben.ac.at

S. Luidold
e-mail: stefan.luidold@unileoben.ac.at

C. Czettl · C. Storf
CERATIZIT Austria GmbH, Metallwerk-Plansee-Str. 71, 6600 Reutte, Austria
e-mail: christoph.czettl@ceratizit.com

C. Storf
e-mail: christian.storf@ceratizit.com

© The Minerals, Metals & Materials Society 2017

H. Kim et al. (eds.), *Rare Metal Technology 2017*,
The Minerals, Metals & Materials Series,
DOI 10.1007/978-3-319-51085-9_16

Introduction

Depending on the impact onto the binder metal and carbide, three main routes can be distinguished regarding the hard metal recycling techniques. Direct processes disintegrate the substrate so the carbide(s) and the binder are mixed in the reclaimed powdery product. The chemical composition of the treated feedstock remains basically unchanged, although minimal process-related contaminations occur. Additionally, the low consumption of energy and therefore the high cost efficiency are beneficial. [1, 2] Within the direct group the zinc process displays the main representative. For instance, in 2000 about 25% of the cemented carbide scrap passed through such facilities in the USA [3] and at least 10% of the raw material demand in Europe [4] was supplied by the same.

Indirect techniques decompose the scrap chemically in their elementary components where these are separated. The process scheme is similar to the primary route and accordingly extensive. For tungsten this implies dissolution, leach purification, crystallization, calcination, reduction and finally carburation in order to receive WC "tungsten carbide". The process's main advantage comprises that the reclaimed powders have the same quality as primary material. [2, 5] Two commercial practices exist inside this category, an oxidation process with chemical digestion or an alkaline salt fusion method. According to Shedd [3] a share of about 35% of the hard metal scrap in the USA is handled by these methods, with an even higher amount of 85% [5] in Germany. Hence, the last class, the semidirect methods, only plays a minor role in today's recycling market. According to this recycling path, the present coatings as well as the binder metals are leached and leave behind a WC skeleton which can be pulverized more easily. Consequently, no extensive procedure should theoretically be necessary for obtaining a reusable pure WC powder. Also, common hydrometallurgical techniques retrieve the binder metal. [5–7]

However, several issues are reported in literature, such as partial oxidation of WC [6–8], slow dissolution of the binder [6, 8], non-removable films [9] and in addition references including powdered hard metal scraps [10] or at least an accompanying mechanical treatment [8, 9, 11]. Notably, the mechanical pulverization of cemented carbides causes troubles, since this category of materials serves for machining others. Thus, this report examines the leaching kinetics of a cutting insert in HCl (hydrochloric acid) and H_2O_2 (hydrogen peroxide) to provide the basis for overcoming the mentioned issues. In order to achieve this, a design of experiments was utilized to obtain statistically relevant data together with an adequate sample preparation. Altogether a simplification and elimination of side effects allowed a proper analysis of the influence of temperature and molar concentrations on the leaching behavior. Moreover, an empirical model equation of the reaction rate was developed to describe the lixiviation of cobalt out of the hard metal substrate.

Materials and Methods

The cutting inserts applied in these experiments were kindly supplied by CERATIZIT Austria GmbH and complied with the properties according to Table 1. As specified in ISO 1832, the mentioned geometry "SNUN120412" relies on a quadratic base with 12 mm comprising 1.2 mm rounded corners and exhibits a height of 4 mm.

A fundamental part of the experiments included the preparation of the test specimen since it eased the subsequent compilation of the model equation. In order to keep the reaction interface almost constant, the cutting insert was hot-embedded in the resin "Struers ClaroFast". The pretreatment of the samples consisted of grinding and polishing, here the last step comprised a 1 μm diamond finish. A drill hole through the resin permitted the central placement in the reaction vessel by a glass bar. During the experiments the thermoplast remained stable in the applied corrosive media. No preferential leaching appeared at the juncture between the cutting insert and the mounting resin which could have affected the results. Corresponding to Fig. 1, the prepared sample only allowed the leaching in one direction.

The experiments were executed in a 1000 ml double-walled borosilicate reaction vessel with a clamp fixed cap which enabled various access possibilities via four glass joints 29/32 and one 14/23 according to DIN 12242. This included a reflux condenser, a purging nitrogen supply, a specimen holder and a sampling port as well. Furthermore, a Pt100 sensor close to the specimen allowed an adequate temperature control. A magnetic stirring device at about 200 min^{-1} maintained a uniform intermixing of the liquid. The implanted design of experiments provided

Table 1 Attributes of the used cutting inserts

Grain size	Medium
Geometry	SNUN120412—ISO 1832
Delivery state	Sintered, not machined
Co	6.0%
WC	93.4%
Composite carbides	0.6%

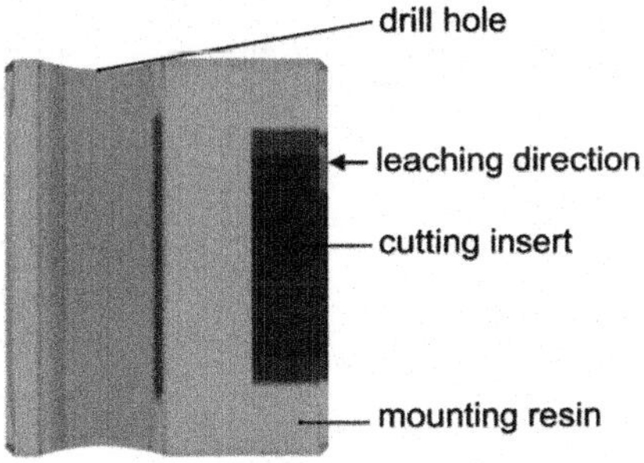

Fig. 1 Specimen for the investigation of the reaction rate

reliable results. Therefore, the software Modde 11 supported the creation of the CCF design as well as its final evaluation. Altogether the investigation compromised of 17 tests with the variable parameters in the same order as reported in Table 2. At the beginning 500 ml of leachant were heated to the desired temperature and subsequently the experiment started by the immersion of the specimen. The monitoring of the leaching process took place by extracting 2.5 ml samples after 5, 10, 20 and 30 min. An ICP-MS realized the analysis according to ÖNORM EN ISO 17294-2. A multi-element standard solution with a mass concentration of $\beta = 10$ mg/l for cobalt served for the calibration in the range of 0.50–100 µg/l. For the analysis a dilution of all samples occurred in following ratios: 1:100, 1:1000, 1:10,000.

Additionally, a mathematic correction considered the binder metal removed by sampling before the linear regression analysis. For the assessment of the reaction process, a rate law is needed. Many diffusion controlled reactions obey the parabolic law. The general form complies with Eq. (1), based on Habashi. [12]

$$k\left[\frac{\text{mg}}{\text{l}\cdot\text{min}^{0.5}}\right]\cdot t^{0.5}\left[\text{min}^{0.5}\right] = y\left[\frac{\text{mg}}{\text{l}}\right] \tag{1}$$

The factor k represents the reaction rate coefficient and t the elapsed time of the ongoing reaction, which together results in the concentration of the leached metal in the solution. As a result, some experiments disclosed a deviation from the parabolic

Table 2 Results of the regression analysis for the k- and n-factor with corresponding coefficient of determination in relation to the variable parameters of the CCF design

No.	T (°C)	HCl (mol/l)	H_2O_2 (mol/l)	$k\left(\frac{\text{mg}}{\text{l}\cdot\text{min}^n}\right)$	n (–)	R^2
01	80	2.00	1.00	2.83	0.57	0.9885
02	30	2.00	1.00	2.62	0.55	0.9975
03	80	2.00	4.00	1.73	0.52	0.9941
04	55	1.25	2.50	4.33	0.46	0.9940
05	80	0.50	4.00	3.58	0.67	0.9925
06	30	1.25	2.50	4.14	0.39	0.9937
07	55	1.25	2.50	3.97	0.53	0.9980
08	30	0.50	1.00	4.17	0.41	0.9748
09	55	1.25	1.00	4.22	0.50	0.9969
10	55	1.25	2.50	3.92	0.53	0.9996
11	55	2.00	2.50	4.20	0.53	0.9969
12	80	1.25	2.50	5.85	0.46	0.9978
13	55	0.50	2.50	4.19	0.47	0.9965
14	30	0.50	4.00	3.01	0.48	0.9970
15	30	2.00	4.00	3.45	0.45	0.9987
16	80	0.50	1.00	6.87	0.43	0.9965
17	55	1.25	4.00	4.28	0.52	0.9887

law. Therefore, a modified empirical law according to Eq. (2) was implemented that applies the power factor (n) as a second parameter.

$$k\left[\frac{\mathrm{mg}}{\mathrm{l} \cdot \mathrm{min}^{\mathrm{n}}}\right] \cdot t^{n}[\mathrm{min}^{\mathrm{n}}] = y\left[\frac{\mathrm{mg}}{\mathrm{l}}\right] \tag{2}$$

A simple logarithm permits a linear regression analysis which was executed by the function "LINEST" of Microsoft Excel 2010.

Results and Discussion

In order to compile the model equation, the determination of the reaction rate coefficient k and the power factor n in dependence of the variable parameters was conducted. The coefficient of determination indicates that most of the variance is accounted for in accordance with the underlying variables k and n as listed in Table 2.

The Fig. 2 reveals the relationship between the reaction rate coefficient and the power factor at different experimental conditions. The two not specified parameters in each diagram exhibit an intermediate adjustment in terms of the experimental boundaries. Surprisingly, the acid- and oxidant concentrations reduce the k-factor at higher amounts, while n is only slightly affected. In terms of temperature,

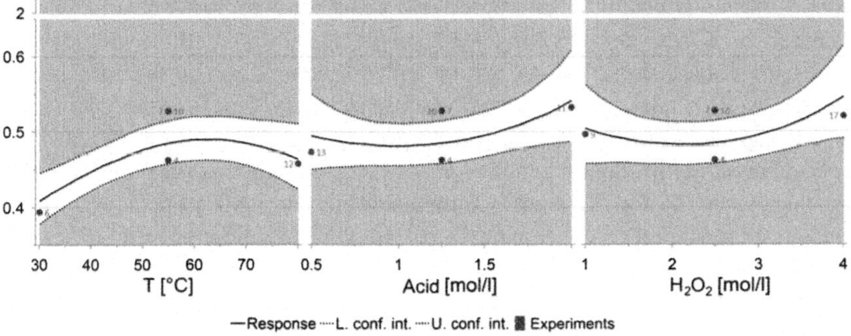

Fig. 2 Prediction plot with 95% confidence interval estimation for the k- and n-coefficients in dependence of a variable parameter

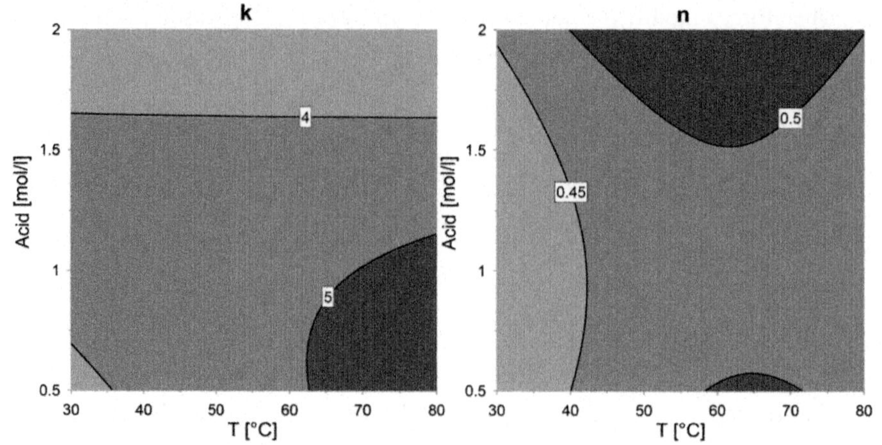

Fig. 3 Response contour plot for the *k*- and *n*-coefficients at 2.5 mol/l H_2O_2 in reference to temperature and acid concentration

n increases up to a peak at about 60 °C, whereas the reaction rate coefficient rises only insignificantly.

The contour diagram in Fig. 3 reports the attainable values at 2.5 mol/l H_2O_2 and variable temperatures as well as acid concentrations. Corresponding to the illustration, the power factor lies around 0.5 for a large area, but can deviate noticeably depending on the parameters.

Conclusion

In agreement with the current investigation, the leaching of a cutting insert consisting of cemented carbide in an aqueous media of HCl and H_2O_2 follows the classical law of diffusion depending on acid, oxidant concentration and temperature adjustment. Nevertheless, a certain deviation of the power factor occurs, which may be caused by a complex leaching mechanism. Possible reactions include the dissolution of cobalt, complex formation and diffusion through the remaining WC skeleton. Additionally, the WC itself may loosen and therefore the diffusion zone does not grow according to the proceeding cobalt dissolution. A further side reaction includes the decomposition of the oxidant into oxygen and water. This report discloses the fundamental parameters which are able to positively affect the leaching characteristics of cemented carbides. The obtained results indicate a substantial relation within the investigated system.

Acknowledgements The financial support by the Austrian Federal Ministry of Science, Research and Economy and the National Foundation for Research, Technology and Development is gratefully acknowledged.

References

1. B. Kieffer, E. Lassner, Tungsten recycling in todays environment. BHM **139**(9), 340–345 (1994)
2. B. Zeiler, Recycling von Hartmetallschrott. Paper presented at the Hagener symposium Pulvermetallurgie 1997, Hagen, 13–14 Nov 1997
3. K.B. Shedd, Tungsten recycling in the United States in 2000. http://pubs.usgs.gov/circ/circ1196-R/
4. T. Karhumaa, M. Kurkela, Review of the Hardmetal recycling market and the role of the zinc process as a recycling option. Paper presented at the 18th international Plansee seminar 2013, Reutte, 03–07 June 2013
5. G. Gille, A. Meier, Recycling von Refraktärmetallen, in *Recycling und Rohstoffe*, ed. by K.J. Thomé-Kozmiensky (TK-Verlag, Neuruppin, 2012), pp. 537–560
6. R. Schiesser, Wertstoffrecycling wolframhältiger Sekundärrohstoffe. Ph.D. thesis, TU Wien, 2003
7. H. Meyer, Verfahren zur Rückgewinnung von Ta(Nb)C-WC-TiC-Mischungen (1979), p. 10
8. W.M. Shwayder, Method of disintegrating sintered hard carbide masses (1969) p. 3
9. N. Kopylov, B. Polyakov, Method of disintegrating sintered hard metal carbide bodies and recovering hard metal carbide powder (1999), p. 20
10. S. Gürmen, S. Stopic, B. Friedrich, Recovery of submicron cobalt-powder by acidic leaching of cemented carbide scrap. Paper presented at the European metallurgical conference, Dresden, 18–21 Sept 2005
11. T. Kojima et al., Recycling process of WC-Co cermets by hydrothermal treatment. J. Mater. Sci. **40**(19), 5167–5172 (2005)
12. Fathi Habashi, Extractive Metallurgy, *General principles*, vol. 1 (Gordon and Breach, New York, 1969)

A New Two-Stage Aluminothermic Reduction Process for Preparation of Ti/Ti-Al Alloys

Kun Zhao and Naixiang Feng

Abstract This work presents a two-stage aluminothermic reduction process for preparing Ti and Ti-Al alloys using Na_2TiF_6. Al-Ti master alloy and pure cryolite as co-products could be obtained. After the first stage reduction, O content of the metal production (particle size of less than 74 μm) was below about 0.35 wt%. Ti (IV), Ti (III) and Ti (0) existed in the Ti-containing cryolite, and the content was about 3 to ~10 wt%. After secondary reduction, Ti content of the clean cryolite was reduced to 0.002 wt%. The Al-Ti master alloy obtained by secondary reduction was composed of Al and $TiAl_3$. A cyclical production process is founded by Al-Ti master alloy returned to the next first and secondary reduction process as reductant, in which Ti and Al are almost 100% recyclable.

Keywords Titanium · Ti-Al alloys · Aluminothermic reduction · Cryolite · Cyclical production

Introduction

Titanium and titanium alloys, as important structural materials besides steel and aluminum, have attracted increasing attentions owing to their superior performances such as light weight, high strength, and high corrosion resistance [1–3]. In recent years, Ti-Al alloys have been considered as top candidate for excellent engineering materials used in aerospace and automotive fields due to their high specific strength and stiffness, high strength retention and high creep resistance at high temperature [4–6].

Currently, Ti metal is produced by a mature commercial technology named as Kroll process, and Ti alloys are usually prepared by blending of pure titanium and metallic elements under high temperature. Kroll process mainly involves the problems of the high-temperature chlorination of rutile or ilmenite, the purification of

K. Zhao · N. Feng (✉)
School of Metallurgy, Northeastern University, Shenyang 110819
People's Republic of China
e-mail: fengnaixiang@163.com

© The Minerals, Metals & Materials Society 2017

H. Kim et al. (eds.), *Rare Metal Technology 2017*,
The Minerals, Metals & Materials Series,
DOI 10.1007/978-3-319-51085-9_17

$TiCl_4$, the reduction of $TiCl_4$ with magnesium and the retreating of $MgCl_2$ melt to magnesium, etc., is a cumbersome process with heavy environmental burdens [7–11]. Therefore, Ti metal and its alloys are relatively expensive, restricting their applications in different fields.

In short, developing a shorter, simpler and higher economic benefit process for preparing titanium and titanium alloys is a significant class in the field of metallurgy as before.

To date, the existing approaches on producing titanium mainly focus on two methods, metallothermic reduction and electrochemical reduction. Depending on its short process flow and low pollution, electro-deoxidation process is the key research direction in the field of titanium metallurgy in the past two decades, in which titanium powder is obtained as the product usually. Nevertheless, electro-deoxidation processes exhibit a common problem on the poor efficiency and troublesome subsequent treatment [12–16]. Besides, aluminothermic reduction method has been reported to recovery of titanium from TiO_2, K_2TiF_6 and Na_2TiF_6 [17–20]. Most of these methods also attributing to low product quality ($\leq$97.01 wt%) and high cost, embodied in the low recovery and the polluted by-product due to the generated of lower valency compound of titanium.

The aim of this study is to introduce a high productivity and pollution-free recycling metallurgical two-stage reduction process for preparing Ti and Ti-Al alloys (e.g. Ti_3Al, TiAl and $TiAl_3$) using industrial Na_2TiF_6 and aluminum powder. Clean cryolite, which can be used in aluminum electrolysis production, as co-product can be obtained after reduction process.

Experimental

Raw Materials

The main raw materials of experiment are Na_2TiF_6 ($\geq$98%, particle size of less than 74 μm), aluminum powder ($\geq$99.9%, particle size of less than 74 μm). The oxygen content of Na_2TiF_6 and Al powders is 0.34 and 0.09 wt% respectively.

Experimental Apparatus and Methods

In the first stage, the general reactions for synthesizing Ti/Ti-Al alloys (Ti_3Al, TiAl and $TiAl_3$) are as follows:

$$12Na_2TiF_6 + 16Al = 12Ti + 3Na_3AlF_6 + 3Na_5Al_3F_{14} + 4AlF_3 \quad (1)$$

$$12Na_2TiF_6 + 20Al = 4Ti_3Al + 3Na_3AlF_6 + 3Na_5Al_3F_{14} + 4AlF_3 \quad (2)$$

$$12Na_2TiF_6 + 28Al = 12TiAl + 3Na_3AlF_6 + 3Na_5Al_3F_{14} + 4AlF_3 \quad (3)$$

$$12Na_2TiF_6 + 52Al = 12TiAl_3 + 3Na_3AlF_6 + 3Na_5Al_3F_{14} + 4AlF_3 \quad (4)$$

In order to make a more uniform composition, Na_2TiF_6 and Al powder were mixed 12 h by high-energy ball milling under a rotation speed of 250 r/min. Then the mixture was uniaxially pressed in a hydraulic press with 40 MPa to make piece with 20 mm in diameter and 40 mm in thickness. The pieces were sintered 2 h at 1100 °C under high pure argon atmosphere, and then, the distillation process started and lasted 2 h in a vacuum of $\sim$0.1 Pa. After cooling to room temperature, the product was taken out from the crucible, and the crystallized product was collected from the condenser.

In the second stage, the said distillated product was mixed with Al powder under a mass radio to be 10:3. The mixture was sintered for 2 h at 1100 °C in an argon protection environment to complete the second stage reduction. Finally, several characterization methods were applied to describe the phases and structures of the different products.

Analysis Methods

The crystal structures of reduction products were identified using X-ray diffraction (XRD) measurements with a Cu-Kα characteristic ray and energy-dispersive spectrometry (EDS). The morphologies of the reduction products were observed using scanning electron microscopy (SEM). X-ray photoelectron spectroscopy (XPS) was used to characterize the form of Ti ions existence in the crystallized product. Inductively coupled plasma-atomic emission spectrometry (ICP-AES) was used to examine the impurities contents in the final product and co-product.

Results and Discussion

Characterization of the Products Obtained by the First Stage Reduction

Figure 1a shows the original form picture of the reduction product, exhibiting a structure of gray-black sponge foam. Figure 1b illustrates XRD profiles of the different reduction products obtained with different proportions of raw materials.

After the first reduction, the single phase of Ti, Ti_3Al, TiAl, and $TiAl_3$ are identified under different proportions of raw materials. O content of the product in powder (particle size of less than 74 μm) is about 0.19–0.35 wt%. Different intermetallic phases have formed due to the different Al proportions. The SEM

Fig. 1 The original form picture (**a**) of the reduction product, and XRD patterns (**b**) of reduction products prepared with different ratio of Al (Al: Na_2TiF_6 is 1#—4:3, 2#—5:3, 3#—7:3, 4#—13:3)

micrographs of different metal products are shown in Fig. 2. It can seen that the metal product displays a loose sponge-like structure.

The component result analyzed by EDS is shown in Table 1, which illustrates that Ti and Al contents in metal products are close to the target product. In addition, O content analyzed by an oxygen-nitrogen analyzer of Ti, Ti_3Al, TiAl and $TiAl_3$ product (particle size of less than 74 μm) obtained by this method was about 0.35, 0.24, 0.13, 0.22 wt%, respectively.

Figure 3 shows Ti 2p photoelectron spectra of the distillated product. It reveals that Ti presented in the distillated product as three kinds of valence state, which was Ti (0), Ti (IV) and Ti (III) respectively. The inset of Fig. 3 evidenced that the distillated product was black.

Titanium element contained should be the major reason for the blackened distillated product obtained after the first stage reduction process. For convenience, the black distillated product is defined as Ti-containing cryolite.

Table 2 shows the chemical compositions of the Ti-containing cryolite fabricated by the first reduction. The mainly elements were F, Na and Al contained in the distillated product. Besides those, a little O, Fe, Si and S also can be found from it, which is come from the raw material. It is worth noticed that Ti was detected approximately 5.29 wt%.

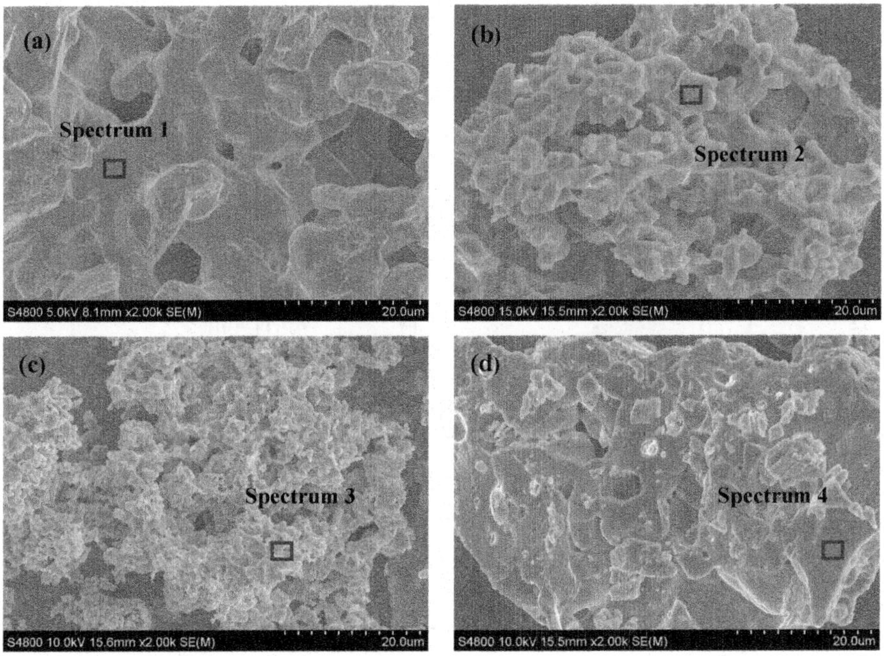

Fig. 2 The SEM micrographs of different metal products after the first stage reduction (**a** Ti, **b** Ti_3Al, **c** TiAl, **d** $TiAl_3$)

Table 1 The component result analyzed by EDS

Mass percent	Ti element	A-element
1	99.48	0.52
2	84.61	15.39
3	65.08	34.92
4	36.89	63.11

Characterization of the Products Obtained by the Second Stage Reduction

As seen from the picture shown in Fig. 4, Ti-containing cryolite was bleached, and a metal ingot was obtained after the secondary reduction. Ti content of the bleached cryolite measured by ICP-AES was merely 0.002 wt%.

Figure 5a shows the micrograph of the metal ingot obtained from the secondary reduction. It demonstrates that there are new phase in two shapes existed in matrix, granular and slender needles likes. An evident boundary can be seen between the new phases in different shapes. The EDS patterns of the new phases and matrix are shown in Fig. 5b, indicating that the new phases in granular and slender needles likes are all $TiAl_3$, and the matrix is associated to Al phase. For convenience, the metal ingot is named as Al-Ti master alloy.

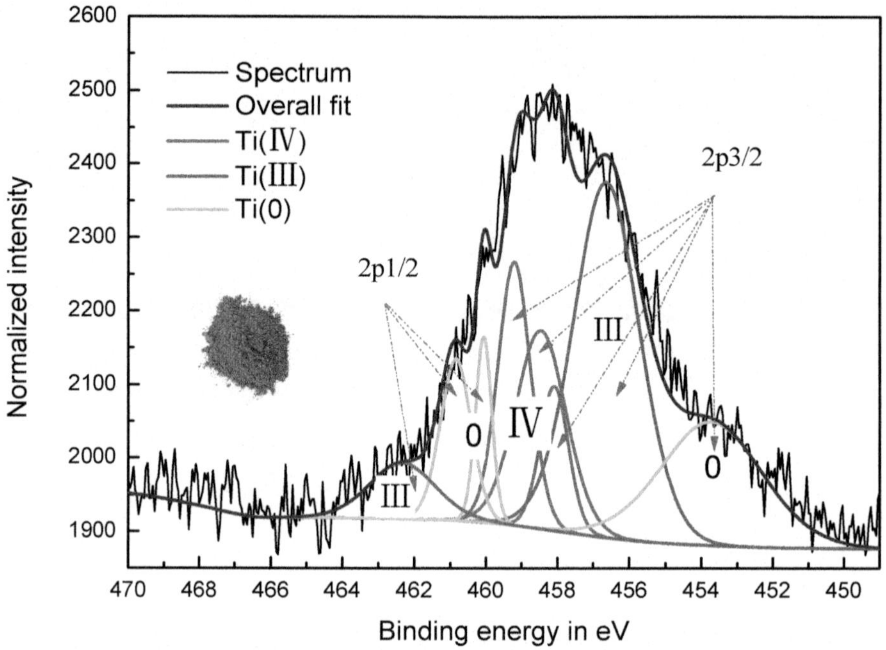

Fig. 3 The XPS trace of the distillated product, and the inset displays its original form photo

Table 2 XRF data of the product attached on the condenser

Element	F	Na	Al	Ti	O	Fe	Si	S	Other
Mass %	52.65	24.38	15.03	5.29	2.33	0.08	0.12	0.02	0.10

Fig. 4 The photo of the crucible section obtained after the secondary reduction

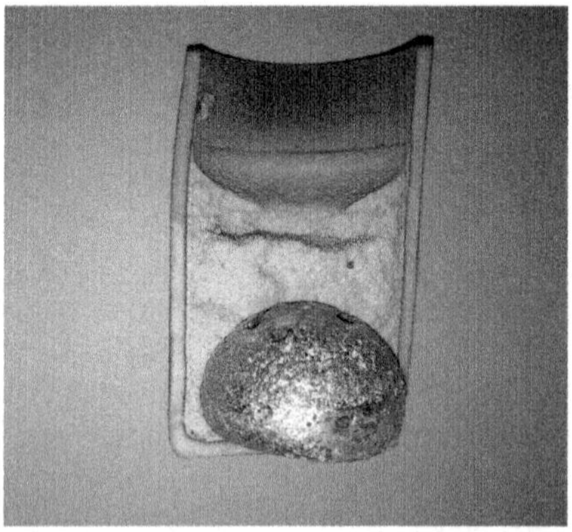

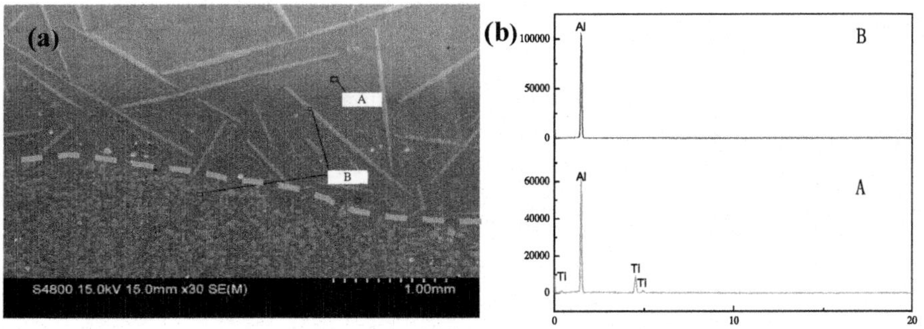

Fig. 5 The SEM image (**a**) and EDS maps (**b**) of the metal ingot obtained after the secondary reduction

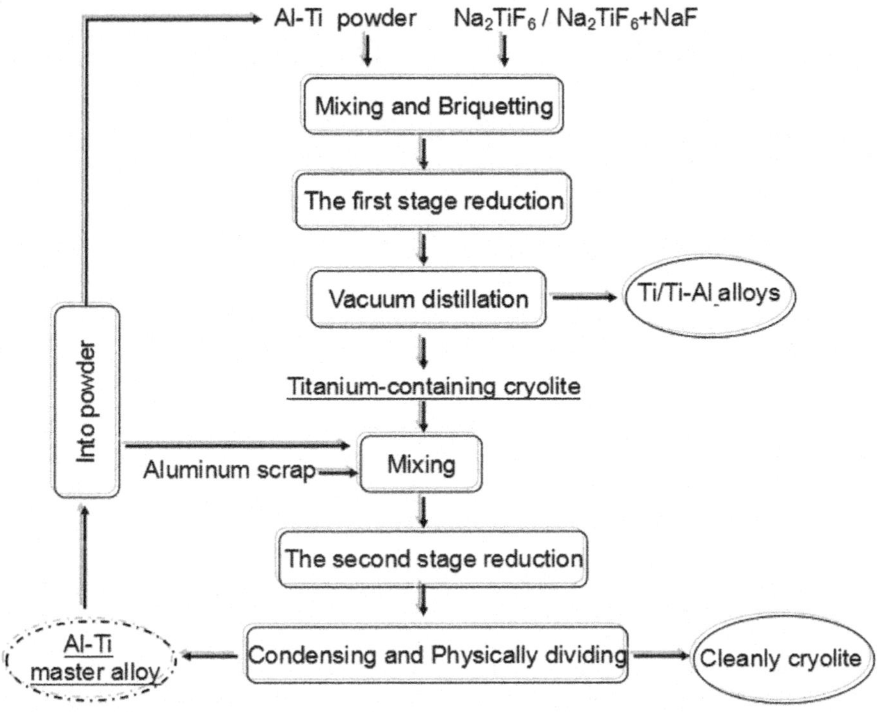

Fig. 6 The flow of the two-stage reduction process

At last, the two-stage reduction process was completed with Ti/Ti-Al alloys, cleanly cryolite, and Al-Ti master alloy as final products. It's worth mentioning that Al-Ti master alloy can be returned to the next first and secondary reduction process and used as reductant. As a result, a cyclical production process is founded without any pollutions and useless products. The flow of this consecutive two-stage reduction process is shown in Fig. 6.

Conclusion

In summary, a two-stage aluminothermic reduction process for preparing Ti/Ti-Al alloys was introduced in this paper. Cleanly cryolite and Al-Ti master alloy could be obtained as co-products. A cyclical production process is founded by Al-Ti master alloy returned to the next first and secondary reduction process as reductant. The new route for production Ti/Ti-Al alloys can be regarded as an environment-friendly metallurgy process, which also brings an exceedingly positive forecast for widespread use of Ti/Ti alloys.

References

1. K. Zhao, Y.W. Wang, J.P. Peng, Y.Z. Di, K.J. Liu, N.X. Feng, Formation of Ti or TiC nanopowder from TiO_2 and carbon powders by electrolysis in molten NaCl–KCl. RSC Adv. **6**, 8644–8650 (2016)
2. B. Wang, K.R. Liu, J.S. Chen, Reaction mechanism of preparation of titanium by electro-deoxidation in molten salt. Trans. Nonferrous Met. Soc. China **21**, 2327–2331 (2011)
3. D.S.M. Vishnu, N. Sanil, L. Shakila, R. Sudha, K.S. Mohandas, K. Nagarajan, Electrochemical reduction of TiO_2 powders in molten calcium chloride. Electrochim. Acta **159**, 124–130 (2015)
4. D.B. Lee, S.W. Woo, Effect of Cr, Nb, Mn, V, W and Si on high temperature oxidation of TiAl alloys. Met. Mater. Int. **11**, 141–147 (2005)
5. S.W. Kim, J.K. Hong, Y.S. Na, J.T. Yeom, S.E. Kim, Development of TiAl alloys with excellent mechanical properties and oxidation resistance. Mater. Des. **54**, 814–819 (2014)
6. H. Jabbar, J.P. Monchoux, M. Thomas, F. Pyczak, A. Couret, Improvement of the creep properties of TiAl alloys densified by spark plasma sintering. Intermetallics **46**, 1–3 (2014)
7. W.J. Kroll, The production of ductile titanium. Trans. Am. Electrochem. Soc. **78**, 35–47 (1940)
8. G.Z. Chen, D.J. Fray, T.W. Farthing, Direct electrochemical reduction of titanium dioxide to titanium in molten calcium chloride. Nature **407**, 361–364 (2000)
9. M.V. Ginatta, Why produce titanium by EW. JOM **52**, 18–20 (2000)
10. R.O. Suzuki, Calciothermic reduction of TiO_2 and in situ electrolysis of CaO in the molten $CaCl_2$. J. Phys. Chem. Solids **66**, 461–465 (2005)
11. A.D. Hartman, S.J. Gerdemann, J.S. Hansen, Producing lower-cost titanium for automotive applications. JOM **50**, 16–19 (1998)
12. T. Uda, T.H. Okabe, Y. Waseda, K.T. Jacob, Contactless electrochemical reduction of titanium (II) chloride by aluminum. Metall. Mater. Trans. B **31**, 713–721 (2000)
13. D.J. Fray, G.Z. Chen, Reduction of titanium and other metal oxides using electro-deoxidation. Mater. Sci. Technol. **20**, 295–300 (2004)
14. K. Ono, R.O. Suzuki, A new concept for producing Ti sponge: calciothermic reduction. JOM **54**, 59–61 (2002)
15. S.Q. Jiao, H.M. Zhu, Electrolysis of Ti_2CO solid solution prepared by TiC and TiO_2. J. Alloys Compd. **438**, 243–246 (2007)
16. D. Slawomir, N. Małgorzata, P. Marek, C. Stanisław, B. Jerzy, A simple method of synthesis and surface purification of titanium carbide powder. Int. J. Refract. Met. Hard Mater. **38**, 87–91 (2013)
17. A.R. Kamali, H. Razavizadeh, M. Hadavi, A new process for Titanium Aluminides production from TiO_2. J. Mater. Sci. Technol. **23**, 367–372 (2007)

18. M. Maeda, T. Yahata, K. Mitugi, T. Ikeda, Aluminothermic reduction of titanium oxide. *Mater. Trans. JIM*, **34**, 599–603 (1993)
19. Y. Chen, D.D.L. Chung, In situ Al-TiB composite obtained by stir casting. J. Mater. Sci. **31**, 311–315 (1996)
20. J.D. Donaldson, C.P. Squire, The transfer of titanium and boron to aluminium master alloys via Na_2TiF_6 and $NaBF_4$. J. Mater. Sci. **13**, 421–426 (1978)

Study on Pre-reduction Mechanisms of Chromium Ore Pellets in SRC Process

Pei-xiao Liu, Yan-xiang Li and Han-jie Guo

Abstract The thermodynamic analysis on chromium ore pellets in solid reduction chromium process is carried out, the result of which is validated by pre-reduction experiment of cold briquetted chromium ore. Polarization Microscope and SEM are employed to observe microstructure and element composition. The thermodynamic analyses indicate that the Cr_2O_3 will be reduced to metallic chromium in solid reaction with the ratio of CO and CO_2 pressure reaching 2017 at 1200 °C. The experimental results agree well with the thermodynamic analyses.

Keywords Solid reduction chromium process · Ferrochrome · Thermodynamic analysis · Lithofacies analysis

Introduction

Nowadays mining yield of chromium ore for smelting ferrochrome is about 20 million tons per year. The fine chromium ores account for about 80%. The poor permeability of burden, irregular furnace operation and slag boil are caused by the use of fine chromium ore in the process of smelting ferrochrome. Therefore, the fine chromium ore must be handled before they are into submerged arc furnace. The pre-process to fine chromium ore has been an important topic for ferroalloy worker all the time.

P. Liu (✉) · Y. Li · H. Guo
Department of Metallurgical and Ecological Engineering,
University of Science and Technology Beijing, Beijing 100083, China
e-mail: liupeixiao1978@126.com

Y. Li
e-mail: liyanxiang@metal.ustb.edu.cn

H. Guo
e-mail: guohanjie@ustb.edu.cn

© The Minerals, Metals & Materials Society 2017

H. Kim et al. (eds.), *Rare Metal Technology 2017*,
The Minerals, Metals & Materials Series,
DOI 10.1007/978-3-319-51085-9_18

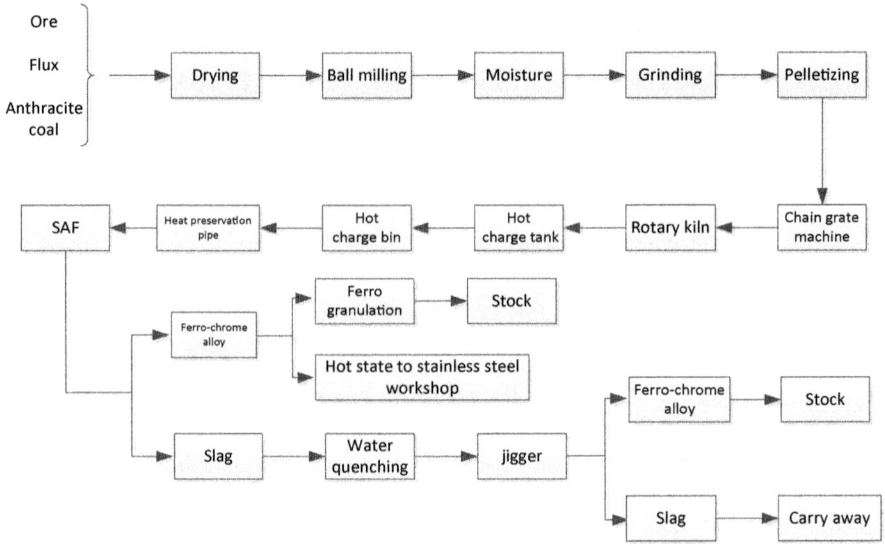

Fig. 1 SRC pre-reduction process for high carbon ferrochrome production

Solid reduction chromium (SRC) was developed by Showa Denko and its industrial production first began in 1970. This process was also called preheating pellets and process flow diagram is shown in Fig. 1.

The pre-reduction pellets were manufactured as follows:

1. The size of ground fine chromium ore below 200 meshes account for 70–80%.
2. The pellets were prepared from ground fine chromium ore and binder in the cylinder ball making machine.
3. The pellets were dried in grate and the strength of them should be improved.
4. The pellets were roasted in rotary kiln under reducing atmosphere. There is a layer of hard shell formed, which is beneficial in the strength improvement. The pellets from kiln could be hot charged into submerged arc furnace.

As for the fundamental research about the SRC process, only NEUSCHUTZ has made a kinetic analysis. In this paper, the pre-reduction mechanism of chromium pellets in the SRC process was studied.

Thermodynamics Analysis of Reduction Reaction in SRC Process

It is obvious that the success of SRC process depends on the degree of pre-reduction of chromium pellets in kiln, which concerns the effect of whole process. The reaction equations of chromium pellets are shown in (1) to (4).

$$2/3Cr_2O_3 + 2C = 4/3Cr + 2CO \qquad (1)$$
$$\Delta G^{\theta} = 123{,}970 - 81.22T, \quad T_s = 1523\ K$$

$$2/3Cr_2O_3 + 26/9C = 4/9Cr_3C_2 + 2CO \qquad (2)$$
$$\Delta G^{\theta} = 114{,}410 - 83.05T, \quad T_s = 1373\ K$$

$$2/3Cr_2O_3 + 18/7C = 4/21Cr_7C_3 + 2CO \qquad (3)$$
$$\Delta G^{\theta} = 115{,}380 - 82.09T, \quad T_s = 1403\ K$$

$$2/3Cr_2O_3 + 54/23C = 4/69Cr_{23}C_6 + 2CO \qquad (4)$$
$$\Delta G^{\theta} = 118{,}270 - 81.75T, \quad T_s = 1403\ K$$

It could be concluded from Eq. (1) to (4) that the initial temperature of Cr_2O_3 reduction reaction is about 1523 K. If the reduction product is Cr_3C_2, initial temperature is about 1373 K. All these reactions are between pure substances. While the existing form of Cr in chromium ore is magnesia-chrome spinel and its molecular formula is $(Mg, Fe)(Cr, Al)_2O_4$. In these conditions, the initial temperature of Cr_2O_3 reduction reaction is higher than that illustrated in Eq. (1)–(4).

There are some iron oxides in magnesite chrome ore, which also could react with C and it was illustrated in Eq. (5).

$$Fe_xO_y + y\,C = xFe + y\,CO \qquad (5)$$

The CO was also generated when the iron oxides were reduced to Fe. There are a mount of chromium ores in pellets after the reduction reaction. The CO gas could not be oxidized to CO_2 with the excess reducing agent. Iron oxide and Cr_2O_3 could coexist in magnesia-chrome spinel and the high density CO would form around or in the magnesia-chrome spinel crystal.

Cr_2O_3 was reduced by CO could illustrated in Eq. (6)–(8).

$$C + 0.5O_2 = CO, \quad \Delta G^{\theta} = -116{,}300 - 83.9T \qquad (6)$$

$$C + O_2 = CO_2, \quad \Delta G^{\theta} = -395{,}350 - 0.54T \qquad (7)$$

$$2Cr + 1.5O_2 = Cr_2O_3, \quad \Delta G^{\theta} = -1{,}120{,}200 + 255.45T \qquad (8)$$

$$CO + 0.5O_2 = CO_2, \quad \Delta G^{\theta} = -278{,}372 - 83.9T \qquad (9)$$

Equation (10) could be obtained from above Equations.

$$3CO + Cr_2O_3 = 3CO_2 + 2Cr, \quad \Delta G^{\theta} = 285{,}084 - 3.75T \qquad (10)$$

Table 1 The ratio of CO and CO_2 pressure at beginning of reduction with different temperature

T/K	t/°C	ΔG^0	$\Delta G^0/3RT$	P_{CO}/P_{CO_2}
1273	1000	280,310	8.8283	6825
1373	1100	279,935	8.1744	3549
1473	1200	279,560	7.6092	2017
1573	1300	279,185	7.1159	1231
1673	1400	278,810	6.6816	798
1773	1500	278,435	6.2963	543
1873	1600	278,060	5.9521	385
1973	1700	277,685	5.6428	282
2073	1800	277,310	5.3633	213
2173	1900	276,935	5.1096	166
2273	2000	276,560	4.8782	131

According to the gibbs free energy of ideal gas Equation, Eq. (10) could be expressed as follows.

$$\Delta G = \Delta G^{\theta} + 3RT\ \ln\frac{P_{CO_2}}{P_{CO}} \tag{11}$$

Another formula could be concluded from Eq. (11) when reduction reaction begin.

$$\Delta G^{\theta}/3RT = \ln\frac{P_{CO_2}}{P_{CO}} \tag{12}$$

If reduction reaction happens, the relationship of temperature and P_{CO}/P_{CO_2} is illustrated in Table 1 and Fig. 2.

It could be concluded that P_{CO}/P_{CO_2} reduce gradually with the increase of temperature. This value is 2017 when the temperature is 1473 K. The Cr_2O_3 could be reduced at this temperature with the excess carbon, which could meet the thermaldynamic requirement in and around the magnesia-chrome spinel crystal. The reducing condition would improve with the increase of temperature, which would be improved obviously when the temperature is 1673 K.

Experiment Analysis

Cold briquetted samples containing carbon were made from South Africa chromite concentrates to investigate the solid pre-reduction at 1000 and 1200 °C in which graphite carbon acted as the reducing agent. Lithofacies samples were made after pre-reduction, and their microstructures and element compositions were analyzed through Leica large polarizing microscope and scanning electron microscopy (SEM).

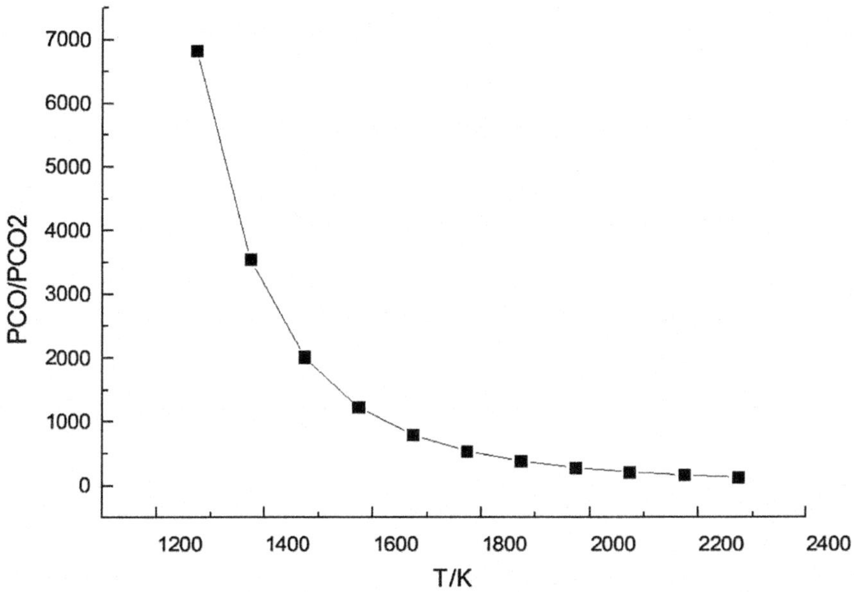

Fig. 2 The relationship between P_{CO}/P_{CO_2} and temperature at the beginning of reduction reaction

Results of lithofacies identification and SEM analysis showed that

(1) Solid reductions were taken place in different degrees for samples at 1000 and 1200 °C. The reduction of Fe was taken place along the edges and cleavage fissures. Magnesiochromites were reduced into chromites within the cleavage fissures. Magnesium and aluminium contents decreased and iron content increased.
(2) Magnesiochromite grain edges were broken after calcinations at 1000 °C. Few of chromite grains were intergrown. And there was some metallic iron reduced from chromite particles or at magnesiochromite grain edge.
(3) Samples after solid reduction at 1200 °C were similar to that at 1000 °C. The reduction of Fe took place along the edges and cleavage fissures, and magnesiochromites were reduced into chromites. Iron content increased and magnesium and aluminium contents decreased.
(4) Much more aluminium chromium slag and some molten slag phase were formed after solid reduction at 1200 °C, and some magnesiochromite were bonded by the molten slag phase.
(5) There were chromium oxide particles wrapped by metallic chromium in the samples after solid reduction at 1200 °C.

Conclusions of Reduction Mechanism of SRC Process

Thermodynamic calculation results show that the reduction of Fe was taken place along the edges and cleavage fissures after calcinations at 1000 °C. Few of chromite grains were intergrown, and there was some metallic iron reduced from chromite particles or at magnesiochromite grain edge. While at 1200 °C with the CO and CO_2 ratio up to 2017, the reduction of Fe took place along the edges and cleavage fissures, and magnesiochromites were reduced into chromites. Magnesium and aluminium contents decreased and iron content increased. There were chromium oxide particles wrapped by metallic chromium in the samples. It was implied that the solid reduction of Cr_2O_3 happened and metallic Cr was generated. Experimental verification was achieved about the generation of metallic Cr under the pre-reduction of chromite concentrates cold briquetted pellets at high temperature. In practice, temperature is mostly controlled above 1400 °C. SRC solid pre-reduction is thermodynamically feasible according to the thermodynamic calculation and reducing roasting experiments. The roasting temperature can be lowered depending on the working condition.

Smelting operations show that comparing with the traditional lump ore or sintering method, cold charge of SRC pre-reduction pellets could save more than 30% of electricity, reduce the amounts of coke and fume and improve the productivity of electric furnace and the recovery of chromium. Additionally, it is helpful for the total enclosing of electric furnace, the recovery of waste heat and the decrease of pollution.

SRC pellets pre-reduction process is advanced ferrochrome production process technology. And it will be the development direction of ferrochrome production technology under the present domestic and foreign raw materials and product markets.

References

1. N. Ralph, A review of the deposits and beneficiation of lower-grade chromite. J. S. Afr. Inst. Min. Metall. **8**, 205–226 (1982)
2. Q.X. Zhang, Ferro-Alloys. **2**, 33–39 (2000)
3. W. Dai, L. Shu, *Ferroalloys Metallurgical Engineering* (Metallurgical Industry Press, Beijing, 1999), p. 195
4. J.H. Zhou, *Ferroalloys Production Technology* (Science Press, Beijing, 1991), p. 77
5. Y. Otani,K. Ichikawa, Manufacture and use of prereduced chromium-ore pellets. Paper presented at the 1st international ferro-alloys congress, Johannesburg, pp. 31–37, 22 Apr 1974
6. C.T. Bi, Ferro-Alloys. **2**, 38–43 (1992)
7. X.G. Zhao, Ferro-Alloys. **5**, 6–10 (1998)
8. R.T. Li, J.S. Chen, Zhejiang Metallurgy. **3**, 1–6 (1987)
9. D. Neuschutz,Kinetic aspects of chromite ore reduction with coal at 1200 to 1550 °C. Paper presented at the 6th international Ferro-Alloys Congress, Cape Town, pp. 65–70, 8 Mar 1992

Recovery of Valuable Metals from High-Content Arsenic Containing Copper Smelting Dust

Xuepeng Li and Dachun Liu

Abstract Smelting-acid leaching process is a way to treat high-content arsenic containing copper. However, smelting process makes a low recovery of valuable metals and acid-leaching process makes large amounts of arsenic-iron slag. Arsenic sulphuration volatilization at a low temperature was studied in this paper and valuable metals were recovered with different methods. Arsenic recycled as As_2O_3 was separated from other valuable metals selectively and calcines were treated by acid pressure leaching. Indium and copper in leaching liquor were made into sponge indium/copper after the processes of extraction-re-extraction-displacing process (ERD process) and displacing process. $ZnSO_4 \cdot 7H_2O$ was collected after concentrating while recovering zinc. Metals, such as tin, bismuth and lead, were recycled as raw materials for tin metallurgy.

Keywords Copper smelting dust · Arsenic · Indium · Acid pressure leaching

Introduction

Copper smelting dust comes from copper smelting process. The main treatments are "reducing smelting-leaching" process and "hydrometallurgy" process. For the first process, metals, such as bismuth, lead and tin, are mainly reduced to alloy and arsenic, zinc, indium, and copper are mainly volatilized to dust. The main problem of this process is the low recovery of valuable metals [1–6]. The second process treats copper smelting dust by leaching, arsenic, zinc, indium, copper are leached to liquid and lead, bismuth, tin are left in slag for the pyrometallurgical processes. Moreover, arsenic is neutralized in the form of arsenic-iron slag by alkali and copper, indium, are replaced as sponge copper and indium by zinc powder. The

X. Li (✉) · D. Liu
Kunming University of Science and Technology, Kunming, Yunan, China
e-mail: 791392138@qq.com

D. Liu
e-mail: 1280919970@qq.com

© The Minerals, Metals & Materials Society 2017

H. Kim et al. (eds.), *Rare Metal Technology 2017*,
The Minerals, Metals & Materials Series,
DOI 10.1007/978-3-319-51085-9_19

main problems of this process are large amounts of arsenic and arsenic-iron waste which is hard to treat and the big risk of toxic AsH_3 gas in the process of replacing Cu and In [7–15]. Thus, it is very important to eliminate the influence of arsenic in the treatment copper dust. Arsenic sulphuration volatilization at low temperature was studied in this paper and valuable metals were recovered with follow-up process, such as acid pressure leaching.

Experimental

Materials and Procedure

Elements content and semi-quantitative analysis of copper smelting dust are shown in Tables 1 and 2. Vulcanizater is sulfur (saled), and catalysator is manganese dioxide (AR).

As shown in Tables 1 and 2, valuable metals are widely present in varying amounts in the copper smelting dust. The content of arsenic is high to 15.32 wt%, which is mainly in forms of As_2O_3. The phases of Pb, Cu and Zn are mainly oxide, sulfate and sulfide, respectively.

A muffle furnace was employed for the arsenic sulphuration volatilization at a low temperature, and the procedures are as follows: First, mixing copper smelting dust and sulfur evenly and reserving the mixture for 3 h at 350 °C; second, raising to the desired temperature.

A 5 L pressure reactor was used for the leaching process. Two steps were taken for the process: First, mixing calcines and H_2SO_4; then, putting the mixture in the pressure reactor with a certain pressure of oxygen.

Table 1 Competent of high-content arsenic copper smelting dust

Element	As	Zn	Pb	Bi	In	Cu	Sn
Pct (%)	15.32	12.52	17.46	3.16	0.125	3.12	3.82

Table 2 Semi-quantitative analysis of high-content arsenic copper smelting dust

Phase	$PbSO_4$	$ZnSO_4 \cdot H_2O$	ZnS	As_2O_3	$Cu_2As_2O_7 \cdot 3H_2O$
Pct (%)	30	25	6	18	3
Phase	$Cu_2O(SO_4)$	$Bi_2O_3(SO_4)$	SnO_2	Others	
Pct (%)	3	8	5	2	

Experimental Theory

A. Theory of arsenic sulphuration volatilization

The reaction between As_2O_3 and sulfur shows in Eq. (1):

$$As_2O_3 + 4.5S = As_2S_3 + 1.5SO_2(g) \quad (1)$$

The analysis above shows that As_2O_3 can be vulcanized to As_2S_3 under a low temperature which is not stable at this temperature rang (boiling point of 707 °C). On the other hand, As_2S_3 will be oxidized to As_2O_3 in the process of dust collection (as shown in Eq. (2)). The other valuable metals, such as indium, lead, tin, zinc and copper, are nearly non-volatile under that condition. Thus, it is available in theory for, selective separation of arsenic from the copper smelting dust (Fig. 1).

$$As_2S_{3(g)} + 4.5O_{2(g)} = As_2O_{3(g)} + 3SO_{2(g)} \quad \Delta G_T^0 = -1240.3 + 0.12T(\text{kJ/mol}) \quad (2)$$

B. Theory of pressure leaching

The phases of valuable metals remained in calcines are mainly oxide, sulfate and sulfide after the removal of arsenic by low-temperature vulcanized vitalization. Pressure leaching can increase the leaching rate of indium, copper and zinc while lead, tin and bismuth nearly can not be leached. The main reactions that take place during the leaching by H_2SO_4 can be expressed as follows:

$$In_2O_3 + 3H_2SO_4 = In(SO_4)_3 + 3H_2O \quad (3)$$

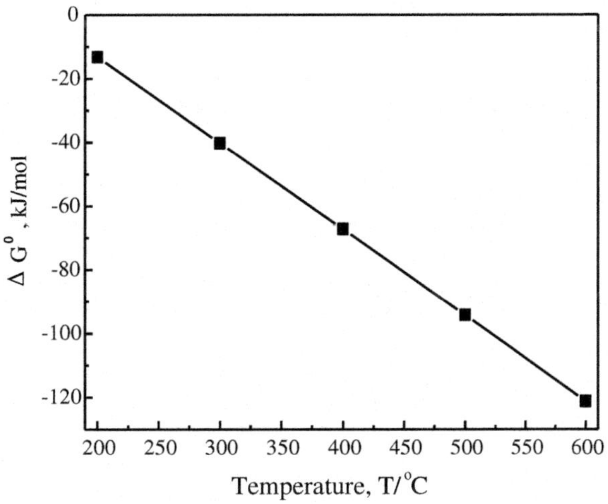

Fig. 1 Relationship between ΔG^0 and temperature of Eq. (1)

$$In_2S_3 + 4.5O_2 + H_2SO_4 = In_2(SO_4)_3 + H_2O + S \tag{4}$$

$$2CuS + O_2 + 2H_2SO_4 = 2CuSO_4 + 2H_2O + 2S \tag{5}$$

$$2CuS + 2H_2SO_4 + O_2 = 2CuSO_4 + 2H_2O + 2S \tag{6}$$

$$4Cu^+ + O_2 + 4H^+ = 4Cu^{2+} + 2H_2O \tag{7}$$

$$2ZnS + O_2 + 2H_2SO_4 = 2ZnSO_4 + 2H_2O + 2S \tag{8}$$

$$ZnO + H_2SO_4 = ZnSO_4 + H_2O \tag{9}$$

Compared with the atmospheric pressure leaching, pressure leaching can increase the partial pressure of oxygen to oxidize the sulfate enough and can increase the activity of valuable metals in liquid, that is to say, pressure leaching can increase the leaching rate of valuable metals. Take Eq. (4) as an example, its equilibrium constant is as shown in Eq. (10).

$$K_c = \frac{a_{In_2(SO_4)_3} a_{H_2O} a_8}{a_{In_2S_3} a_{H_2SO_4} \left(\frac{P_{o_2}}{P^0}\right)^{4.5}} = \frac{a_{In_2(SO_4)_3} (P^0)^{4.5}}{a_{H_2SO_4} (P_{O_2})^{4.5}} \tag{10}$$

Because K_c is a constant value at a certain temperature, the activity of $In_2(SO_4)_3$ in liquid will increase when the partial pressure of oxygen increased. It means the content of $In_2(SO_4)_3$ will be increased by pressure leaching. It is the same to other sulfates.

Results and Discussions

Low-Temperature Vulcanized Vitalization Experiments

High-content arsenic containing copper dust was vulcanized for 3 h at 350 °C and then volatilized in certain temperature in a muffle furnace. Evaporation temperature and amount of vulcanizater are decided in this part.

A. **Effect of evaporation temperature on evaporation rate**

As shown in Fig. 2, the evaporation rate of arsenic increases while temperature increasing. The suitable vitalization temperature is 630 °C through the experiments. Under that conditions the evaporation of arsenic is 91.39% and the content of arsenic in calcines is reduced to 1.1%, meanwhile, the evaporation rates of Pb, Zn, In, Cu, Bi, Sn are 5, 6, 5.65, 2.31, 2.21, 5.51%, respectively.

Fig. 2 The effect of vitalization temperature on the evaporation rate of arsenic and other valuable metals, vulcanizater of 13 wt%, vulcanized for 3 h at 350 °C

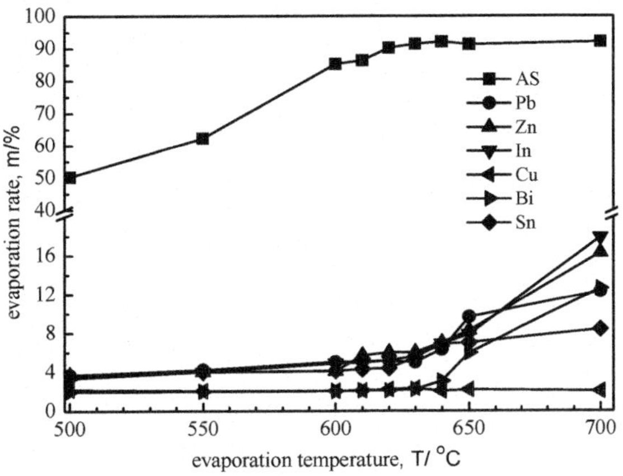

B. **Effect of amount of vulcanizater on evaporation rate**

As shown in Fig. 3, the evaporation rate of arsenic increases while the addition amount of vulcanizater increasing, meanwhile, the evaporation rates of other metals are very low. The better amount of vulcanizater addition is 13 wt% through the experiments (Fig. 4).

It was indicated that the selectively removal of arsenic by low-temperature vulcanized vitalization is feasible through the experiments. The determined conditions are below: the amount of vulcanizater addition is 13 wt%; vulcanized for 3 h in 350 °C then volatilized for 3 h at 630 °C. Under those conditions, the evaporation rate of arsenic can reach to 91.39%, the content of arsenic remained in calcines was reduced to 1.1%. Meanwhile, the evaporation rates of Pb, Zn, In, Cu, Bi, Sn are 5, 6, 5.65, 2.31, 2.21, 5.51%, respectively.

Fig. 3 The effect of the addition amount of vulcanizater on the evaporation rate of arsenic and other valuable metals, vulcanized for 3 h in 350 °C then volatilized for 3 h at 630 °C

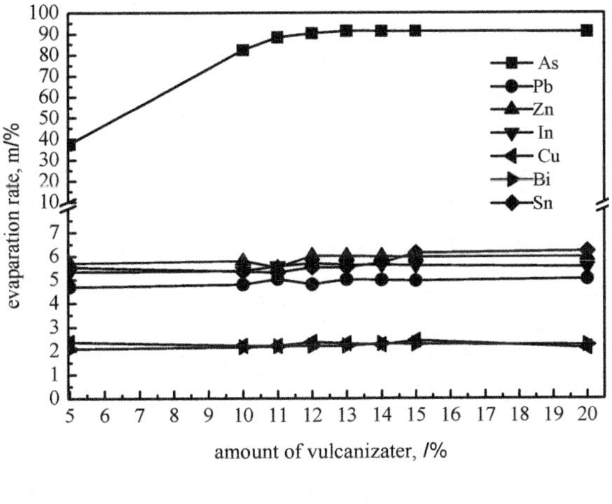

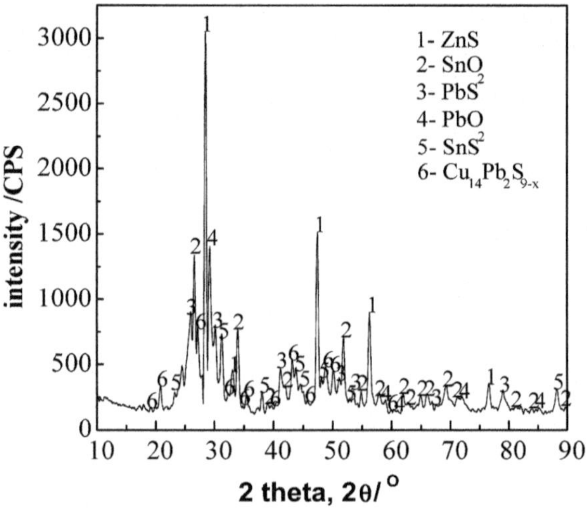

Fig. 4 The XRD analysis of the calcines. It shows the elements Zn, Cu, Pb, Sn are in the forms of oxide and sulfide

Pressure Leaching Experiments

Calcines are treated by H_2SO_4 pressure leaching. The influence of leaching temperature, sulfuric acid concentration and liquid-solid ratio (L/S) are discussed in this part.

A. **Effect of leaching temperature on leaching rate**

As shown in Fig. 5, the leaching rates for elements, such as Cu, Zn and In, are 80, 75, 35% respectively at 90 °C under atmospheric pressure. Meanwhile, Pb, Bi and Sn can barely be leached. The leaching rates of Cu, Zn and In has been raised dramatically by pressure leaching. However, the leaching rates for Cu, Zn and In reached to 98.99, 90.2 and 96.5% at 150 °C. While the changes for Pb, Bi and Sn were not obviously at the same temperature. Thus, the leaching temperature can be taken as 150 °C.

B. **Effect of H_2SO_4 content on leaching rate**

As shown in Fig. 6, the leaching rates reached to 98.99, 90.2 and 96.5% for Cu, Zn and In when H_2SO_4 concentration is 150 g/L. The leaching rates are barely changed with the increase of sulfuric acid concentration for Pb, Bi and Sn. Thus, 150 g/L of the sulfuric acid was taken as one of the leaching parameters.

C. **Effect of L/S on leaching rate**

As shown in Fig. 7, the leaching rate of Cu, Zn and In raises when L/S increased. When L/S is 4:1, the leaching rates of Cu, Zn and In are 96.27, 94.6 and 88.6%, respectively. Meanwhile, Pb, Bi and Cu can barely be leached. Although the leaching rates are a bit lower than L/S of 5:1 and 6:1, the ratio of 4:1 is suitable for

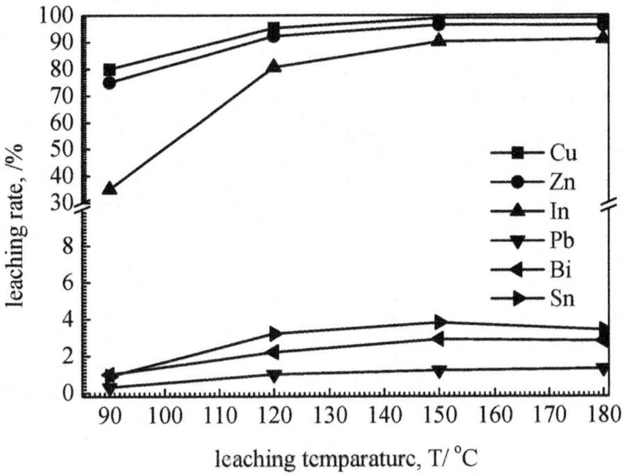

Fig. 5 The influence of temperature on the leaching rate of valuable metals. The parameters for such experiments are below: H_2SO_4 concentration is 150 g/L, L/S is 6:1, leaching time is 2 h, the pressure in pressure reactor is 0.7–0.8 MPa, the partial pressure of oxygen is about 0.3–0.4 MPa

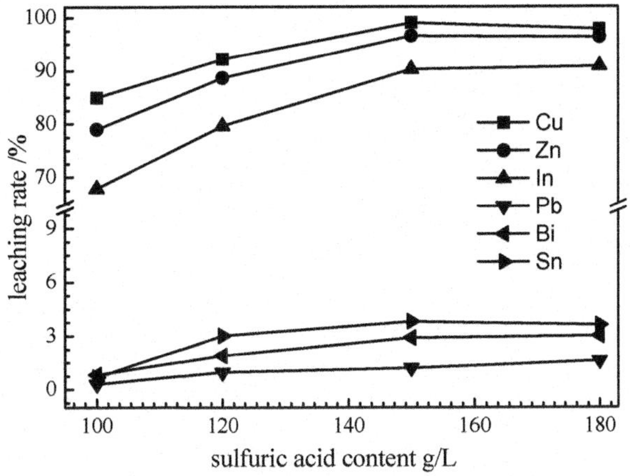

Fig. 6 The concentration influence of H_2SO_4 on the leaching rates of valuable metals. The parameters for such experiments are below: leaching temperature is 150 °C, L/S is 6:1, leaching time is 2 h, the pressure in pressure reactor is about 0.7–0.8 MPa, the partial pressure of oxygen is about 0.3–0.4 MPa

saving the leaching volume. Thus, L/S value of 4:1 can be taken as one of the leaching parameters.

D. **Effects of leaching time on leaching rate**

As shown in Fig. 8, the leaching rate of Cu, Zn and In raises when leaching time increased. The leaching rate of Cu, Zn and In almost stay stable after 2 h. Therefore, the optimization time for the leaching experiments is 2 h.

E. **Treatment on leaching liquid**

After acid leaching, Cu, Zn and In were almost leached to leaching liquid. Table 3 shows the components of one-step leaching liquor and residue.

As shown in Table 3, Sn, Pb, Bi mainly remained in the residue. On the other hand, Zn, Cu and In were leached. Sn, Bi, Pb mostly remained in residue can be used as raw materials for tin metallurgy. The component of five-time-step leaching liquid is shown in Table 4.

Fig. 7 The influence of L/S on the leaching rates of valuable metals while leaching temperature is 150 °C, H_2SO_4 concentration is 150 g/L, leaching time is 2 h with 0.7–0.8 MPa in pressure reactor, the partial pressure of oxygen is about 0.3–0.4 MPa

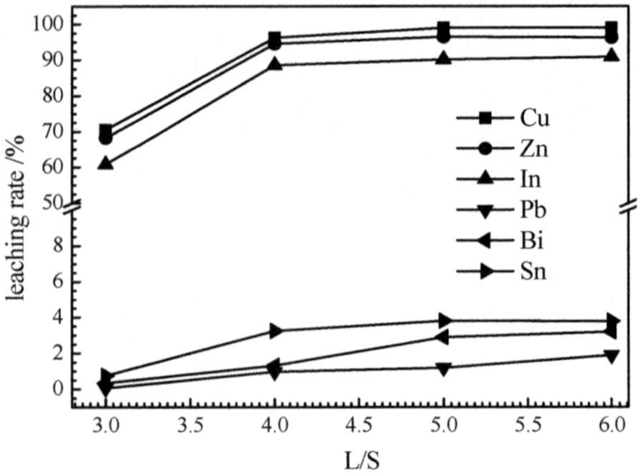

Fig. 8 The influence of leaching time on the leaching rates of valuable metals, leaching temperature is 150 °C, H_2SO_4 concentration is 150 g/L, L/S is 4:1 with 0.7–0.8 MPa in pressure reactor, the partial pressure of oxygen is about 0.3–0.4 MPa

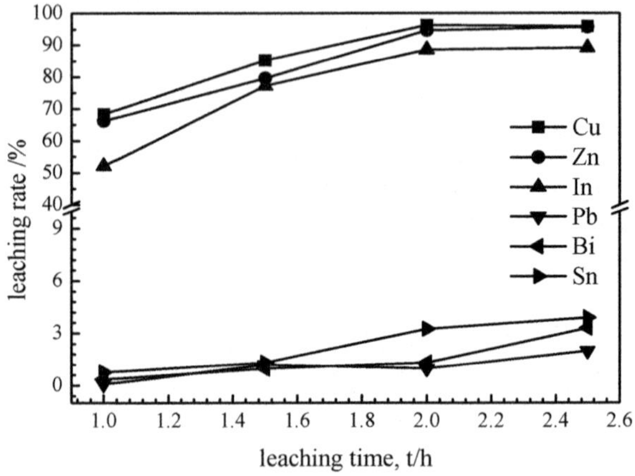

Table 3 Component of one-step leaching liquor and leaching slag

	Zn	Cu	In	Pb	Bi	Sn	As
Leaching liquor (g/L)	11.3	2.35	0.21	0.08	0.12	0.37	0.15
Residue (%)	1.0	0.07	100 g/t	25.45	6	6.89	1.89

Table 4 Component of five-time-step leaching liquor and leaching residue

	Zn	Cu	In	Pb	Bi	Sn	As
Leaching liquor (g/L)	53.9	11.95	1.2	0.38	0.72	1.02	0.56

As shown in Table 4, Zn, Cu and In were concentrated after five-time-step leaching. Indium was extracted by di(2-ethylhexyl) phosphoric acid (P204) then re-extracted and replaced by zinc power to sponge indium. Copper was replaced by zinc power to sponge copper. Because of the low content of As, there was no AsH_3 gas while displacing Cu through instrument test. Zn was made to $ZnSO_4$ $7H_2O$ by concentrating.

After the experiments above, the valuable metals mentioned above in high-content arsenic containing copper dust were comprehensive recovered.

Conclusions

As was selectively separated by volatile sulfide at low temperature under the conditions, the addition amount of vulcanizater is 13 wt%, vulcanized for 3 h at 350 °C then volatilized for 3 h at 630 °C. Under those conditions, the evaporation rate of arsenic is 91.39%, the content of arsenic remained in calcines was reduced to 1.1%. The influence of As was eliminated for it was selectively removed and recycled. Thus, the risk of toxic AsH_3 gas in displacing process and big amount of arsenic-iron slag were avoided. The evaporation rates of Pb, Zn, In, Cu, Bi and Sn are 5, 6, 5.65, 2.31, 2.21, 5.51% respectively.

The calcines were leached by pressure leaching, the leaching rates of Cu, Zn, In were 96.27, 94.6, 88.6% respectively under the conditions concentrate of H_2SO_4 150 g/L, L/S 4:1, leaching time 2 h, pressure for 0.7–0.8 MPa. Sn, Bi and Pb stayed remained in leaching residue and can be used as raw materials for tin metallurgy. The elements component of Cu, Zn, In in leaching liquid is up to 11.95, 53.9, 1.2 g/L respectively after they were leached for 5 times. Sponge indium and copper were obtained which were replaced by zinc power. For element zinc, it was made into $ZnSO_4$ $7H_2O$ by concentrating.

References

1. W. Chen et al., Study on the new process for treating copper converter dusts. Non-ferrous Min. Metall. **3**(19), 45–47 (2003)
2. Y. Xu et al., A new process for comprehensive utilization of high-arsenic dust from Isa furnace. China Nonferrous Metall. **5**, 16–18 (2005)
3. H. Li, The process practice for recycle of lead, copper and bismuth from the fume of the copper converter. Yunnan Metall. **2**(40), 52–56 (2011)
4. Z. Liu, Analysis and suggestion on comprehensive recovery process of copper smelting dust. China Nonferrous Metall. **5**, 44–48 (2015)
5. S. Ruan, Y. Lu, On comprehensive recovery of valuable metals from ESP dust in copper smelting. Nonferrous Smelting **6**, 41–44 (2003)
6. Z. Yu, Comprehensive utilization of copper converter dust. Nonferrous Smelting (1), 37–40 (1997)
7. T. Li, Process and practice of hydrometallurgical treatment of as-enriched dust. China Nonferrous Metall. **5**, 11–14 (2015)
8. J.-J. Ke, R.-Y. Qiu, Recovery of metal values from copper smelter flue dust. Hydrometallurgy **12**(2), 217–224 (1984)

9. Z. Xu et al., Pressure leaching technique of smelter dust with high-copper and high-arsenic. Chin. J. Nonferrous Metals **18**(1), S59–S63 (2008)
10. M. Li, Comprehensive recovery of valuable elements in copper smelting process. China Nonferrous Metall. **2**, 71–73 (2014)
11. Z. Wang et al., The new technics processing of recovering valuable metals by using hydrometallurical process to treat the copper dust. Hunan Nonferrous Metals **6**(26), 20–23 (2010)
12. L. Guan et al., Leaching process for comprehensive recovery of valuable metals from copper smelter dust with low indium. Chin. J. Rare Metals **1**(32), 88–93 (2008)
13. A.J. Monhemius, P.M. Swash, Removing and stabilizing as from copper refining circuits by hydrothermal processing. JOM **9**, 30–33 (1999)
14. Y. Liao et al., Research on kinetics of leaching of arsenic from dust containing high arsenic. J. Sichuan Univ. (Eng. Sci. Ed.) **3**(45), 200–206 (2015)
15. A.A. Samuel, S. Ake, Selective leaching arsenic and antimony from a tetrahedrite rich complex sulphide concentrate using alkaline sulphide solution. Miner. Eng. **23**, 1227–1236 (2010)

Sulfuric Acid Leaching of Mechanically Activated Vanadium–Bearing Converter Slag

Junyi Xiang, Qingyun Huang, Xuewei Lv and Chenguang Bai

Abstract The extraction of vanadium from mechanically activated converter slag was studied in dilute sulfuric acid solution. The effects of roasting and leaching parameters were investigated. The results show that Ca/V mole ratio, roasting temperature, leaching temperature, S/L ratio and leaching time significantly affected the dissolution of vanadium while only roasting time has relatively minor effects under the conditions in this work. The over-high roasting temperatures and Ca/V mole ratios performed negative effects on the dissolution of vanadium. The optimum conditions for roasting was found to be 800 °C, 60 min and 1:1 Ca/V mole ratio, and that for leaching conditions was pH 2.5, 60 °C, 20:1 L/S ratio,—150 mesh and 40–80 min. More than 90% of vanadium was leached from the converter slag under these optimum conditions.

Keywords Vanadium · Converter slag · Mechanical activation · Hydrometallurgy · Leaching

Introduction

Vanadium is defined as a strategic metal as it causes noticeable improvements in physical and chemical properties for steels and alloys [1, 2]. Vanadium originates from primary sources such as phosphate rock, titaniferous magnetite, and uraniferous sandstone and siltstone, in which it constitutes less than 2% of the host rock [3]. Vanadium-bearing titanomagnetite ores and their vanadium rich slags are the major sources for the production of vanadium pentoxide [4, 5]. China ranks third in

J. Xiang · X. Lv (✉) · C. Bai
School of Materials Science and Engineering,
Chongqing University, Chongqing 400044, China
e-mail: lvxuewei@163.com

Q. Huang
School of Materials and Metallurgical Engineering, Chongqing University
of Science and Technology, Chongqing 401331, China

© The Minerals, Metals & Materials Society 2017

H. Kim et al. (eds.), *Rare Metal Technology 2017*,
The Minerals, Metals & Materials Series,
DOI 10.1007/978-3-319-51085-9_20

vanadium in the world, right after Russia and South Africa. The major producers of vanadium-bearing titanomagnetite in China are Panzhihua Iron and Steel Group Corp., followed by Chengde Iron and Steel Group Co. Ltd. [6]. Vanadium-bearing titanomagnetite ores contain about 0.1–0.3% V_2O_5 by weight. After reduction of titanomagnetite ore in a traditional blast furnace, vanadium stays dissolved in molten pig iron. The molten pig iron is next oxygen lanced in a converter which caused the vanadium transfer to the slag phase [7, 8]. The vanadium content in the converter slag may range from 12 to 25% V_2O_5 by weight [9]. More than 1.2 million tons of vanadium-bearing steel slag are produced each year in China [10].

Sodium salt roast process is the most mature and commercial process to extract vanadium from converter slag [11]. It involves the oxidation roasting of converter slag with the addition of sodium chloride or sodium carbonate to from three kinds of water soluble vanadate, leaching the roasted slag with water, purification of aqueous solution, and precipitation steps. Unfortunately, it brings a series of environment problems, like emission of corrosive gases (Cl_2, HCl, SO_2 and SO_3), the recovery of sodium salt, and producing huge quantities annually of waste water contain sodium salt [12]. What's more, the fusion agglomeration of the slag at high roasting temperature leads to ring formation in rotary kiln as the low melting point of sodium salts.

To minimize the emission of waste gas, waste water and waste solid, adapt the stricter policies on environmental protection, it has led to extensive studies conducted on other way to cleaner production. Calcification roasting process is a promising way [13]. Instead of sodium salts, lime or limestone are used as additives during calcification roasting process to transform vanadium-bearing spinels into acid soluble calcium vanadates. Leaching of calcification-roasted slag with sulfuric acid to produce soluble $(VO_2)_2SO_4$ is followed by purification, precipitation and roasting. The calcification roasting is environment friendly due to the roasting with lime will not emit corrosive gases. It has been reported that the recovery of vanadium is significantly influenced by the roasting and the followed leaching steps [11, 14]. In our study, we found that mechanical activation pretreatment for vanadium-bearing converter slag can significantly accelerate the oxidation roasting step and increase the leaching efficiency in the following leaching step. In this study, leaching ratio of vanadium from mechanical activated converter slag with calcification roasting followed by sulfuric acid leaching is investigated. The effect of roasting and leaching parameters on oxidation and leaching ratio of vanadium were studied.

Experimental

Samples for this study were taken from the converter slag disposal site in Panzhihua Iron and Steel Group Corp. As shown in Fig. 1, the converter slag was crushed to a particle size finer than 125 μm (120 US mesh) through a series of jaw crushers and

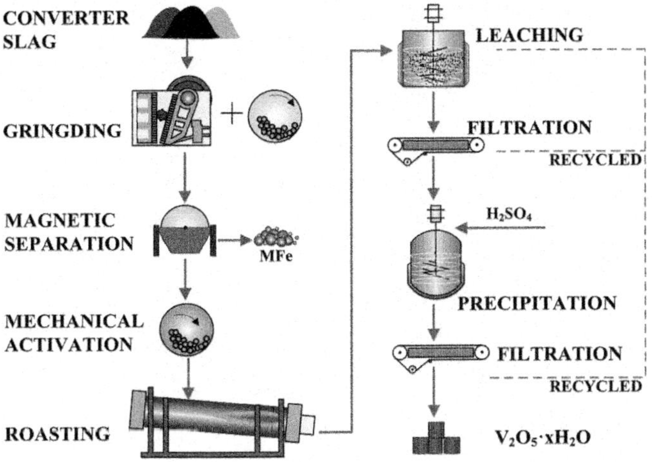

Fig. 1 Calcification roasting-acid leaching process for vanadium recovery from converter slag

Table 1 Chemical analysis of the slag, wt%

V_2O_5	TFe	TiO_2	SiO_2	MnO	CaO	MgO	Al_2O_3	Cr_2O_3
15.29	31.00	14.38	14.50	7.32	2.57	2.92	3.47	1.43

ball mills. Then separated the magnetic iron through a magnetic separation. The non-magnetic part was used to recovery of vanadium. The chemical analysis of the non-magnetic part was shown in Table 1.

The ore was mixed with a determined amount of calcium carbonate, then introduced into a planetary ball mill (Retsch PM 100, Germany) for mechanical activation. The mole ratio of vanadium-to-calcium in the mixture was maintained at 0.5, 1, 1.5 and 2 in the separate experiments. The mechanical activation experiments were conducted at a ball to ore weight ratio of 5:1, rotation rate of 400 rpm, and lasting 80 min. The SEM image of the activated sample is shown in Fig. 2. The activated sample had extremely fine particles with an average diameter of about 1 μm. The activated slag was then loaded into an alumina crucible and introduced inside a muffle furnace once reached the desired temperature. The roasting process was carried at 750, 800, 850, and 900 °C for 2 h in separate experiments. The mixture was stirred at 20 min intervals in order to inhibit agglomeration. After roasting, the crucible was removed from the furnace and cooled down to room temperature in air. The roasted slag was milled and screened by sieves of different meshes.

The roasted slag was leached in sulfuric acid solution at a certain temperature, agitation speed, S/L ratio and time. A 500 mL three-necked round-bottomed flask equipped with a water condenser, a thermometer and a pH meter. The flask was placed in a thermostatically maintained water bath and the reaction mixture was

Fig. 2 SEM image of the mechanical activated sample

stirred by a magnetic stirrer at 150 rpm. Sulfuric acid concentration of 15% was used to maintain the acidity of the leaching solution at 2.5 ± 0.2 in the entire leaching process. After a specific leaching time the slurry was filtered and the leaching residue was dried. The concentration of vanadium in the solution was analyzed using inductively coupled plasma optical emission spectrometry (ICP-OES) by an ICAP 6000 series instrument (Thermo Fisher Scientific, USA).

Results and Discussion

The effects of roasting parameters including Ca/V mole ratio, roasting temperature and holding time on the leaching ratio of vanadium was shown in Fig. 3. The effect of Ca/V mole ratio was determined by varying the Ca/V mole ratio in the range of 0.5–2.0 using 800 °C roasting temperature, 2 h holding time, 50 °C leaching temperature, 60 min leaching time and 20:1 L/S ratio. It was shown that the leaching ratio of vanadium increased first with the increase of Ca/V mole ratio reached a maximum then decreased with the further increase of Ca/V mole ratio. The optimum Ca/V mole ratio was 1:1.

The effect of roasting temperature was determined by varying the roasting temperature in the range of 750–900 °C and keeping the other conditions same as mentioned above. The leaching ratio of vanadium directly increased when increased the leaching temperature from 750 to 800 °C. Further increasing temperature from 800 to 900 °C decreased the leaching ratio gradually.

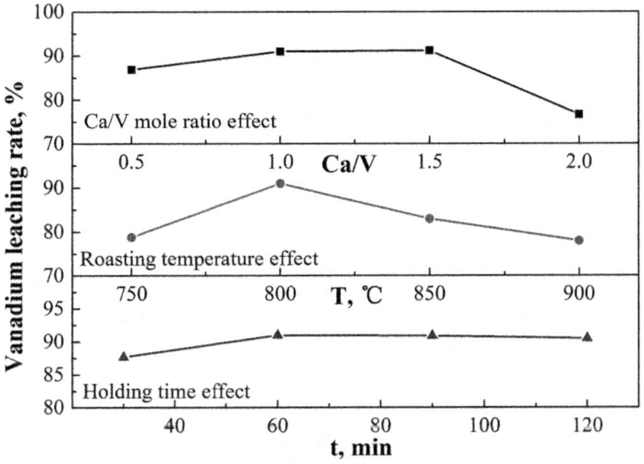

Fig. 3 Effects of Ca/V mole ratio, roasting temperature and holding time on the leaching of vanadium

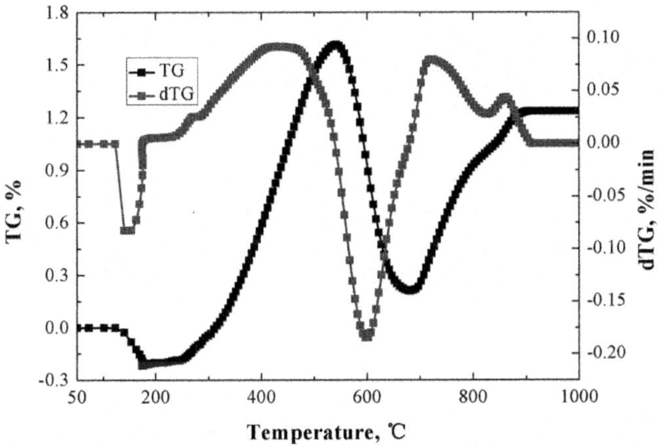

Fig. 4 TG-dTG diagram of the mechanically activated sample

The TG-dTG experiment was conducted under the condition of Ca/V mole ratio of 1:1, air flow rate of 20 ml/min and heating rate of 10 K/min. As shown in Fig. 4, the sample weight increases by 1.82% in the temperature range of 200–540 °C, then dramatically decreases by 1.4% in the temperature range of 540–683 °C, and then increases by 1.02% in the temperature range of 683–907 °C. The weight decrease may be due to the decomposition of calcium carbonate, or perhaps the reaction of vanadium pentoxide with calcium carbonate.

$$V_2O_5 + CaO = CaV_2O_6 \quad (1)$$

$$V_2O_5 + 2CaO = Ca_2V_2O_7 \quad (2)$$

$$V_2O_5 + 3CaO = Ca_3V_2O_8 \quad (3)$$

$$V_2O_5 + CaCO_3 = CaV_2O_6 + CO_2 \quad (4)$$

$$V_2O_5 + 2CaCO_3 = Ca_2V_2O_7 + 2CO_2 \quad (5)$$

$$V_2O_5 + 3CaCO_3 = Ca_3V_2O_8 \quad (6)$$

$$CaCO_3 = V_2O_5 + CaO \quad (7)$$

ΔG^θ−T diagram for reactions (1)–(7) at P_{O2} = 1 atm were shown in Fig. 5. It was shown that the ΔG^θ values for reaction (4)–(6) are negative which indicate that the reactions between $CaCO_3$ and V_2O_5 could happen without the decomposition of $CaCO_3$. The initial decomposition temperature of $CaCO_3$ was 860.9 °C at CO_2 partial pressure of 1 atm. As shown in Fig. 6, the initial decomposition temperature directly decreased with decreasing CO_2 partial pressure. The decomposition temperature is less than 629.19 when CO_2 partial pressure lower than 0.01 atm.

The effect of holding time was determined by roasting samples for a total of 2 h and keeping other conditions same as mentioned above. The increase in holing time increased the leaching ratio of vanadium in first 60 min then almost unchanged with the extent of time. Therefore, the subsequent roasting tests were performed at 800 °C, 1:1 Ca/V mole ratio and 1 h holding time.

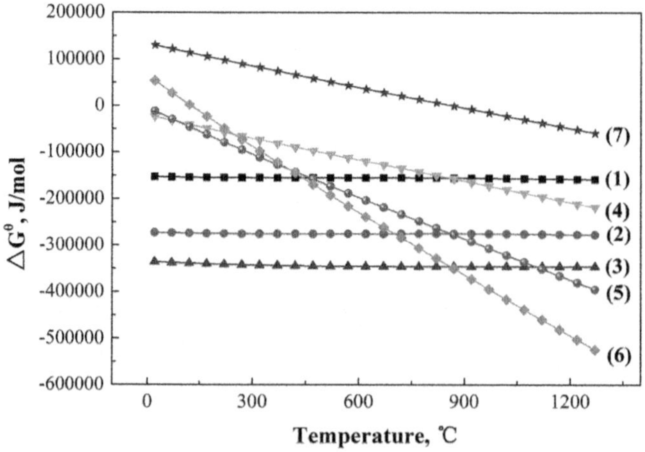

Fig. 5 ΔG^θ−T diagram for reactions in oxidation process at PO_2 = 1 atm

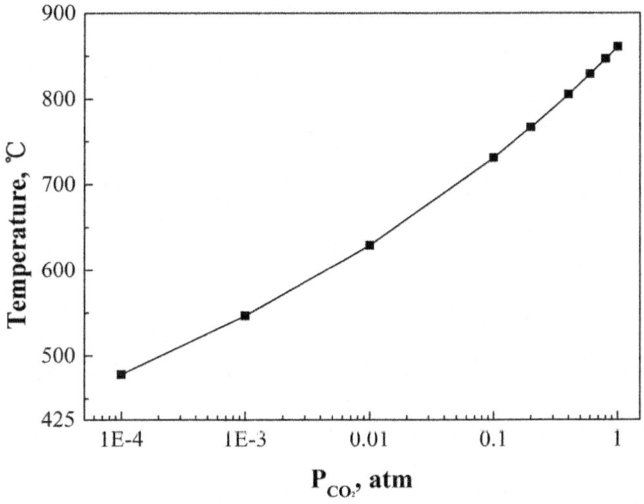

Fig. 6 The initial decomposition temperature of calcium carbonate at different CO_2 partial pressure

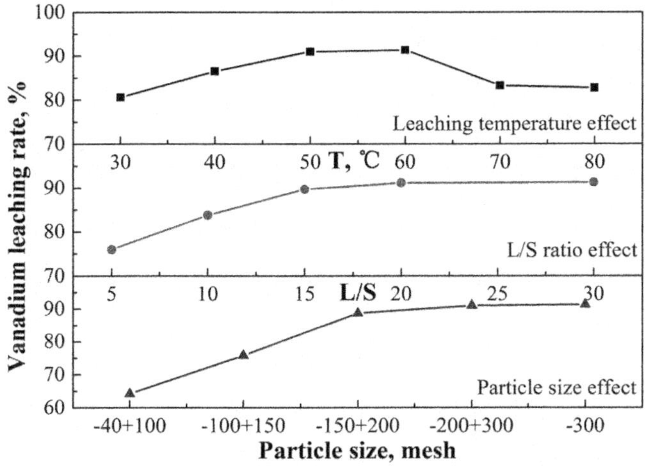

Fig. 7 Effects of leaching temperature, L/S ratio and particle size on the leaching of vanadium

The effect of leaching temperature on the leaching ratio of vanadium was determined by varying the temperature in the range of 30–80 °C using 800 °C roasting temperature, 1:1 Ca/V mole ratio, 1 h holding time, 60 min leaching time and 20:1 L/S ratio. As shown in Fig. 7, the leaching ratio of vanadium gradually increased with increasing leaching temperature till it reached around 50 °C and then decreased with the further increasing temperature. Increasing of leaching temperature lead to a fast diffusion of ions in the solution, and then accelerated the leaching rate. However, excessive temperature also leads to a fast formation of calcium sulfate ($CaSO_4$).

The calcium sulphate stayed attached to the surface of the samples, increased the thickness of product layer, and further makes diffusion difficult rough the product layer. Moreover, the decrease in solubility of V^{5+} at higher temperature possible due to the hydrolysis of dissolved V^{5+} to solid state as $V_2O_5{\cdot}xH_2O$. Therefore, the subsequent leach tests were performed at 50 °C leaching temperature.

The effect of L/S ratio on the leaching ratio of vanadium was studied by varying L/S ratio in the range of 5–30 and keeping other conditions same as mentioned above. The leaching ratio of vanadium increased from approximately 75% to about 90% with the increase in L/S ratio from 5 to 20, indicating that higher L/S ratio is required to enhance the leaching of the vanadium.

The effect of particle size on the leaching ratio of vanadium was determined by using 5 different kinds of particles, −40 + 100 mesh, −100 + 150 mesh, −150 + 200 mesh, −200 + 300 mesh and −300 mesh and keeping other conditions same as mentioned above.

The leaching ratio of vanadium increased from about 65% to about 90% with the decrease in particle size from −40 + 100 mesh to −150 + 200 mesh. The leaching ratio slightly increased with the further decreasing particle size. Finer particle size give higher vanadium leaching, but also increase the burden of filtration. Therefore a moderate particle size of −150 mesh was chosen as the optimum particle size in the current study.

The effect of leaching time studied by leaching samples with a particle size finer than 150 mesh for a total of 120 min using 20 L/S ratio at 50 °C (Fig. 8). It is shown that the leaching ratio of vanadium dire rapidly increased to about 85% in the first 20 min, then gradually increased to about 92% after 120 min of leaching.

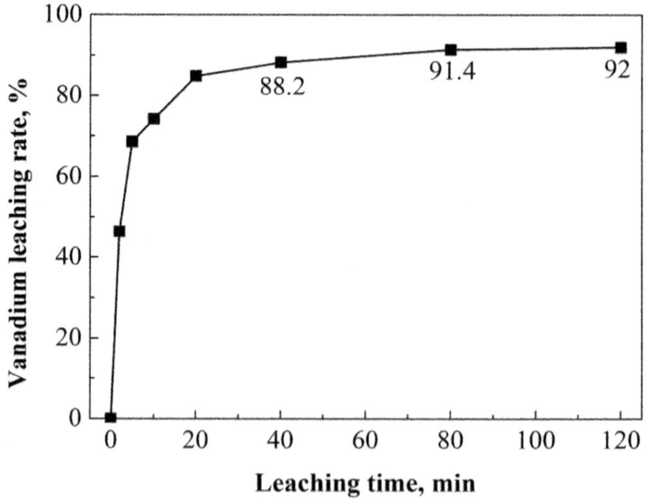

Fig. 8 Effects of leaching time on the leaching of vanadium

Conclusions

The following conclusions can be drawn from the present investigation:

(1) Both roasting and leaching parameters have a significant effect on the ratio of vanadium from the converter slag. The best roasting conditions for mechanically activated samples was 800 °C, 60 min and 1:1 Ca/V mole ratio. Higher roasting temperatures and Ca/V mole ratios caused a negative effect on the dissolution of vanadium as the generation of liquid phase during the roasting process.

(2) The leaching of vanadium was found to be a rapid process and about 80% of leaching was completed within the first 10–20 min of leaching. The optimum leaching conditions was 50 °C, 20:1 L/S ratio, −150 mesh and 40–80 min. High temperature is unfavorable to the leaching of vanadium due to the fast formation of calcium sulfate which will increase in the thickness of product layer.

(3) More than 90% of vanadium was leached from the converter slag under the optimum conditions.

Acknowledgements This work was supported by the Natural Science Foundation of China [grant numbers 5140, 4047] and the Basic and Frontier Research Program of Chongqing [grant numbers cstc2014jcyjA50011].

References

1. R. Moskalyk, A. Alfantazi, Processing of vanadium: a review. Miner. Eng. **16**, 793–805 (2003)
2. Z. Liu et al., Hydrometallurgical leaching process intensified by an electric field for converter vanadium slag. Hydrometallurgy **155**, 56–60 (2015)
3. USGS, USGS mineral commodity summaries, US Geological Survey (2016)
4. S.M.J. Mirazimi, F. Rashchi, M. Saba, Vanadium removal from roasted LD converter slag: optimization of parameters by response surface methodology (RSM). Sep. Purif. Technol. **116**, 175–183 (2013)
5. Z.-H. Wang et al., Research and prospect on extraction of vanadium from vanadium slag by liquid oxidation technologies. Trans. Nonferrous. Met. Soc. China **24**, 1273–1288 (2014)
6. P.R. Taylor et al., Extractive metallurgy of vanadium-containing titaniferous magnetite ores: a review. Miner. Metall. Process. **23**, 80–86 (2006)
7. H.-Y. Li et al., Selective leaching of vanadium in calcification-roasted vanadium slag by ammonium carbonate. Hydrometallurgy **160**, 18–25 (2016)
8. I.N. Monakhov et al., The flow of vanadium-bearing materials in industry. Metallurgist **48**, 381–385 (2004)
9. M. Aarabi-Karasgani et al., Leaching of vanadium from LD converter slag using sulfuric acid. Hydrometallurgy **102**, 14–21 (2010)
10. H. Yang, L. Jing, B. Zhang, Recovery of iron from vanadium tailings with coal-based direct reduction followed by magnetic separation. J. Hazard. Mater. **185**, 1405–1411 (2011)

11. Z. Yang et al., Leaching kinetics of calcification roasted vanadium slag with high CaO content by sulfuric acid. Int. J. Miner. Process. **133**, 105–111 (2014)
12. G. Zhang et al., Effects of microwave roasting on the kinetics of extracting vanadium from vanadium slag. JOM **68**, 577–584 (2015)
13. J. Zhang et al., Mechanism of vanadium slag roasting with calcium oxide. Int. J. Miner. Process. **138**, 20–29 (2015)
14. J. Zhang et al., Effect of acid leaching on the vanadium leaching rate in process of vanadium extraction using calcium roasting. J. Northeas. Univ. (Nat. Sci.) **35**, 1574–1577 (2014)

Present Status and Development of Comprehensive Utilization of Vanadium-Titanium Magnetite

Shiju Zhang, Songli Liu, Wenhui Ma, Kuisong Zhu, Li Cao and Yongnian Dai

Abstract There is an abundance of vanadium-titanium magnetite in Panzhihua, China, which contains more than 8.73×10^8 tons of TiO_2 accounting for 90.6% of the national reserves. To fully utilize Panzhihua titanium resources, many processes were proposed. The development and utilization of vanadium-titanium magnetite resource have extremely vital significance. However, its current use of technology is inadequate. This article is a review on techniques to make comprehensive use of vanadium-titanium magnetite, especially in the titanium extraction, and the existing problems are also introduced. The authors predict its possible tendency of development of comprehensive utilization of vanadium-titanium magnetite in the future.

Keywords Utilization of vanadium-titanium magnetite · Panzhihua · Status and development

---

S. Zhang · W. Ma (✉) · Y. Dai
Faculty of Metallurgical and Energy Engineering, Kunming University of Science and Technology, Kunming 650093, Yunnan, People's Republic of China
e-mail: mwhsilicon@126.com

S. Zhang · S. Liu · K. Zhu
Resources and Environmental Engineering College, Panzhihua University, Panzhihua 617000, Sichuan, People's Republic of China

W. Ma · Y. Dai
State Key Laboratory of Complex Nonferrous Metal Resources Cleaning Utilization in Yunnan Province, Kunming University of Science and Technology, Kunming 650093, Yunnan, People's Republic of China

W. Ma · Y. Dai
Engineering Research Center for Silicon Metallurgy and Silicon Materials of Yunnan Provincial Universities, Kunming University of Science and Technology, Kunming 650093, Yunnan, People's Republic of China

L. Cao
Materials Science and Engineering College, Xihua University, Chengdu 610039, Sichuan, People's Republic of China

© The Minerals, Metals & Materials Society 2017

H. Kim et al. (eds.), *Rare Metal Technology 2017*,
The Minerals, Metals & Materials Series,
DOI 10.1007/978-3-319-51085-9_21

Introduction

The main products of vanadium-titanium magnetite beneficiation are titanium iron concentrate (iron ore) containing about 50% of titanium in V-Ti magnetite, ilmenite concentrate containing about 50% of titanium, and cobalt sulfide concentrate [1–4].

Titanium iron concentrate is mainly used for iron making through blast furnace process, which produces a large amount of titanium-containing blast furnace slag with a titanium content of 20%–30 wt%. So far, more than 70 million tons of the titanium-containing blast furnace slag has been produced in Panzhihua and there are no effective ways to make use of it.

Ilmenite concentrate is mainly used to produce titanium-rich materials. Current industrial practices for this purpose mainly include the Electric Smelting Method, the Reduction Roast Method, Choice Chlorination Method, and Acid Leaching Method. However, none of these methods are economically effective for preparation of high quality titanium-rich materials. Impurities such as Ca and Mg, especially those in Panzhihua vanadium-titanium magnetite cannot be removed effectively using the Electric Smelting Method and the Reduction Roast Method. Development of the Choice Chlorination Method is still in a preliminary stage in China while the Acid Leaching Method causes many environmental problems. At present, most titanium white is produced using the traditional sulfuric acid method in China, which generates a lot of "three wastes" (waste gas, waste water and waste residues). With the improvement of the national environmental protection system, it is highly demanded to improve the traditional sulfuric acid method and speed up the industrialization of the Choice Chlorination process.

There is an abundance of vanadium-titanium magnetite in Panzhihua, which contains more than 8.73×10^8 tons of TiO_2 accounting for 90.6% of the national reserves. However, its current use of technology is inadequate. To fully utilize Panzhihua titanium resources, many processes were proposed. The development and utilization of vanadium-titanium magnetite resource have extremely vital significance. This article is a review on techniques to make comprehensive use of vanadium-titanium magnetite, especially on the titanium extraction. Existing problems are also introduced and possible development tendencies are predicted.

Status and Development of Utilization Technique for Titanium Iron Concentrate

The main component of titanium iron concentrate is magnetite. For titanium iron concentrate in Panzhihua, the content of iron is about 50%, and the content of TiO_2 is about 12%. Three process methods, namely the Blast Furnace iron making process, the rotary kiln-electric furnace process, and the reduction-magnetic separation process, dominate the titanium iron concentrate utilization industry in China.

Blast Furnace Iron Making Process

Blast furnace iron making process is one of the earliest methods to utilize titanium iron concentrate. In the process, almost all of the titanium goes into the blast furnace slag, vanadium is reduced into the molten iron. After iron extraction, it produces a large quantity of titanium-containing blast furnace slag with a TiO_2 content of 20%–25 wt%. Over the years, the slag accumulated has reached an enormous amount of 70 million tons. There have been a lot of research focusing on the comprehensive utilization of slag, but none of this research has been successful in developing an effective process. Given the importance of the blast furnace iron making process to the metallurgical industry, it is not very economical and environmentally friendly for the comprehensive utilization of titanium iron concentrate as a substantial amount of titanium is left wasted in the titanium-bearing blast furnace slag which is harmful to the environment [5–8].

Non-blast Furnace Iron Making Process

The method of extracting vanadium and titanium from titanium iron concentrate is referred to as the non-blast furnace iron making process, such as electric furnace smelting process and reduction-grinding separation process. Electric furnace smelting process is a method of separation of titanium and iron by electric furnace. It is principally similar to the blast furnace iron making process but is simpler in operation. Nevertheless, it's not as widely used because of the (high) viscosity of titanium slag. In reduction-grinding separation process, the ferriferous oxides are reduced in solid phase by the reductant of coal firstly, but the titanium remains oxide, then ferrous micro-bead is separated from the products by magnetic separation. Although titanium and iron can be separated from titanium iron concentrate in solid state in the reduction-magnetic separation process without formation of foaming slag, high requirement of iron particles in the (subsequent) reduction process constitutes a major restraint to the wide application of the non-blast furnace iron making process [9, 10].

Status and Development of Utilization Technique on Ilmenite Concentrate

Ilmenite concentrate with TiO_2 grade of about 45% is the main raw material for preparation of titanium dioxide and sponge titanium. As the Ti content in the titanium concentrate is not high enough for production of Ti or Ti alloy, further enrichment is desired to transform the titanium concentrate into high titanium-containing slag and rutile. The enrichment treatment can also reduce the production cost of the final product. There are many concentration methods, three of which will be review as below.

Electric Smelting Method

Putting ilmenite concentrate and coke in electric arc furnace and smelting at temperature above 1350 °C, ferriferous oxides are reduced to liquid iron settling in the furnace, and TiO_2 is enriched in the slag floating on liquid iron meanwhile. Impurities cannot be removed perfectly with this method. The content of inorganic impurities in ilmenite concentrate, especially non-ferrous impurities, has an important influence on titanium slag grade. The titanium slag grade prepared from Panzhihua titanium concentrate through the electric furnace smelting method is not very high ($\sim$wt 75%) due to the presence of alkaline earth metal oxides in the titanium concentrate, e.g., MgO, CaO, and Al_2O_3. Further enrichment processes are essential to prepare high titanium slag, but these processes, e.g., the sulfuric acid process, may have a bad impact on the environment. The Electric Smelting method features less pollutants generated and simple operation and it is suitable for hydropower rich areas [11].

Choice Chlorination Method

Different substances in ilmenite concentrate have different thermodynamic qualities. Under certain condition (temperature at about 900–1100 °C), iron oxides will react with Cl_2 to form $FeCl_3$ which can be removed easily. The process has the advantages of simple process, easy production and low power consumption. However, the process is difficult to solve the problems caused by impurities such as $MgCl_2$ and $CaCl_2$, and the corrosion problems caused by chlorine and hydrogen chloride as well [11].

Reduction Process

Researches on reduction process are relatively early. Most of the processes are carbon reduction methods wherein titanium concentrate mixed with carbon powder is reduced. Product of the reaction in air is a solid solution of TiC and TiNO. Using the reduction product as raw material, $TiCl_4$ can be synthesized through chlorination. The main problem of this method is that the reduction temperature is high and the production capacity is low.

There are many wet metallurgical utilization methods to enrich titanium, such as acid Leaching Method, reduction-hydrochloric acid solution, etc. These methods have common disadvantages: difficulties in dealing with acid leaching waste and iron products, inefficiencies, and serious corrosion of equipment [12, 13].

Status and Development of Utilization Technique on Titanium-Bearing Blast Furnace Slag

Almost all of titanium iron concentrates use in blast furnace and most of TiO_2 go into blast furnace slag. Iron concentrate used in blast furnace iron making process has produced almost 70 million tons titanium-containing blast furnace slag with TiO_2 content of 20%–25 wt% in Panzhihua. Comprehensive utilization of titanium-bearing blast furnace slag has great practical significance. Many scholars at home and abroad have done a great deal of researches on the slag.

Technique on Non-Titanium Extraction

Techniques on non-titanium extraction are mainly found in the following applications: cement admixture or concrete aggregate; materials of sanitary porcelain, glazed tiles, ceramic tiles or floor tiles; preparation of cast stone, alkali resistant glass fiber or mineral wool. It is a great waste of titanium resources that titanium in the slag is not effectively used with these techniques. The blast furnace slag is directly used for the production of building materials and other products without extracting valuable elements in it. These processes are poor in economic efficiency, and their products are of low relative added values. Along with the development of titanium and its alloy industry, comprehensive utilization of titanium in blast furnace slag has become the consensus of researchers. Non-Titanium extraction technologies have been gradually phased out [1, 14].

Technique on Titanium Extraction

The blast furnace slag in Panzhihua is an important titanium resource, which contains rich titanium and can be exploited comprehensively. Universities and scientific research institutions at home and abroad have conducted a lot of researches on titanium extraction from the slag.

1. Preparation of TiO_2 by sulfuric acid leaching of Titanium-bearing Blast Furnace Slag: This method produces titanium white from blast furnace slag with sulfuric acid leaching, hydrolyzing, extraction and precipitating separation. To produce a single ton of TiO_2, it takes about 6 tons of concentrated sulfuric acid, etc. The recovery rate of the Titanium is up to 73.4%, but this process consumes large

amounts of sulfuric acid and produces a large number of residual acid leaching liquid and dregs which are difficult to use and cause environment pollution. In addition, the production efficiency of the process is not high [15].

2. Preparation of Ti-Si-Al alloy: The process produces Ti-Si-Al alloy from blast furnace slag by reduction with 75% ferrosilicon (lime as a flux). The purity of product in this process is not high, and there is still a lot of residual titanium remaining in slag. The cost of this process is relatively high as well [16].
3. Alkali-Treatment Process: This method is to separate titanium from the slag by NaOH. The consumption of alkali in this process is large (one ton of slag using about 200–250 kg of NaOH). If considering the problem of sodium salt recovery, the process would be further complicated and costs would be greatly increased. Moreover, the enrichment of titanium is not good, and alkali treatment of blast furnace slag at high temperature will produce more serious air pollution [17].
4. Selective Enrichment of perovskite from Titanium-bearing Blast Furnace Slag: This process makes titanium components dispersed in the slag transfer and concentrate in the designed mineral phase-perovskite phase, and separates perovskite phase from the modified slags by beneficiation. The advantages of this process are low environmental pollution, low cost and high processing capacity. The perovskite is of different and non-uniform size, and accretes with spinel. This above feature makes the valuable single mineral-perovskite difficult to liberate [18, 19].
5. High-temperature Carbonization and Low-temperature Chlorination: TiC is prepared by the carbon thermal reduction method from the slag at high temperature (1600–1700 °C) and distributed diffusely in products. After crushing and magnetic separation, TiC will be enriched in slag. Then, the slag containing TiC can produce $TiCl_4$ by reaction of chlorination at low temperature. It is very difficult to dissociate and enrich TiC from products. Serious corrosion of equipment occurs and a large number of dilute hydrochloric acid is produced in chlorination process [1, 20].

Status and Development of Utilization Technique on High Titanium Slag

The grades of Ti-rich material from ilmenite concentrate in Panzhihua are around 75% because of lots of alkaline earth metal oxides such as MgO, CaO and Al_2O_3 contained in ilmenite concentrate. The Ti-rich material is only suitable for production of titanium dioxide as a raw material in sulfuric acid leaching process instead of chlorinating process. To prevent serious environmental problems created by sulfuric acid method, Ti-rich material needs to be further enriched to high titanium slag with grades higher than 90 wt% which is suitable for preparation of pigment and sponge titanium [1, 19].

Conclusion

Based on the introduction and analysis of these technologies on comprehensive utilization of vanadium-titanium magnetite in Panzhihua, all of these technologies are restricted by the problems of technology advantages, industrialization, energy saving, environmental protection, etc. The abundant situation of titanium mineral resources in Panzhihua is not commensurate with titanium industry and preparation technology.

So far, in Panzhihua, the titanium iron concentrate is mainly used in blast furnace to produce vanadium-bearing hot metal, and the ilmenite concentrate is mainly used in electric arc furnace to produce titanium slag, and the blast furnace slag is lack of the rational industrial scheme to recovery titanium.

Vanadium and titanium iron concentrate is one kind of iron, vanadium, titanium multiple-element symbiotic composite ore, which has extremely high comprehensive utilization value. Practice on the development and utilization of vanadium-titanium magnetite resources in Panzhihua is very significant. Although a great deal of research has been done, comprehensive utilization technique on vanadium-titanium magnetite has not been able to make a breakthrough progress.

Efficient preconcentration and extraction of titanium from titanium bearing blast furnace slag, titanium concentrate, rich titanium material or high titanium slag, and changing iron making process for the main method of utilization of titanium iron concentrate are the focuses and also challenges in the research and development of comprehensive utilization of vanadium-titanium magnetite. With the social development in recent years, the energy and environmental problems get more and more concerned, extracting hydrogen from coke oven gas in BF iron making processes and preparing Ti-rich material or synthetic rut by Hydrogenous Reduction will be probably a good approach.

References

1. S. Liu, S. Yang, Present state and perspectives of complex utilization on Panzhihua BF slag. Light. Met. **7**, 48–50 (2007)
2. Y. Yang, Titanium: the metal giant in the new millennium. Eng. Sci. **4**(3), 21–31 (2002)
3. C. Bai, Study on some physical chemistry problems of blast furnace slag-bearing titania. Dissertation, Chongqing University, 2003
4. X. Zou, *Study on direct selective extraction of* $TiM_x(M = Si, Fe)$ *alloys from Ti-bearing compound ores*. Dissertation, Shanghai University, 2012
5. U.S. Geological Survey, *Mineral commodity summaries 2009* (USGS, Washington, 2009), pp. 178–179
6. Z. Yuan, C. Xu, S. Zheng, J. Zhou, Comprehensive utilization of titanium resources in Panzhihua. Mod. Chem. Ind. **23**(5), 1–4 (2003)
7. G. Deng, X. Wang, X. Che, Today and tomorrow of titanium industry. Iron Steel Vanadium Titanium **24**(1), 1–7 (2003)
8. X. Li, J. Pu, The latest developments of integrated utilization on Panzhihua high titanium-bearing bf slag. Iron Steel Vanadium Titanium **32**(2), 10–14 (2011)

9. M. Chen, The process of direct reduction of vanadium titanium magnetite rotary kiln. Sichuan Metall. **4**, 11–17 (1992)
10. D. Zhu, Y. Guo, Innovative process for comprehensive utilization of vanadium-bearing titanomagnetite concentrate. Multipurp Utilization Miner. Res. **2**, 16–20 (1999)
11. G. Deng, Present situation and future development of titanium rich materials. Titanium Ind. Prog. **4**, 1–5 (2000)
12. F. Joseck, M. Wang, Y. Wu, Potential energy and greenhouse gas emission effects of hydrogen production from coke oven gas in U.S. steel mills. Int. J. Hydrog. Energ. **33**, 1445–1454 (2008)
13. X. Si, X. Lu, C. Li, S. Guo, W. Ding, Experimental studies on oxidation of Panzhihua ferrous powder. J. Cent. South Univ. (Sci. Technol.) **42**, 56–61 (2011)
14. S. Zhu, L. Lu, Developments of study on comprehensive utilization of Ti-bearing blast furnace slag and selective separation technology of Ti component. Express Inf. Mining ind. **24**(3), 9–11 (2008)
15. Q. Chen, Scale-up experiment on TiO_2 and Sc_2O_3 recovery from B.F. slag at pangang. Iron Steel Vanadium Titanium **3**, 64–68 (1995)
16. Z. Li, C. Xu, Z Li, Y. Zhou, The study on smelting Ti-Si ferroalloy by DC electro thermal process using PISC blast furnace titaniferous slag. J. Chongqing Univ. (Nat. Sci. Ed.) **19**(4), 82–86 (1996)
17. Z. Zhou, B. Zhang, Z. Zhu, A test of titania separation from high titania bearing blast furnace slag. Iron Steel Vanadium Titanium **4**, 35–38 (1999)
18. T. Lou, Y. Li, L. Li, Z. Sui, Study of precipitation of perovskite phase from the oxide SLA. Acta Metall. Sin. **36**(2), 141–144 (2000)
19. Z. Sui, Z. Guo, L. Zhang, L. Zhang, M. Wang, T. Lou, G. Li, Green separation technique of Ti component from Ti-bearing blast furnace slag. J. Mater. Metall. **5**(2), 93–97 (2006)
20. Y. Xiong, B. Liang, C. Li, Extraction and separation of Titanium from air-cooled Ti-bearing blast furnace slag. Chin. J. Process Eng. **8**(6), 1092–1097 (2008)

Review of TiO_2-Rich Materials Preparation for the Chlorination Process

Songli Liu, Li Cao, Kuisong Zhu, Shiju Zhang and Kui He

Abstract Chlorination process is a clean and efficient way to manufacture of titanium dioxide pigment. However, the limited natural rutile promotes the investigation on the upgrading of ilmenite ore into titanium-rich materials. The principle and producing situation of some production methods of TiO_2-rich materials preparation for the chlorination process are discussed. The advantage and disadvantage of the Reduction smelting produces titaniferous slag, Carbothermal reduction-nitridation, Acid leaching, and Reductive leaching produce synthetic rutile are also summed in this paper. This paper presents a review of these technologies and carbothermal reduction-nitridation method is recommended to deal with Panzhihua ilmenite ore to meet the needs of the titanium dioxide pigment production by chlorination process.

Keywords Rich titanium materials · Chlorinate · Carbothermal reduction-nitridation

S. Liu (✉)
Machinery and Electrical Engineering, Yangtze Normal University,
Fuling 408100, Chongqing, China
e-mail: 393106756@qq.com

S. Liu · L. Cao · K. Zhu · S. Zhang · K. He
Resources and Environmental Engineering, Panzhihua University,
Panzhihua 617000, Sichuan Province, China

S. Liu · L. Cao
Materials Science and Engineering, Xihua University,
Chengdu 610039, Sichuan Province, China

K. Zhu · S. Zhang
Faculty of Metallurgical and Energy Engineering, Kunming University of Science and Technology, Kunming 650093, Sichuan Province, China

© The Minerals, Metals & Materials Society 2017

H. Kim et al. (eds.), *Rare Metal Technology 2017*,
The Minerals, Metals & Materials Series,
DOI 10.1007/978-3-319-51085-9_22

Introduction

Panzhihua, China is considered a significant resource of ilmenite, and makes up a large proportion (35%) of the world's total reserves [1]. Unfortunately, Panzhihua ilmenite has high contents of calcium and magnesium. The main component of Panzhihua ilmenite is $FeTiO_3$, the main features of which are (1) its structure is dense, optionally poor, and difficult to separate TiO_2 to other constituents; (2) the content of TiO_2 is low compared to rutile. Panzhihua ilmenite has a high impurity content, particularly MgO content [2].

Although ilmenite can be directly used as the raw material such as titanium dioxide or titanium sponge, it is lengthy, costly and produces large amounts of by product, which results in some environmental problems [3]. The technology in China is facing increasing pressure from government. Therefore, the development of chlorination technology is imperative. The production of Chlorinated titanium dioxide and titanium sponge require high-quality titanium-rich materials, therefore the preparation of material for Panxi ilmenite mainly is used in chlorinated titania process. Preparing chlorinated rich titanium materials not only for China's titanium industries has a decisive significance, but also for the development of metallurgical technology has great significance.

Preparation of Rich Titanium-Rich Materials

Reduction Smelting

The reduction smelting of ilmenite is in the 1600–1800 °C, then titanium enriches in slag phase, so we call it as titanium slag and high content of titanium is called high titanium slag. Reduction smelting which includes open arc furnace, semi-closed arc furnace and enclosed arc furnace is a commonly used method. The main reactions of the reduction smelting are as follows [4]:

$$FeTiO_3 + C = Fe + TiO_2 + CO \quad (1)$$

$$2FeTiO_3 + 3C = 2Fe + Ti_2O_3 + 3CO \quad (2)$$

$$3FeTiO_3 + 4C = 3Fe + Ti_3O_5 + 4CO \quad (3)$$

$$2FeTiO_3 + C = FeTi_2O_5 + 2Fe + CO \quad (4)$$

$$Fe_2O_3 + C = 2Fe + 3COFe \quad (5)$$

Relatively speaking, the reduction smelting technology is relatively mature technology and suit large scale production, but it has high power consumption than per ton high titanium slag (including TiO_2 72–92%). Power consumption is about

2200–3500 kW-h and 80 kg graphite electrode [5]. Reduction smelting dislodge not iron impurity ability is not remarkably, and can not get high-grade titanium-rich material.

Preparation of Synthetic Rutile by Slag Enrichment

1. Hydrochloric acid direct leaching slag Process

The main component of titanium slag is anosovite which is inert to inorganic acid, so it is difficult to direct product synthetic rutile by using hydrochloric acid in the atmospheric pressure [6]. Using hydrochloric acid pressure leaching, oxidation roasting and reduction of acid leaching, which grade of TiO_2 is higher than normal pressure leaching, but the effect is not obvious, can not meet the production requirements [7].

2. Titanium slag pretreat in high temperature—leaching Process

UGS as a representative of the QIT company, QIT roast the slag which was after reduction smelting in a high temperature (950–1100 °C), then hydrochloric acid pressure leaching in the 150 °C. If the SiO_2 of extraction product content is high, we can reduce the content by alkali leaching. After calcining obtain the titaniferous slag (TiO_2 > 95%) which is called UGS [8]. In spite of the process length and high energy consumption, QIT's use of a large airtight furnace which capacity is 250 kt/a, rich ore resources and cheap power resources is beneficial from the economy of scale.

3. Titanium slag adding additive calcination modification—pickling Process

Many domestic and foreign scholars have done a lot of research on slag modified and make slag structure and phase composition change by adding modifying agent and calcining. The impurity phase of modified titanium slag can easy to remove through chemical method or physical method and obtain high-quality synthetic rutile [9].

Adding alkali metal ion to calcine can improve reaction activity and facilitate reduction effect. Domestic and international's study mainly focus on adding Na_2CO_3 calcining which can destroy the structure and morphology of anosovite in order to speed up the rate of the reaction product of leaching, acid leaching to improve the impurity [1].

The study found that adding suitable amount of iron powder after melting salt activation treatment and leaching with low concentration of hydrochloric acid in the atmospheric pressure activation of high titanium slag can dissolve impurities such as Fe, Si, Al of the titaniferous slag, adjust pH make metatitanic acid precipitation to wash out. The TiO_2 content of synthetic rutile which was prepared by calcining can reach more than 96% [10].

While it is possible to add additives and calcining to obtain a high-grade synthetic rutile, but some additives are expensive and a larger amount of its industrial production needs further study.

Reduction-Rusting Process

The Australian National Institute of Chemistry researched and developed the reduction-rusting process which can make synthetic rutile be produced on a large scale. With the reduction-rusting process, the ferric oxide in ilmenite is reduced to iron, which gathers rust in water to obtain the rich material containing titanium dioxide. Then the material is washed by dilute acid or water, filtrated and dried. Finally the synthetic rutile which is used for produce $TiCl_4$ material can be produced [11]. This synthetic rutile is an excellent source of chloride titanium dioxide production. Domestic and international research on the reduction of corrosion focused on shortening the corrosion time and the most studied is the most obvious effect of corrosion was studied. Becher method using ammonium chloride solution as corrosion liquid and NH_{4+} NH_4^+ is mainly used as a buffer to prevent the "in situ corrosion" occurs [12]. Becher method cannot remove other non-ferrous impurities and the reaction time is long, so it is only suitable for the treatment of high-grade ilmenite slag.

China began to study on the reduction-rusting process technology which is low yields and TiO_2 grade of less than 90% since the early 1970s to product synthetic rutile. Our technology remain has low productivity, long time to rust and corrosion product instability issues [13].

Hydrochloric Acid Leaching

The hydrochloric acid leaching method generally requires different pretreatment according to different ilmenite. The weathered ilmenite sands due to the poor acid-soluble of Fe_2O_3 usually use weak reduction ilmenite to improve the leaching rate and the effect.

American beauty base company is using the circulation of BCA hydrochloric acid leaching method to weak reduction by dilute hydrochloric acid pressure leaching method. Synthetic rutile containing TiO_2 content is 94% was obtained from 54 to 65% TiO_2 titanium iron ore [14].

Hydrochloric acid leaching plays an increasingly important role in the leaching of the metallic iron from the reduction products; the hydrochloric acid is preferred in leaching because it has some advantages, such as fast leaching, remarkable impurity removal and acid regeneration technology that can usefully remove residual iron and other impurities from ilmenite to form synthetic rutile [15].

Carbothermal Reduction-Nitridation

Titanium resources can selectively be chlorinated at temperatures below 200 °C if the TiO_2 in the feed is nitrided before chlorination. Although the SiC can direct chlorination, it can control Si not change to SiC before the nitriding process. Ca, Mg and other impurities are same as TiO_2 which is need a higher temperature to nitride [16]. The mainly reaction of the Carbothermal reduction-nitridation as follow [17]:

$$FeTiO_3 + (4 - x)C + \frac{x}{2}N_2 = TiC_{1-x}N_x + Fe + 3CO$$

X is between 0 and 1.

Ananthapadmanabhan and Taylor [18] synthesized titanium nitride in a thermal plasma reactor from ilmenite ore concentrate using methane and ammonia as the reactive gases. The product has been characterized by X-ray diffraction and SEM and results show that the amount of methane remarkably affects the phase composition of the product.

Welham and Llewellyn [19] synthesized single crystal TiC or TiN powders from ilmenite or TiO_2 within a single low temperature stage. The titaniferous powders were ball milled for 100 h in a laboratory scale with magnesium powder and either graphite or nitrogen as reductant.

In recent years, the scholars has made a lot of attempts and efforts in the preparation of carbo-reduction Ti (C, N) [20], but carbo-reduction process of iron reduction has not been clear enough. The process is not systematic. The pure of Ti (C, N) is not enough, the particle size is too small, or limited production, the subsequent separation of Ti (C, N) and purification of study still blank.

Conclusions

How to utilize the rich reserves of the low grade ilmenite containing high calcium and magnesium in Panxi area to prepare high quality titanium concentrates is the key to promote the development of titanium industry. At present, the methods of preparing titanium-rich materials can not economically and efficiently prepare high-quality titanium-rich materials suitable for chlorination process. Reduction smelting produce titaniferous slag and Reductive leaching produce synthetic rutile method cannot remove the impurities such as calcium and magnesium which must use acid leaching to remove, this not only increased the process and improve the cost, but also bring huge harm to ecology environment. The Acid leaching methods can prepare the suitablc for the high quality of rich titanium chloride process, but there is the use of iron is bad and has same environmental problems.

Carbothermal reduction-nitridation of ilmenite can selectively be chlorinated at low temperatures. In the low temperature the impurities such as calcium and magnesium are not molten so can't affect the chloride process, which realize the selective chlorination and reduce the chlorine consumption in chloride process.

In light of the development trend of international titanium materials, the carbothermal reduction-nitridation of ilmenite technology was recommended for possessing Panzhihua natural resource.

References

1. C. Li, B. Liang, H.Y. Wang, Preparation of synthetic rutile by hydrochloric acid leaching of mechanically activated Panzhihua ilmenite. Hydrometallurgy **91**(1), 121–129 (2008)
2. L. Wang, H. Wang, Distribution and production status of titanium resources in China. Chin. J. Rare Met. **28**(1), 265–267 (2004)
3. A. Adipuri, G. Zhang, O. Ostrovski, Chlorination of titanium oxycarbide produced by carbothermal reduction of rutile. Metall. Trans. B **39B**, 23–34 (2008)
4. G. Deng, Rich titanium material production present situation and the future development. Prog. Titanium Ind. (4), 1–5(2000)
5. W. Mo, D. Lu, *The Titanium Metallurgy*, vol. 6 (Metallurgical Industry Press, Beijing, 1979), pp. 123–163
6. W. Jiang, X. Jiang, S. Wang, Study on preparing artificial rutile from high titanium slag. Nonferrous Met. (Extr. Metall.), **3**, 008(2012)
7. Y.-F. Guo, C.-M. Xiao, T. Jiang, Enrichment of moderate and low grade titania feedstock by activated roasting-acidic leaching. Chin. J. Nonferrous Met. **15**(9), 1446 (2005)
8. T. Lasheen, Soda ash roasting of titania slag product from Rosetta ilmenite. Hydrometallurgy **93**(3), 124–128 (2008)
9. B. Zhong, T. Xue, Enrichment of low grade reduced titanium slag by H_3PO_4 activation roasting and acid leaching. Chin. J. Process Eng. **13**(3), 378–384 (2013)
10. Y. Fan, X. Jiang, Experimental study on preparation of synthetic rutile by titania slag. Min. Metall. **21**(4), 50–53 (2013)
11. J. Farrow, I. Ritchie, P. Mangano, The reaction between reduced ilmenite and oxygen in ammonium chloride solutions. Hydrometallurgy **18**(1), 21–38 (1987)
12. E. Kumari, K. Bhat, S. Sasibhushanan, Catalytic removal of iron from reduced ilmenite. Miner. Eng. **14**(3), 365–368 (2001)
13. Y. Yang, T. Lei, Titanium concentrates production methods and suggestions for development. Yunnan Metall. **35**(1), 41–44 (2006)
14. W. Cheng, Artificial rutile production method (BCA). Paint Coat. Ind. **3**, 009 (1980)
15. M.G. Shahien, M.M.H. Khedr, A.E. Maurice, Synthesis of high purity rutile nanoparticles from medium-grade Egyptian natural ilmenite. Beni-Suef Univ. J. Basic Appl. Sci. **4**(3), 207–213 (2015)
16. D.S. van Vuuren, G.T. Tshilombo, Nitriding of ilmenite and high-grade slag fines. J. S. Afr. Inst. Min. Metall. **111**(3), 173 (2011)
17. P. Fusheng, T. Aitao, Z. Likui, *Reaction synthesis of titanium carbonitride and its composites*, vol. 3 (Chongqing university Press, 2005)
18. P.V. Ananthapadmanabhan, P.R. Taylor, W. Zhu, Synthesis of titanium nitride in a thermal plasma reactor. J. Alloy. Compd. **287**(1), 126–129 (1999)
19. N.J. Welham, D.J. Llewellyn, Formation of nanometric hard materials by cold milling. J. Eur. Ceram. Soc. **19**(16), 2833–2841 (1999)
20. A. Adipuri, G. Zhang, O. Ostrovski, Chlorination of titanium oxycarbonitride produced by carbothermal nitridation of rutile. Ind. Eng. Chem. Res. **48**(2), 779–787 (2008)

Part IV
Poster Session

Adsorbents for Selective Recovery of Heavy Rare Earth Elements

Takeshi Ogata, Hirokazu Narita and Mikiya Tanaka

Abstract There is an increasing attention to recover dilute rare earth elements (REEs) from acidic solutions in order to meet their demands and stabilize their supply chain. Adsorption is suitable for recovering target metal ions when their concentrations are low. Many adsorbents for REEs have been studied; nevertheless, there are no adsorbents in practical use. We have developed adsorbents consisting of silica gel particles modified with diglycolamic acid groups. The adsorbent has the ability to selectively adsorb RE ions from solutions containing high concentrations of base metal ions in a low pH region, and the adsorption capacities of the adsorbent for the heavy REEs were higher than those for the light ones. Moreover, this adsorbent meets the requirements for a practical adsorbent, such as high durability, easy desorption, and high adsorption rate. We therefore conclude that the adsorbent can contribute the production of heavy REEs from underutilized resources.

Keywords Rare earth · Adsorption · Separation · Recovery

Introduction

Rare earth elements (REEs)—a group of 17 elements consisting of scandium, yttrium, and the lanthanoids—are used in cutting-edge technologies such as electric cars, smart phones, and wind turbines. A U.S. Department of Energy report states that in the medium term (2015–2025), REEs such as yttrium, neodymium, europium, terbium, and dysprosium are important elements for the clean energy industry; and, except for Nd, all of these elements are middle and heavy REEs [1]. The demand for heavy REEs is increasing, but most heavy REEs are produced from

T. Ogata (✉) · H. Narita · M. Tanaka
Environmental Management Research Institute, National Institute of Advanced Industrial Science and Technology (AIST), 16-1, Onogawa, Tsukuba, Ibaraki 305-8569, Japan
e-mail: t-ogata@aist.go.jp

© The Minerals, Metals & Materials Society 2017

H. Kim et al. (eds.), *Rare Metal Technology 2017*,
The Minerals, Metals & Materials Series,
DOI 10.1007/978-3-319-51085-9_23

ion-adsorption-type ores in a single location, southern China [2]. Therefore, the development of new sources of heavy REEs is important to ensure a secure supply of these elements.

We focus on underutilized resources of heavy REEs such as low-grade ores [3–6] and REE–containing End-of-Life products [7, 8]. Development risks can be avoided through recovery of REEs in the form of byproducts from the present industrial processes; furthermore, REEs can potentially be obtained at low cost. However, REEs only exist in low concentrations in such underutilized resources; what is more, base metals such as iron and aluminum coexist in high concentrations. Adsorption is an effective hydrometallurgical technology for recovering target metal ions when their concentrations are low. Many adsorbents for REEs have been studied, but none so far has been appropriate for practical use.

With the goal of designing a practical adsorbent, we considered (i) ligands that selectively adsorb REEs, (ii) supports to immobilize these ligands, and (iii) methods to introduce ligands onto supports. (i) We focused on diglycolamide compounds that are tridendate oxygen donor ligands as candidates with selectivity towards REEs. Diglycolamide compounds have been studied during nuclear energy research, where they are employed as extractants in solvent extraction processes; their application in the separation of actinides and lanthanoids has been investigated [9]. Narita and Tanaka, for example, synthesized *N,N′*-dimethyl-*N,N′*-di-*n*-octyl-diglycolamide as an extractant for separating REEs and base metals; they succeeded in the selective extraction of REEs at relatively high pH [10]. Further, rather than considering diamide compounds, Naganawa et al. synthesized dioctyldiglycol amic acid (DODGAA), as an analogue to *N,N,N′,N′*-tetraoctyl diglycolamide [11]; this extractant has selectivity for REEs in the relatively weak acid range of pH 2–4 [12]. Recovery of low concentrations of REEs is typically carried out in the dilute acid range (pH of ca. 1–2); for this reason, we selected diglycolamic acid as the adsorbent ligand. (ii) For supports on which to immobilize ligands, factors such as swelling, chemical and physical stability and cost need to be considered. Silica gel and zeolites are widely used as inorganic supports, while polystyrenes and polyacrylates are widely employed as organic supports. Adsorbents are generally spherical in shape. We used spherical silica gel particles with a low degree of swelling, and good chemical resistance, and mechanical strength. (iii) Broadly speaking, there are two methods employed for immobilizing ligands on supports: physical and chemical methods. The physical immobilization method imparts the ligands with a relatively high degree of freedom and for this reason, high selectivity on par with that of the solvent extraction method can be expected; however, physically immobilized ligands tend to leak out from their supports due to the physical nature of the bond and in many cases it is difficult to use such materials for an extended period of time. On the other hand, chemically immobilized ligands are chemically bonded to the support; therefore degradation is limited. We introduced the diglycolamic acid ligand to the surface of a silica gel support via chemical bonding (silane coupling). This method of introducing a ligand to a support is expected to facilitate rapid adsorption as well as extended utilization. The EDASiDGA (3-(ethylenediamino)propyl silica gel bearing

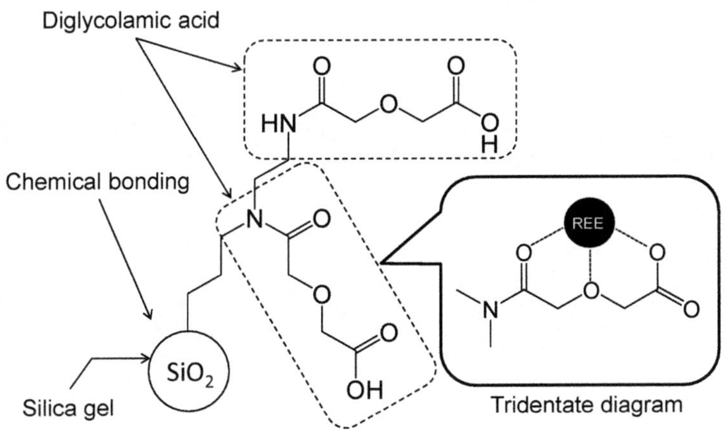

Fig. 1 Chemical structure of EDASiDGA adsorbent

immobilized diglycolamic acid ligands) adsorbent shown in Fig. 1 was developed based on such a guiding principle [13–16].

In the present study, we performed column experiments to clarify the feasibility of EDASiDGA for practical use. In addition, we evaluated the selective adsorption of heavy RE ions from a simulated solution containing dilute RE ions and high concentrations of base metal ions at low pH (pH 1.0).

Experimental

Silica gel particles modified with diglycolamic acid were prepared as described in our previous paper [13, 14]. Briefly, diglycolic anhydride and 3-(ethylenediamino) propyl silica gel were added to dichloromethane and allowed to react at 298 K for 3 days, at which point the particles were filtered off, washed with dichloromethane and ethanol to remove unreacted starting materials, and rinsed again with water. The washed and rinsed particles were dried under vacuum.

Column adsorption experiments were carried out with the EDASiDGA adsorbent: An 8 mm internal diameter glass column was packed with 1.0 g of EDASiDGA (bed volume: 1.86 cm^3) that was then washed with a hydrochloric acid solution adjusted to pH 1.0. An appropriate quantity of adsorption test solution was then fed through the packed column and the effluent was collected. Then, a hydrochloric acid solution (pH 2.0) was fed as a cleaning solution, followed by a 1 M sulfuric acid solution as an eluent and the eluates were collected. Metal ion concentrations in each of the eluate fractions collected were measured using an inductively coupled plasma spectrometer (Shimadzu, ICPE-9000). All column experiments were conducted at room temperature (291 ± 2 K).

Results and Discussion

To investigate the adsorption and desorption rates of the EDASiDGA adsorbent, dysprosium recovery from 1 mM dysprosium chloride solution at pH 1.0 (adjusted by using HCl) was evaluated at various flow rates. We set the space velocity (*SV*) from 5 to 100 h^{-1}. Here, the *SV* is defined as the flow rate/the volume of the adsorbent bed in the column; *SV* indicates how many bed volumes of the test solution can be treated per hour. The results are shown in Fig. 2, in which the ratio of the concentration of dysprosium ion in the effluent (*C*) to its initial concentration (C_0) is plotted versus bed volumes. When the dysprosium solution was fed through the column as the adsorption test solution, the dysprosium concentration of the effluent was below the detection limit for all *SV* tested. After breakthrough, the concentration of dysprosium in the effluent rapidly increases and approaches the initial concentration. Furthermore, major differences in *SV* were not apparent for the breakthrough curve, and an acceptable breakthrough curve were obtained regardless of the *SV* values. When taking into consideration the fact that commercially available adsorbents are utilized at *SV*s of 5–40 h^{-1} [17], our results show that the adsorption rate of EDASiDGA is sufficiently high for practical application. The dysprosium that was adsorbed were quantitatively desorbed with ease using 1 M sulfuric acid solution and no tailing of the desorption curve was apparent. These findings clearly show that the adsorbed REEs are easily desorbed and recovered using relatively low concentrations of acid.

A test solution was prepared to simulate conditions of low concentrations of REEs and high concentrations of base metals in order to verify the practicality of the adsorbent. The simulated solution was adjusted to pH 1.0 using hydrochloric acid; each REEs concentration was 0.1 mM and each base metal concentration was 0.1 M (one-thousand times that of the REEs). After the test solution was fed, the base metals aluminum, calcium, iron (III), copper, and zinc did not interact with the

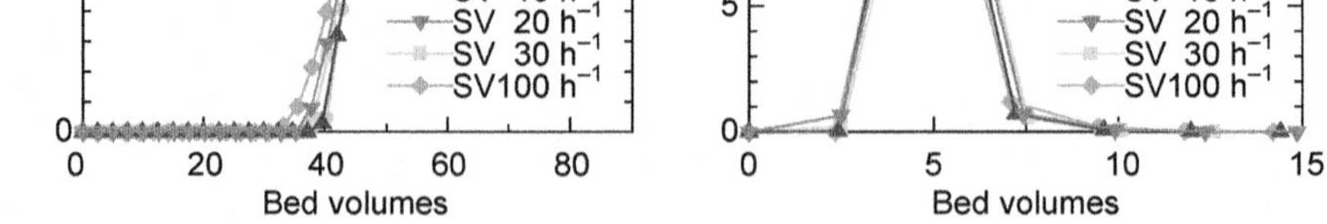

Fig. 2 Space velocity (*SV*) dependency of dysprosium on EDASiDGA. Adsorption solution: aqueous dysprosium solution (initial concentration: 1 mM, pH 1.0); washing solution: HCl (pH 2.0); eluent: H_2SO_4 (1 M)

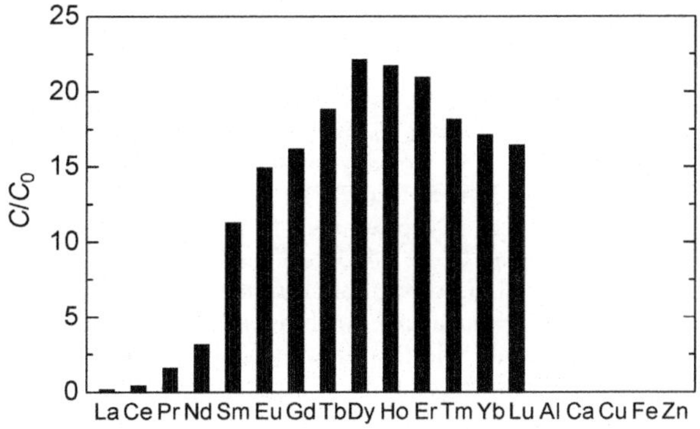

Fig. 3 Ratios of final effluent metal ion concentration (C) versus initial concentration (C_0)

adsorbent in the column and were discharged. Conversely, the rare earth elements were retained in the column. After a certain volume had passed through, rare earth elements started to break through in the order of atomic number (lanthanum, cerium, praseodymium…): this is thought to be due to displacement of light rare earth elements by heavy rare earth elements because of the higher adsorption capacity of the adsorbent for the latter. After heavy rare earth elements such as dysprosium, holmium, and erbium had broken through, the metal ions still adsorbed were desorbed using 1 M sulfuric acid solution. Figure 3 shows the ratio of the concentration of each metal ion in the effluent (C) to its initial concentration (C_0). The REEs alone were recovered, with almost no base metals such as aluminum, calcium, iron (III), copper, and zinc present in the effluent. The findings show that feeding an aqueous test solution to the packed column results in REEs being selectively retained in the column; these are then desorbed and recovered using acid solution of relatively low concentration. Furthermore, selectivity for medium and heavy REEs is higher than for light REEs; dysprosium can be concentrated to more than 20 times its initial concentration.

Conclusions

Targeting the recovery of REEs from underutilized resources such as low-grade ores and wastes, we developed a new adsorbent capable of selectively separating and recovering low concentrations of REEs from high concentrations of base metals. This adsorbent can selectively adsorb medium and heavy REEs, which can then be easily desorbed and recovered using relatively low concentrations of acid. Special mention should be made of the high adsorption capability for heavy REEs that are particularly valued as resources. Adsorption-desorption rates of the adsorbent are more than sufficient to be applied practically, while properties other than adsorption, such as physical stability, compare favorably with commercially available adsorbents. Thus, the absorbent has the potential to be applied commercially and may contribute to the recovery of REEs from underutilized resources.

References

1. *Critical Materials Strategy* (U.S. Department of Energy, 2011)
2. *Rare Earths: Market Outlook to 2020*, 15th edn. (Roskill, 2015)
3. F. Habashi, The recovery of the lanthanides from phosphate rock. J. Chem. Technol. Biotechnol. **35A**, 5–14 (1985)
4. Y. Kato et al., Deep-sea mud in the pacific ocean as a potential resource for rare-earth elements. Nat. Geosci. **4**, 535–539 (2011)
5. Y. Kon et al., Geochemical characteristics of apatite in heavy REE-rich deep-sea mud from Minami-Torishima area, southeastern Japan. Resour. Geol. **64**, 47–57 (2014)
6. T. Ogata et al., Selective recovery of heavy rare earth elements from apatite with an adsorbent bearing immobilized tridentate amido ligands. Sep. Purif. Technol. **159**, 157–160 (2016)
7. M. Tanaka et al., Recycling of rare earths from scrap, in *Handbook on the Physics and Chemistry of Rare Earths*, ed. by J.C.G. Bünzli, V.K. Pecharsky, vol. 43 (Amsterdam, Elsevier, 2013), pp. 159–212
8. K. Binnemans et al., Recycling of rare earths: a critical review. J. Clean. Prod. **51**, 1–22 (2013)
9. S.A. Ansari et al., Chemistry of diglycolamides: promising extractants for actinide partitioning. Chem. Rev. **112**, 1751–1772 (2012)
10. H. Narita, M. Tanaka, Separation of rare earth elements from base metals in concentrated HNO_3, H_2SO_4 and HCl solutions with diglycolamide. Solvent Extr. Res. Dev. Jpn. **20**, 115–121 (2013)
11. H. Naganawa et al., A new "green" extractant of the diglycol amic acid type for lanthanides. Solvent Extr. Res. Dev. Jpn. **14**, 151–159 (2007)
12. Y. Baba et al., Selective recovery of dysprosium and neodymium ions by a supported liquid membrane based on ionic liquids. Solvent Extr. Res. Dev. Jpn. **18,** 193–198 (2011)
13. T. Ogata, H. Narita, M. Tanaka, Immobilization of diglycol amic acid on silica gel for selective recovery of rare earth elements. Chem. Lett. **43**, 1414–1416 (2014)
14. T. Ogata, H. Narita, M. Tanaka, Adsorption behavior of rare earth elements on silica gel modified with diglycol amic acid. Hydrometallurgy **152**, 178–182 (2015)
15. T. Ogata, H. Narita, M. Tanaka, Rapid and selective recovery of heavy rare earths by using an adsorbent with diglycol amic acid group. Hydrometallurgy **155**, 105–109 (2015)
16. T. Ogata, H. Narita, M. Tanaka, Adsorption mechanism of rare earth elements by adsorbents with diglycolamic acid ligands. Hydrometallurgy **163**, 156–160 (2016)
17. Dow Water & Process Solutions Home Page. Product Data Sheet. http://www.dowwaterandprocess.com/en/products/a/amberlite_ir120_h

Behavior of Sec-Octylphenoxy Acetic Acid (CA-12) in Yttrium Recovery from High Concentrated Heavy Rare Earths Mixture

Corradino Sposato, Alessandro Blasi, Assunta Romanelli, Giacobbe Braccio and Massimo Morgana

Abstract The solvent extraction technique is the only industrial way used to separate and purificate yttrium from other rare earths. In order to improve this process several extractants have been tested during last years. In this work, the behavior of sec-octylphenoxy acetic acid (CA-12) in Yttrium recovery from high concentrated heavy rare earth mixture was investigated. Sodium hydroxide solution was used in order to presaponify organic phase composed by CA-12 and Tributyl phosphate (TBP) diluted in kerosene. In the investigated condition, TBP confirmed its role of phase modifier not significantly altering the extraction behavior of CA-12. The CA-12-TBP system showed a high affinity in extraction for lighter rare earths such as Sm, Eu, and Gd, leaving yttrium in aqueous phase. Using a feed concentration ($\sum$ [RE]) of 1 M, organic mixture is capable to extract around the 70% of metals in a single extraction test showing this extraction sequence Y < Lu < Yb < Tm < Er < Ho < Dy < Tb < Gd < Eu < Sm. Moreover for [CA-12] = 1.79 M and [RE] = 2 M the instability of extraction system occurs.

Keywords CA-12 · Solvent extraction · Rare earths · Liquid-liquid extraction

C. Sposato (✉) · A. Blasi · A. Romanelli · G. Braccio · M. Morgana
Italian National Agency for New Technologies, Energy and Sustainable Economic Development, ENEA Research Center of Trisaia, S.S. 106 Ionica, km 419 + 500, 75026 Rotondella, MT, Italy
e-mail: corradino.sposato@enea.it

A. Blasi
e-mail: alessandro.blasi@enea.it

A. Romanelli
e-mail: assunta.romanelli@enea.it

G. Braccio
e-mail: giacobbe.braccio@enea.it

M. Morgana
e-mail: massimo.morgana@enea.it

© The Minerals, Metals & Materials Society 2017

H. Kim et al. (eds.), *Rare Metal Technology 2017*,
The Minerals, Metals & Materials Series,
DOI 10.1007/978-3-319-51085-9_24

Introduction

In last years, heavy rare earths and yttrium are reported as Critical Raw Materials in the Strategic paper of EU community [1]. Yttrium is widely used for luminescent lamps production [2], Light Emitting Diode and other electronical devices [3]. Yttrium is produced from minerals [4] and naphthenic acid is the extractant most used to separate it from other heavy rare earths [5]. Minerals such as loparite, fergusonite, xenotime, monazite and bastnasite are the main source of rare earths, while a lower amount of REs are also contained in rock phosphate which is considered as secondary source for rare-earth elements. Yttrium separation and purification is a difficult task and solvent extraction technique is used to perform it [6–10]. Valiente et al. [7] studied the distribution of yttrium (III) between acidic aqueous chloride solutions and organic solutions of di-(2-ethylhexyl)-phosphoric acid (D2EHPA) dissolved in kerosene. Sposato et al. [11] made a comparison among three different extractant such as (2-ethylhexyl)-mono (2-ethylhexyl) ester phosphonic acid (P507), secondary-octyl phenoxy acetic acid (CA-12)) and bis (2,4,4-trimethylpentyl)phosphinic acid (Cyanex272) in the separation of heavy rare earths from a yttrium-rich nitrate aqueous source. Extraction of yttrium and some trivalent lanthanides from nitrate and thiocyanate solutions using Cyanex 923 in xylene was investigated by Reddy et al. [12]; they studied lanthanides complexes built by TRPO extractant from nitrate and thiocyanate media. Ramachandra Reddy et al. [9] performed studies about extraction of Y (III) from phosphoric acid solutions with TOPS 99 (Talcher Organo phosphorus solvent) as extractant. Gupta et al. [13] studied the solvent extraction and separation of yttrium and lanthanides by using Cyanex 923 from different acid media. Solvent extraction of Sc, Y, La and Gd from hydrochloric acid solutions with bis(2,4,4-trimethylpentyl) monothiophosphinic acid (Cyanex 302), was investigated by Li and coworkers [14]. In a further works [15] Wei Li et al. investigated extraction and separation of yttrium with CA-12) and TBP from other rare earths in chloride medium, TBP was used as phase modifier to achieve fast phase separation and higher stability of organic phase. In present work we study the behavior of CA-12 in Yttrium recovery from a mixture with a high concentration of heavy rare earth.

Experimental

Apparatus

Digital pH meter (Metrohm 780) was used for pH measurements. Automatic titrator (Metrohm 905 Titrando), equipped by an iSolvotrode probe, was used to determinate the molar concentration of CA-12). Inductively coupled plasma optical emission spectrometer (ICP-OES, Perkin Elmer, Optima 8300 model) was used for the analysis of rare earths in aqueous media.

Reagent

CA-12 was kindly provided by Treibacher Industrie and its concentration (3.58 M) was titrated by using standard NaOH. This extractant, used without any further purification, was diluted with kerosene. The kerosene (provided by oleotecnica S.p. A.) is a selected fraction of linear hydrocarbons compounds (C10–C13, aromatic compounds <2%). Pure Sodium hydroxide (provided by Carlo Erba ACS-ISO) was dissolved in demineralized water and used for the pre-saponification of the extractant. TBP >99% (provided by Sigma-Aldrich) was used as phase modifier. Stock solution containing all rare earths and yttrium were prepared dissolving their salts, of analytical grade, in demineralized water (kindly provided by Treibacher Industry) and their concentrations checked by using ICP-OES (Table 1). All other reagents were of analytical grade.

Methods

The experiments were carried out shaking equal volumes of aqueous solution (feed) and CA-12) diluted with kerosene for 40 min with the support of a mechanical shaker at room temperature (298 ± 1 K) to ensure complete equilibration. The laboratory is provided with microclima control. The ingredients in the aqueous phase were analyzed by ICP-OES, and the concentration of rare earths in organic phase was determined by mass balance. The preliminary experiments, by using HCl stripping solution, indicated that the concentration of metals in organic phase by mass balance was consistent. Distribution ratio was obtained as $D = [RE]o/[RE]a$, where 'a' and 'o' denote aqueous and organic phase. The value of pH was determined after extraction and phase separation. An amount of 15 vol.% of Tributyl phosphate (TBP), used as phase modifier, is added to the extractant mixture before kerosene dilution.

Results and Discussion

The Role of the Phase Modifier

CA-12 is an organic carboxylic acid extractant composed by an aromatic 6 carbons-ring with two functional groups in meta position ($-C_8H_{17}$ and–OCH_2COOH). Complexes of CA-12 and rare earths make easily emulsion in alkyl

Table 1 Feed relative concentration of rare earth elements

	Dy	Er	Eu	Gd	Ho	Lu	Sm	Tb	Tm	Y	Yb
Feed (wt%)	9.3	5.4	1.5	11.7	2.3	0.5	4.1	1.7	0.8	58	4.6

solution due to their small solubility, so it is necessary to add a modifier (usually an alcohol like iso-octanol) to avoid it [16]. Li studied the use of aliphatic alcohols as modifiers in CA12 system for long term industrial tests [17]. The concentration of CA-12 with isooctanol as modifier decreases with time degrading the extraction system, probably the esterification of CA-12 and isooctanol occours. Otherwise, the concentration of CA-12-TBP system is constant for long time making this system a better choice [18]. For these reasons we decided to work using CA-12) with TBP as extractant system.

In Fig. 1 is reported the amount of rare earth extracted (calculated as $\sum$RE extracted in organic phase/$\sum$RE in feed * 100) as a function of pH with and without the use of TBP. We observed that the use of a phase modifier does not significantly alter the amount of rare earth extracted from the feed. Separation factors (β) of rare earth elements versus Yttrium, calculated as $D_{\text{of single element}}/D_{\text{of yttrium}}$ are not influenced by presence of TBP (Table 2). Moreover, with the use of TBP we observe a better separation of the phases in term of rate and to avoid emulsification.

CA-12 Behavior for Feed Concentration = 0.2 M

The amount of rare earths extracted (calculated as $\sum$RE extracted in organic phase/$\sum$RE in feed * 100) versus pH at different CA-12) concentration is reported in Fig. 2.

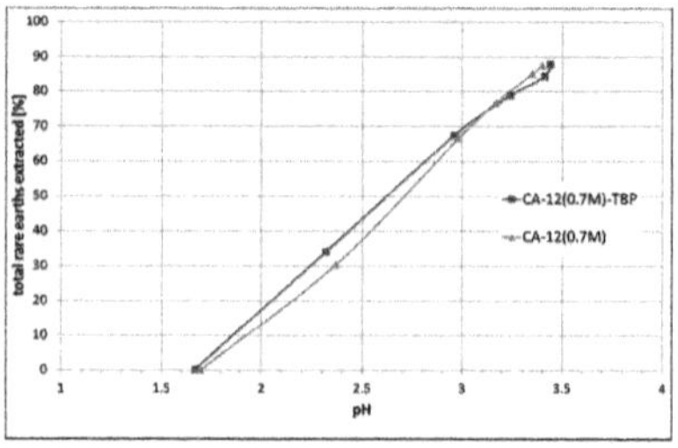

Fig. 1 The effect of phase modifier on the rare earth extracted. [(CA-12)] = 0.7 M, TBP = 15 vol.% (if present), [feed] = 0.2 M

Table 2 Separation factors (β) of mixture rare earths elements [(CA-12)] = 0.7 M with and without TBP

	0.7 M + TBP	0.7 M
Gd	4.26	4.34
Dy	2.45	2.58
Er	1.50	1.53
Sm	8.93	9.03
Yb	1.15	1.07
Ho	2.20	2.27
Tb	4.63	4.75
Eu	7.07	7.19
Tm	1.38	1.39
Lu	0.98	0.96

As expected, at higher extractant concentration, we observe an increased amount of extracted rare earths with the increase of pH. For [CA-12] = 1.79 M, just at pH = 3.4, we can obtain the total extraction of rare earths contained in the feed. In this case, we don't have the possibility to separate rare earths metals but this underline that CA-12 is a strong extractant. The lowest value of pH, for each curves, is reached without adding sodium hydroxide, and this result shows as CA-12 doesn't work as extractant for pH < 2.

Data reported in Figs. 3 and 4 shows the effect of *D* versus pH on the mixture rare earths using the feed at 0.2 M and changing CA-12 concentration.

With pH increasing we observe an improvement of rare earths extraction.

Compared with other rare earths, yttrium and lutetium are the most difficult to be extracted by CA-12 which means that it is possible to be separated from other rare earths. The results shows that the value of D is Y $\leq$ Lu < Yb < Tm < Ho < Er < Dy < Gd < Tb < Eu < Sm. The values of D increase with the increasing of extractant concentration, in particular from 0.5 to 0.7 M concentration

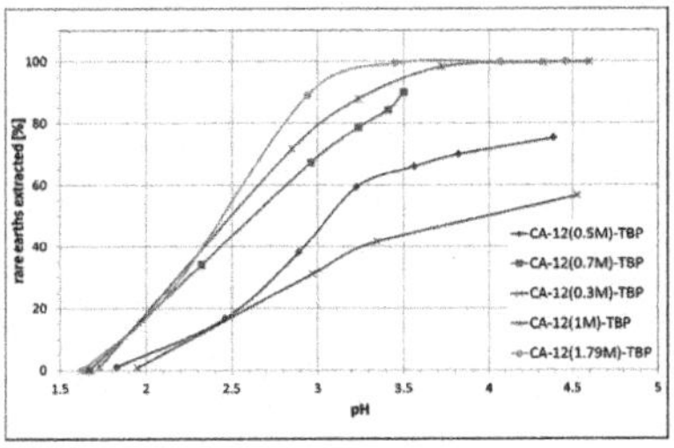

Fig. 2 Total amount of rare earths extracted versus pH at different CA-12-TBP-kerosene solution concentration for [feed] = 0.2 M

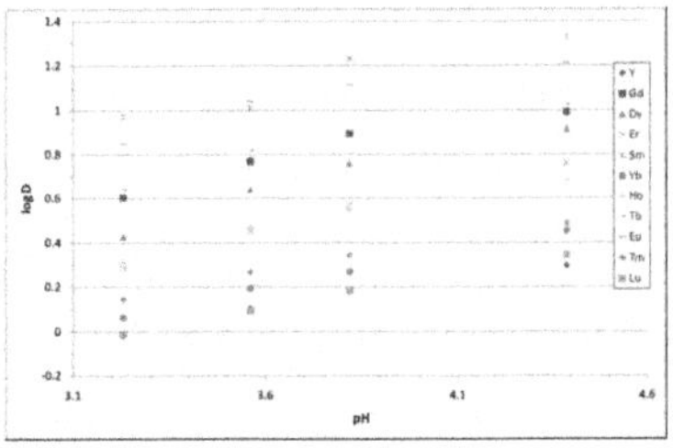

Fig. 3 Relationship between logarithm distribution ratio D and pH of aqueous phase: [(CA-12)] = 0.5 M; TBP = 15%; [feed] = 0.2 M

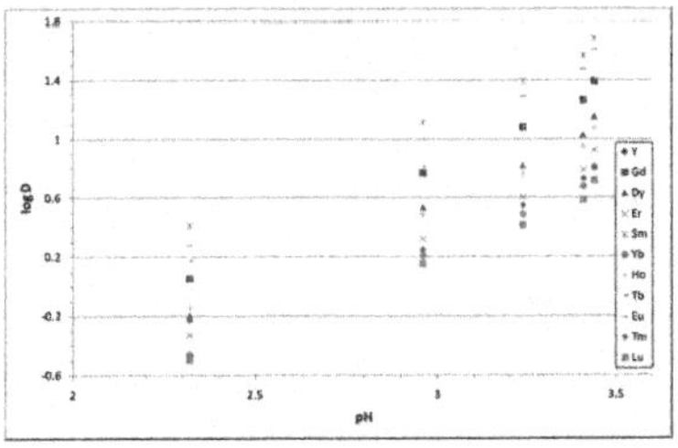

Fig. 4 Relationship between logarithm distribution ratio D and pH of aqueous phase: [(CA-12)] = 0.7 M; TBP = 15%; [feed] = 0.2 M

Table 3 Separation factors (β) of mixture rare earths elements

	CA-12 molar concentration			
	0.3 M	0.5 M	0.7 M	1 M
Gd	4.33	4.67	4.26	3.96
Dy	2.98	3.25	2.45	2.68
Er	1.88	2.34	1.50	1.59
Sm	8.56	9.62	8.93	8.36
Yb	1.98	1.20	1.15	1.36
Ho	2.29	2.49	2.20	1.94
Tb	3.81	5.51	4.63	3.57
Eu	7.30	8.25	7.07	7.61
Tm	1.99	1.39	1.38	1.39
Lu	1.62	0.90	0.98	1.08

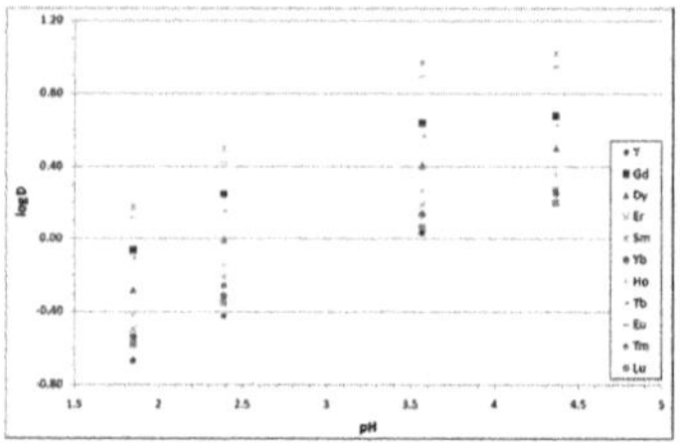

Fig. 5 Relationship between logarithm distribution ratio D and pH of aqueous phase: [(CA-12)] = 1.79 M; TBP = 15%; [feed] = 1 M

of CA-12, at pH = 3.24 this increasing is around 160% (this value is calculated as $(D_{element\ [0.5M]} - D_{element\ [0.7M]}/D_{element\ [0.5M]}) * 100$).

Separations factors (β), reported in Table 3, are calculated as $D_{of\ single\ element}/D_{of\ yttrium}$. The separation coefficient values are nearly constant for all CA-12) concentration. Lutetium and yttrium are the most difficult to be extracted from aqueous phase by using CA-12 as extractant. CA-12 presents the best selectivity for Sm, Eu, Tb and Gd with the higher values of separation factors. CA-12 presents a good affinity for all rare earths and the selectivity increase with the decrease of ionic radio dimension [19], except for an inversion for Holmium/Erbium and Terbium/Gadolinium, probably due to their relative concentrations in feed.

CA-12 Behavior for High Feed (1 and 2 M) Concentration

Data reported in Fig. 5 shows the effect of D versus pH with [feed] = 1 M. The results shows that the value of Distribution factors are Y < Lu < Yb < Tm < Er < Ho < Dy < Tb < Gd < Eu < Sm. Enhancing feed concentration, the selectivity for the extraction increase with the decrease of ionic radio dimension without exception. Moreover, with [CA-12)] = 1.79 M is possible to extract about the 70%

Table 4 Separation factors (β); [CA-12] = 1 M, 1.79 M and [RE] = 1 M

Element	[CA-12] = 1 M	[CA-12] = 1.79 M
Gd	4.3	3.8
Dy	2.8	2.3
Er	1.8	1.4
Sm	7.2	7.1
Yb	1.36	1.2
Ho	2.2	1.6
Tb	3.7	3.2
Eu	6.4	6.3
Tm	1.6	1.23
Lu	1.4	1.05

of all rare earths contained in the feed at pH = 4.3 in a single step. These result underlines that CA-12-TBP solution presents a strong capability to extract rare earth from feed just in a single extraction step.

In the Table 4 are reported the separation factors (β) of rare earth elements versus Yttrium, calculated as $D_{\text{of single element}}/D_{\text{of yttrium}}$.

Best separation factors are for Samarium and Europium, 7.2 and 6.4 respectively, at [CA-12)] = 1 M. The enhancing of extractant concentration decrease separation selectivity.

The possibility to separate yttrium and then the other rare earth became very interesting. In particular, we observe that yttrium is the metals most difficult to be extract, so it is left in aqueous phase to be separated. On the contrary, metals such as Sm, Eu and Gd present a very good affinity to CA-12-TBP system, so they are loaded in organic phase and separated/recovered after stripping by an HCl concentrated solution [15].

Tests carried out with [CA-12)] = 1.79 M and [RE] = 2 M showed a total instability of extraction system. In particular, we observed the impossibility to separate aqueous and organic phase after the shaking of the system and the formation of a gel-phase occurs.

Conclusions

TBP showed to be a good phase modifier with a marginal effect on extractant power of the system. CA-12) + TBP showed a high affinity in extraction of lighter rare earths: the best extracted is Sm while the worst is Y (Y < Lu < Yb < Tm < Er < Ho < Dy < Tb < Gd < Eu < Sm). It resulted to be a very strong extractant system with an interesting potential to separate yttrium from other rare earths. It was impossible to work at feed concentration of 2 M for the incoming instability of the system.

Acknowledgements The authors wish to thank R&D division of Treibacher Industrie, in particular Dr. Stephan Pirker, Dr. Christiane Kartusch and Gunther Bierbaumer for support and help during this work.

References

1. Report on Critical Raw Materials for the EU, May 2014, http://www.amg-nv.com/files/Report-on-Critical-Raw-Materials-for-the-EU-2014.pdf
2. C.H. Yan, J.T. Jia, C.S. Liao, S. Wu, G.X. Xu, Rare earth separation in China. Tsinghua Sci. Technol. **11**, 241–247 (2006)
3. R. Mueller-Mach, G. Mueller, M.R. Krames, H.A. Höppe, F. Stadler, W. Schnick, T. Juestel, P. Schmidt, Highly efficient all-nitride phosphor-converted white LED, Phys. Status solidi (a) (2005) 1–6
4. J.W. Anthony, R.A. Bideaux, K.W. Bladh, M.C. Nichols (ed.), Handbook of mineralogy (Mineralogical Society of America, Chantilly, VA), pp. 20151–1110. http://www.handbookofmineralogy.org/
5. A. Peng, Z.R. Dai, C.X. Wang, Collect. Rare Earth **39** (1982)
6. G.X. Xu, ChY Yuan, *Solvent extraction of rare earths*, 1st edn. (Science Press, Beijing, 1987), p. 89
7. E. Antico, A. Masanaa, M. Hidalgo, V. Salvado, M. Iglesias, M. Valiente, Solvent extraction of yttrium from chloride media by di(2-ethylhexyl)phosphoric acid in kerosene. Speciation studies and gel formation. Anal. Chim. Acta **327**, 267–276 (1996)
8. M.I. Saleh, M.F. Bari, B. Saad, Solvent extraction of lanthanum(III) from acidic nitrate-acetato medium by Cyanex 272 in toluene. Hydrometallurgy **63**, 75–84 (2002)
9. B. Ramachandra Reddy, S. Radhika, B. Nagaphani Kumar, Liquid-liquid extraction studies of trivalent yttrium from phosphoric acid solutions using TOPS 99 as an Extractant, Sep. Sci. Technol. **45**, 1426–1432 (2010)
10. A. Blasi, C. Sposato, G. Devincenzis, P. Garzone, M. Morgana, Definition of the process to separate light rare earths by working with (2-Ethylexyl)-Mono(2-Ethylexyl)ester phosphonic acid (P507) in a mixer settler battery. Rare Metal Technol. **2014**, 197–200 (2014)
11. C. Sposato, A. Blasi, G. Devincenzis, P. Garzone, M. Morgana, Comparison among different extractants, As (2-ethylhexyl)-mono (2-ethylhexyl) ester phosphonic acid (P507), secondary-octyl phenoxy acetic acid (CA-12) and BIS(2, 4, 4-trimethylpentyl)phosphinic acid (CYANEX272), in the separation of heavy rare earths via hydrometallurgical processes. Rare Metal Technol. **2014**, 201–206 (2014)
12. M.L.P. Reddy, R.L. Varma, T.R. Ramamohan, S.K. Sahu, V. Chakravortty, Cyanex 923 as an extractant for trivalent lanthanides and yttrium. Solv. Extr. Ion Exchange **16**, 795–812 (1998)
13. B. Gupta, P. Malik, A. Deep, Solvent extraction and separation of tervalent lanthanides and yttrium using cyanex 923. Solv. Extr. Ion Exchange **21**, 239–258 (2003)
14. D. Wu, C. Niu, D.Q. Li, Y. Bai, Solvent extraction of scandium(III), yttrium(III), lanthanum (III) and gadolinium(III) using cyanex 302 in heptane from hydrochloric acid solutions. J. Alloy. Compd. **374**, 442–446 (2004)
15. Wei Li, Xianglan Wang, ShulanMeng, Deqian Li, Ying Xiong, Extraction and separation of yttrium from the rare earths with sec-octylphenoxy acetic acid in chloride media. Sep. Purif. Technol. **54**, 164–169 (2007)
16. D.Q. Li, S.L. Meng, A process of liquid-liquid extraction and separation of high purity yttrium, CN Patent 99118261.8, 1999
17. D.Q. Li, Progress in separation processes of rare earths, in *Paper presented at The 4th International Conference on Rare Earth Development and Application*, Beijing, 2001, pp. 11–20

18. Y. Wang, W. Liao, D. Li, A solvent extraction process with mixture of CA12 and Cyanex272 for the preparation of high purity yttrium oxide from rare earth ores. Sep. Purif. Technol. **82**, 197–201 (2011)
19. Y. Marcus, *Ion properties* (M. Dekker, New York, 1997)

Preparation of Molybdenum Powder from Molybdenite Concentrate Through Vacuum Decomposition-Acid Leaching Combination Process

Chongfang Yang, Yuezhen Zhou, Dachun Liu, Wenlong Jiang, Fansong Liu and Zewei Liu

Abstract A novel process that the combination of vacuum decomposition and acid leaching was proposed. The influence of distillation times, diameter and thickness of the feeding material on the mass fraction of Molybdenum (Mo) and sulphur (S) in residual was investigated. The influence of leaching temperature, leaching time, concentration of hydrochloric acid solution, and ratio of liquid to solid on leaching rates of Mo and S in acid leaching process was also investigated. Due to the fact that SO_2 emission happening in the long process of traditional molybdenum metallurgy could be avoided in this novel method, this combination process was environmentally-friendly. Molybdenum powder was obtained through handling molybdenite concentrate under the optimal condition parameters, and the mass fraction of Mo reached 98.29 wt%. So it's feasible to produce molybdenum powder from molybdenite concentrate by using this new method.

Keywords Molybdenite concentrate · Vacuum decomposition · Acid leaching · Molybdenum powder

---

C. Yang · Y. Zhou · D. Liu · W. Jiang · F. Liu · Z. Liu
The National Engineering Laboratory for Vacuum Metallurgy, Kunming University of Science and Technology, Kunming 650093, People's Republic of China
e-mail: 476559736@qq.com

C. Yang · Y. Zhou · D. Liu · W. Jiang · F. Liu · Z. Liu
Key Laboratory for Nonferrous Vacuum Metallurgy of Yunnan Province, Kunming 650093, People's Republic of China

D. Liu · W. Jiang (✉)
State Key Laboratory of Complex Nonferrous Metal Resources Clear Utilization, Kunming 650093, People's Republic of China
e-mail: 122397715@qq.com

© The Minerals, Metals & Materials Society 2017
H. Kim et al. (eds.), *Rare Metal Technology 2017*,
The Minerals, Metals & Materials Series,
DOI 10.1007/978-3-319-51085-9_25

Introduction

Molybdenum (Mo) is widely used in special steels, aviation, alloys, chemistry, electronic and medical industries. MoS_2 is the main component of molybdenite which is the raw material to produce molybdenum metal products and chemical products. Molybdenite concentrate treatment can be classified into two major categories, including hydrometallurgy and pyrometallurgy. The common feature of these two methods is that the sulfide ores were converted into oxide or its salts, then the further purification of intermediates was carried out, and finally the Mo metal was obtained by reduction of MoO_2. These methods are widely used by large manufacturers in the world due to low cost and easy operation. But simultaneously there are some problems and the insufficiency, such as long flow sheet, large amount of SO_2 emission and serious environmental pollution [1–5].

To solve the SO_2 pollution problem, the main extraction processes for molybdenite concentrates include lime-roasting and pressure oxidation leaching. The sulfur was converted to calcium sulfate for the lime-roasting, it greatly reduced the emergence of SO_2 gas, but it generated a lot of calcium sulfate slag [6]. The sulfur was converted to sodium sulfate into the solution for the pressure oxidation leaching, this method avoided the problem of SO_2 emission, but led to a high production cost, because of heavy use of sodium hydroxide [7].

Many scholars have made several meaningful researches on vacuum decomposition process of molybdenite concentrate, from economy, working conditions and environmental protection, etc. Chen [8] investigated vacuum decomposition process of analytic grade molybdenum disulfide and molybdenite concentrate respectively, and useful experimental parameters were obtained. Liu et al. [2] simulated the crystal structure of MoS_2 by dynamics simulations, and studied the thermal decomposition of MoS_2, theoretical calculation provided guidance for the experimental results. Wang et al. [9, 10] studied the key steps, morphology and phase evolution of thermal decomposition process of molybdenum concentrate in vacuum, which was verified by vacuum decomposition experiments.

The main purpose of this research is to verify the feasibility of this new method of producing molybdenum powder from molybdenite concentrate, and further investigate the influence of distillation times, material diameter and thickness (in vacuum decomposition process) and leaching temperature, leaching time, concentration of hydrochloric acid, ratio of liquid to solid (in leaching process) on properties of the products.

Theoretical Analysis

Vacuum Decomposition Basis

The possible mechanism of thermal decomposition of MoS_2 was as follows, and the Gibbs free energy for reactions at different pressures were shown in Table 1 [9–11].

$$4MoS_2 = 2Mo_2S_3 + S_2 \quad (1)$$

$$Mo_2S_3 = Mo + 1.5S_2 \quad (2)$$

As can be seen from Table 1 that the initial decomposition temperature decreases monotonously with decreasing pressure. And comparing to reactions happened at atmospheric pressure, whose initial decomposition temperatures at above 2000 K, the temperatures were 1491 and 1575 K when the pressure fell to 10 Pa. In order to investigate the volatilization behaviors of Mo and S during vacuum decomposition process, saturated vapor pressure of Mo, S and MoS_2 were calculated in the Clausius-Clapeyron equation expressed as follows [12]:

$$\lg P^{\theta} = AT^{-1} + B\lg T + CT + D \quad (3)$$

$$\lg P^{\theta} = -AT^{-1} + B \quad (4)$$

where p^{θ} was the saturated vapor pressure, Pa. A, B, C and D were evaporation constants [13]. T was the absolute temperature, K. The melting point and saturated vapor pressure of Mo, S and its compound were shown in Table 2 and Fig. 1.

When the temperature was between melting point (1458 K) and initial decomposition temperature (1491 or 1575 K) of MoS_2 and Mo_2S_3, MoS_2 existed in liquid form in the experimental material. And according to the saturated vapor pressure of MoS_2, it could not evaporate into condensate during the vacuum decomposition process. When the temperature was higher than initial decomposition temperature of MoS_2, it was decomposed into Mo and S_2. Mo could not evaporate into

Table 1 Initial decomposition temperature of MoS_2 and Mo_2S_3 at different pressures

Reaction	Initial decomposition temperature/K			
	101,325 Pa	1000 Pa	100 Pa	10 Pa
$4MoS_2 = 2Mo_2S_3 + S_2(g)$	2053	1728	1601	1491
$Mo_2S_3 = 2Mo + 1.5S_2(g)$	2282	2052	1707	1575

Table 2 Melting points of Mo, S and its compound

	Mo	S	MoS_2
T_M/K	2896	388	1458

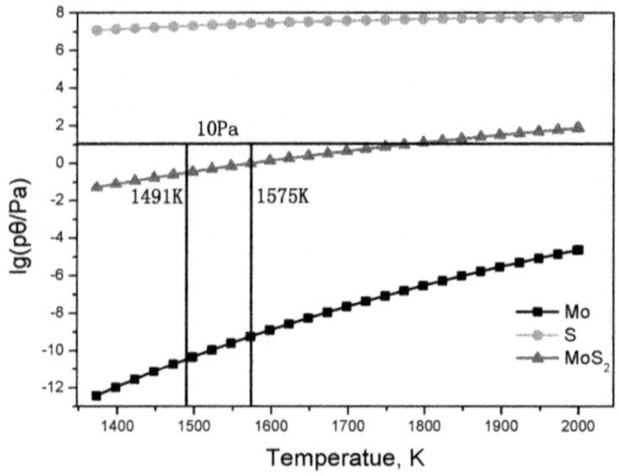

Fig. 1 Relationship between temperature and saturated vapor pressure

condensate because its melting point was much higher than chamber's temperature, and S could easily evaporate into condensate because its saturated vapor pressure was much higher than chamber pressure (5–35 Pa).

The theoretical research shows that vacuum decomposition is feasible to separate Mo and S in molybdenite concentrate. Mo nearly enriched in the residual, and S could easily evaporate into the condensate.

Acid Leaching Basis

The main impurities in crude molybdenum were mainly divided into three parts: acid oxide (SiO_2), alkaline oxides (Al_2O_3, MgO and CaO) and sulfides (such FeS). So the Gibbs free energy of leaching reactions of the alkaline oxides and sulfides in hydrochloric acid solution was shown in Table 3 [14].

As shown in Table 3, alkaline oxides (Al_2O_3, MgO and CaO) could be easily dissolved with hydrochloric acid, and sulfides (such FeS) could be partly dissolved with hydrochloric acid. Consequently, it was feasible to remove impurities and obtain molybdenum powder from crude molybdenum through acid leaching process.

Table 3 Gibbs free energy of leaching reactions

Leaching reaction	T/K	$\Delta G/(kJ\ mol^{-1})$
$Al_2O_3 + 6H^+ = 2Al^{3+} + 3H_2O$	293–373	$\Delta G = 0.3745T - 237.98$
$MgO + 2H^+ = Mg^{2+} + H_2O$		$\Delta G = 0.0477T - 145.51$
$CaO + 2H^+ = Ca^{2+} + H_2O$		$\Delta G = 0.0214T - 184.78$
$FeS + 2H^+ = Fe^{2+} + H_2S$		$\Delta G = -0.2833T + 99.98$

Table 4 Chemical composition of molybdenite concentrate

Element	Mo	S	O	C	Al	Si	Mg	Ca	Fe	Cu	Others
wt%	50.73	35.96	5.61	0.32	4.52	1.29	0.12	0.10	0.78	0.40	0.17

Material

The high grade molybdenite concentrate was selected in this experiment. The chemical composition of molybdenite concentrate was got by chemical titration was shown in Table 4, and XRD pattern of molybdenite concentrate which were used in this research was shown in Fig. 2, respectively.

Apparatus and Methods

The vacuum decomposition equipment is developed independently by the National Engineering Laboratory for Vacuum Metallurgy. Major components of the system include vacuum equipment, heating equipment and Evaporator-Condenser set The schematic diagram of the vertical vacuum distillation furnace is shown in Fig. 3. The ohmic heating method was employed, and the heating temperature range is 300–1973 K. The vacuum level for this equipment is 5 Pa.

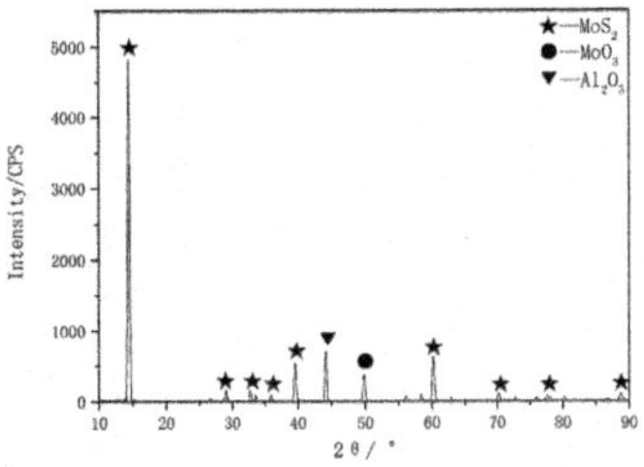

Fig. 2 XRD pattern of molybdenite concentrate

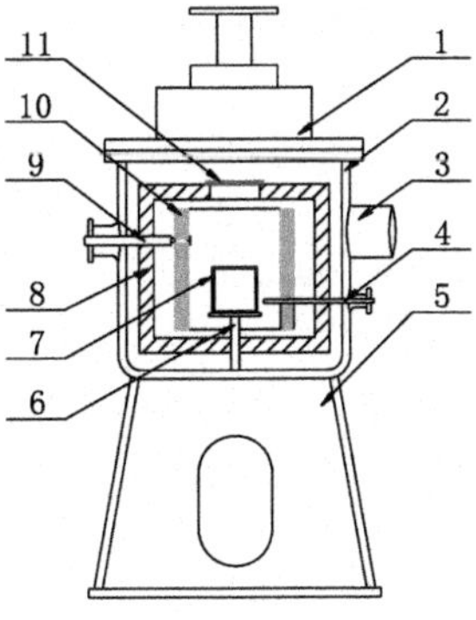

Fig. 3 Schematic diagram of vertical vacuum distillation furnace: *1* Furnace lid; *2* Furnace body; *3* Vacuum pipe; *4* Temperature sensor; *5* Furnace bottom; *6* Crucible holder; *7* Crucible; *8* Heat holding cover; *9* Electrode; *10* Heating unit; *11* Condenser

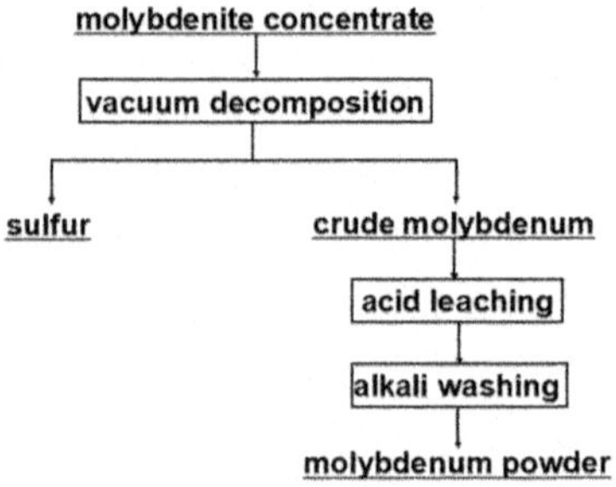

Fig. 4 Main flowsheet of the process

The main flowsheet of the process is shown in Fig. 4. The process mainly includes vacuum decomposition, acid leaching and alkali washing.

Results and Discussion

Effect of Distillation Times

Our previous research work, verified the feasibility of producing crude molybdenum from molybdenite concentrate by vacuum decomposition process. The influence of temperature and heat preservation time on the mass fraction of Mo and S in residual have been obtained [15]. the better condition of temperature and heat preservation time was 60 min and 1823 K, respectively. However, it was only part of the experimental rules with the vacuum decomposition of molybdenite concentrate. So the influence of distillation times on the mass fraction of Mo and S in residual was further investigated, as shown in Fig. 5. As shown in Fig. 5, as the distillation times increased, the mass fraction of Mo in residual increased from 90.17 to 90.88 wt%, while the mass fraction of S decreased from 0.48 to 0.30 wt%, which indicates that the increase of the distillation times was beneficial to decreasing the content of S in the residual, however, the decrease of the mass fraction was too small to be considered. Thus, compared with the influence of temperature and heat preservation time on the mass fraction of Mo and S in residual, the influence of distillation times could be ignored.

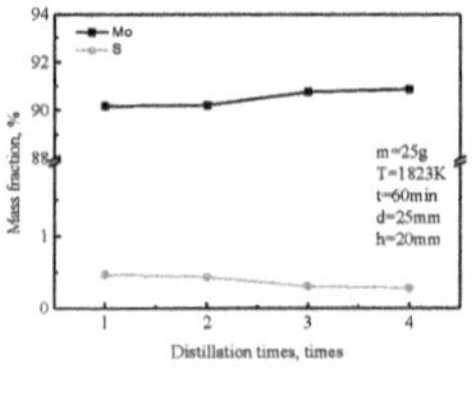

Fig. 5 Influence of distillation times on the mass fraction of Mo and S in residual (d is the material diameter, h is the material thickness)

Effect of Material Diameter and Thickness

To investigate the interplay of diameter and thickness in feeding material, the experiments were carried out. The results influence of diameter and material thickness on the mass fraction of Mo and S in residual, as shown in Fig. 6. Figure 6 shows the mass fraction of Mo decreased, while the mass fraction of S increased, with diameter of feeding material distributed in the range of 20–30 mm. Compared with the influence of the diameter of feeding material, the influence of the material thickness was similar.

As mentioned above, optimal condition parameters of vacuum decomposition experiment can be summarized as follows: experiment was carried under the pressure of 5–35 Pa for 60 min at 1823 K, distillation times was once, the sample mass was 25 g, material diameter and thickness was 25 and 20 mm, respectively. The chemical composition of the residual was got by chemical titration, and the removal rate was calculated from Eq. (5), were shown in Table 5. XRD pattern of residual were shown in Fig. 7.

$$r_V = \frac{m_X - m_y}{m_X} \times 100\% \quad (5)$$

where r_V is the removal rate of element in the vacuum decomposition process, %. m_x is the content of element in the molybdenite concentrate, g. m_y is the content of element in residual.

As shown in Table 5 and Fig. 7, residual was obtained after vacuum decomposition was crude molybdenum. The condensate gathered from the condenser was crude sulfur. The direct recovery rate of Mo was 97.28%, which indicated that a very small amount of Mo was lost, and the possible reason accounting for the phenomenon above could be that the mechanical loss was caused by vacuum system. As to the impurity elements, such as O, Al, Si and Cu, their removal rates were relatively large, ranging from 67.8 to 90.15%. However, the mass fraction of

Fig. 6 Influence of material thickness on the mass fraction of Mo and S in residual

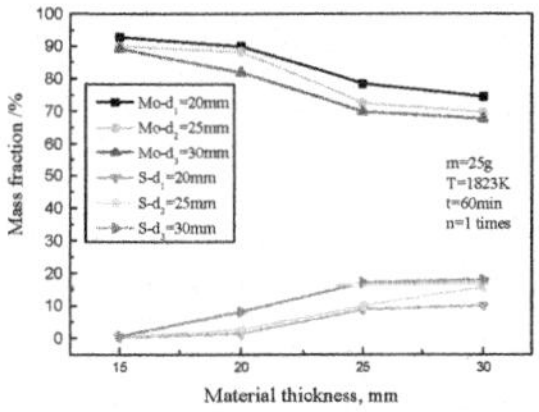

Table 5 Chemical composition and removal rate (r_V)

Element	Mo	S	O	C	Al	Si	Mg	Ca	Fe	Cu	Others
wt%	90.17	0.48	3.30	0.90	3.09	0.30	0.21	0.17	0.98	0.07	0.33
r_V.%	2.72	99.27	67.80	−53.94	62.58	87.27	4.22	6.96	31.23	90.15	–

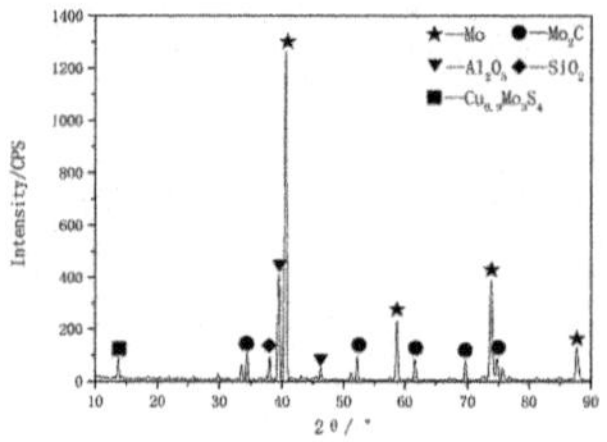

Fig. 7 XRD pattern of residual after vacuum decomposition

C increased from 0.32 to 0.9 wt% in the residual. The reason was the formation of Mo_2C, as shown Fig. 7, because the high-purity graphite crucible used in experiments provided the carbon source and high temperature promoted carbon to participate in the experimental material.

Effect of Temperature

In this work, the crude molybdenum was obtained after vacuum decomposition, the mass fraction of Mo and S was 90.17 and 0.48 wt%, respectively. The crude molybdenum was the experimental materials at the leaching process. Leaching experiments were performed under ordinary pressure, the sample mass was 18.0 g, and the dried powders were smaller than 200 mesh. The influence of leaching temperature on the leaching rate of Mo and S was studied, and the leaching rate of Mo and S in the leaching slag was calculated from Eq. (6) presented in Fig. 8.

$$r_L = \frac{m_i - m_j}{m_i} \times 100\% \tag{6}$$

where r_L is the leaching rate of Mo and S in the leaching process, %. m_i is the content of element in the crude molybdenum, g. m_j is the content of element in leaching slag.

Figure 8 shows when the leaching temperature increased from 333 to 358 K, the leaching rate of S increased from 83.08 to 92.39%, and the leaching rate of Mo ranged from 1.29 to 1.9%. Beyond 358 K, the leaching rate of S increases slowly,

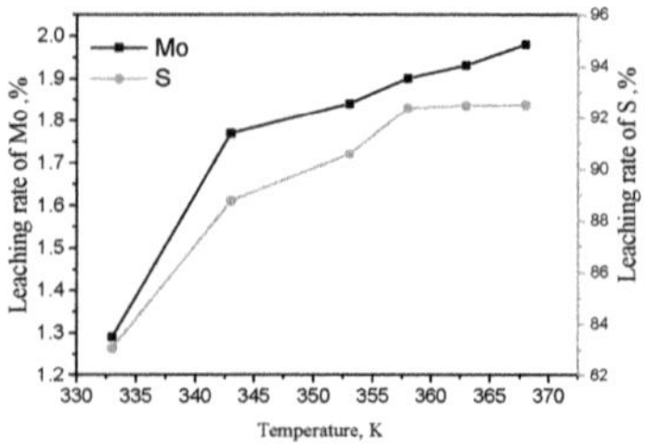

Fig. 8 Influence of temperature on Mo and S leaching (L/S ratio 8 ml/g, 120 min, initial hydrochloric acid concentration 4 mol/L)

but the leaching rate of Mo increases quickly,so a further increase in temperature to 368 K has no significant effect, and further experiments were carried out for 358 K.

Effect of Time

The leaching rate of Mo and S is also influenced by leaching time. The higher leaching rate of S may not be achieved in a short time whilst long time have disadvantages such as higher production cost. So, in this work, an intermediate time range from 15 to 240 min was selected and its influence was studied as shown in Fig. 9. It can be seen the leaching rate of S is increased from 90.48 to 93.59%, and the leaching rate of Mo ranged from 0.78 to 1.93%, when the leaching time increased from 30 to 120 min. Beyond 120 min, the leaching rate of S increases slowly, but the leaching rate of Mo increases quickly, so all further experiments were carried out for 120 min.

Effect of Hydrochloric Acid Concentration

In order to investigate the effect of the hydrochloric acid concentration on the leaching rate of Mo and S, the experiments was carried out at 358 K for 120 min with sample mass was 18.0 g, and the results was shown in Fig. 10. It can be seen from Fig. 10 that the effect of hydrochloric acid concentration under the same

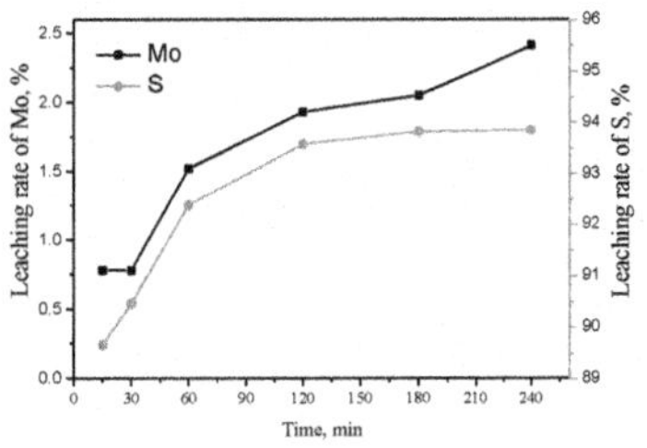

Fig. 9 Influence of time on leaching rate of Mo and S (L/S ratio 8 ml/g, 358 K, initial hydrochloric acid concentration 4 mol/L)

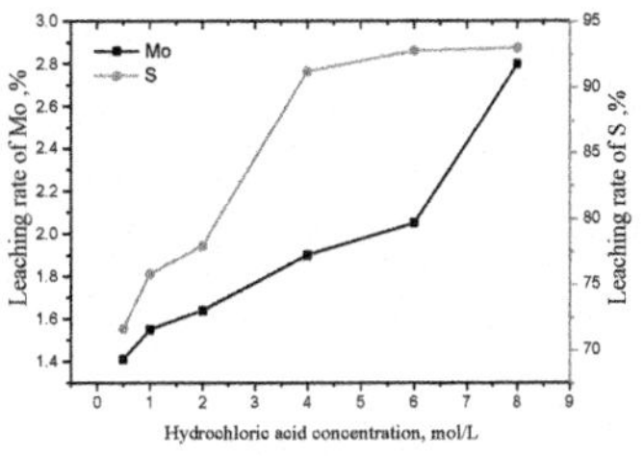

Fig. 10 Influence of hydrochloric acid on leaching rate of Mo and S (L/S ratio 8 ml/g, 120 min, 358 K)

conditions as before. Therefore the optimum hydrochloric acid concentration appears to be 4 mol/L and all further experiments were carried out at this hydrochloric acid concentration.

Effect of Liquid-to-Solid Ratio

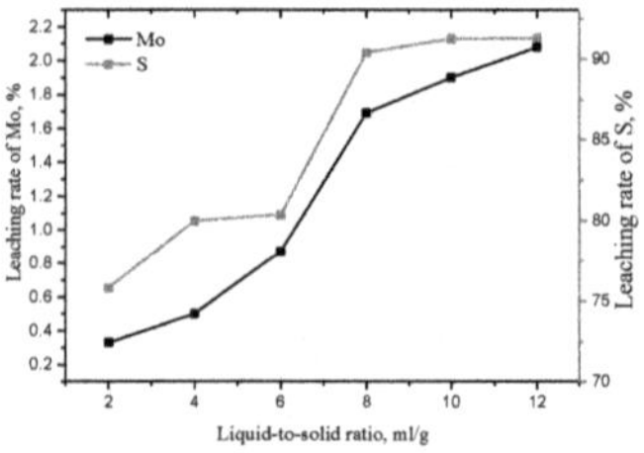

Fig. 11 Influence of liquid-to-solid ratio on leaching rate of Mo and S (initial hydrochloric acid concentration 4 mol/L, 120 min, 358 K)

The ratio of liquid to solid plays an important role in leaching processes. The results of liquid-to-solid ratio shown in Fig. 11 under optimum leaching conditions, indicate that the leaching rate of S increases from 75.83 to 90.38%, the leaching rate of Mo increases from 0.33 to 1.69% as the liquid-to-solid ratio increases from 2 to 8 ml/g. A further increase in liquid-to-solid ratio up to 12 ml/g has not helpful to leaching rate of S, and the S leaching rate can get to 90.38% in 120 min. So 8 ml/g is selected.

Under above optimization experimental conditions, acid leaching slag was obtained. In order to further remove the acid oxides, in this work, the acid leaching slag was handled by alkali washing (1.0 mol/L, NaOH solution). And the slag obtained after alkali washing process was the molybdenum powder. The chemical composition of the leaching slag was got by chemical titration, and the removal rate of element in the molybdenum powder was calculated from Eq. (7) were shown in Table 6.

$$r_R = \frac{m_a - m_b}{m_a} \times 100\% \quad (7)$$

Table 6 Chemical composition and removal rate (r_R) of molybdenum powder

	Mo	S	O	C	Al	Si	Mg	Ca	Fe	Cu	Others
wt%	98.290	<0.005	0.190	0.920	0.180	<0.005	<0.005	0.060	0.240	0.006	0.099
r_R.%	4.86	99.99	98.34	−41.18	98.04	99.81	97.95	70.54	84.89	99.26	–

where r_R is the removal rate of element through vacuum decomposition-acid leaching combination process, %. m_a is the content of element in the molybdenite concentrate, g. m_b is the content of element in the molybdenum powder.

Conclusions

(1) It's feasible to produce molybdenum powder from molybdenite concentrate by using the new method proposed in this research. Above all, this novel process was short flow, environment-friendly compared with the traditional methods.
(2) The crude molybdenum was obtained by vacuum decomposition of molybdenite concentrate. The mass fraction of Mo and S is 90.17 and 0.48 wt%, respectively. The optimal condition parameters are as follows: the pressure of 5–35 Pa for 60 min at 1823 K, distillation times is once, material diameter and thickness is 25 and 20 mm, respectively.
(3) The molybdenum powder was obtained by the acid leaching and alkali washing of the residues of vacuum decomposition. The mass fraction of Mo and S is 98.29 and 0.005 wt% respectively, and removing rate of S reached 99.99%. And the optimal condition parameters are as follows: the leaching temperature, leaching time, concentration of hydrochloric, concentration of NaOH and ratio of liquid to solid is 358 K, 120 min, 4.0, 1.0 mol/L and 8 ml/g, respectively.

Acknowledgements The authors acknowledge the financial support from National Natural Science Foundation of China (NO. U1202271), the Program for Innovative Research Team in University of Ministry of Education of China (NO. IRT1250), and the Program for Innovative Research Team in Nonferrous Metal Vacuum Metallurgy of Ministry of Science and Technology (Grant No. 2014RA4018), and the Cultivating Plan Program for the Technological Leading Talents of Yunnan Province (2014HA003).

References

1. K.X. Jiang, Y.F. Wang, X.P. Zou et al., Extraction of molybdenum from molybdenite concentrates with hydrometallurgical processing. JOM **64**(11), 1285–1289 (2012)
2. D.C. Liu, X.M. Chen, D.J. Li, et al., Simulation of MoS_2 crystal structure and the experimental study of thermal decomposition
3. T.G. Xiang, B.H. Yang, *Metallurgy of molybdenum* (Central South University Press, Changsha, 2009)
4. Q.X. Zhang, Q.S. Zhao, *Metallurgy of molybdenum and tungsten* (Metallurgical Industry Press, Beijing, 2005)
5. E. Blanco, H.Y. Sohn et al., The kinetics of oxidation of molybdenite concentrate by water vapor. Metall. Mater. Trans **38**(4), 689–693 (2007)
6. P.M. Guo, D.G. Wang, P. Zhao. Thermodynamic analysis on non-oxidation of molybdenite roasting process. Nonferrous Metals: Metall. Part (2) **19**, 6–8, (2010)
7. Z.F. Cao, H. Zhong, Z.H. Qiu et al., A novel technology for molybdenum extraction from molybdenite concentrate. Hydrometallurgy **99**, 2–6 (2009)

8. J. Chen, *Studies on Vacuum Decomposition of Molybdenite Concentrate* (Kunming university of science and technology, Kunming, 2009), pp. 16–49
9. L. Wang, P.M. Guo, J.M. Pang et al., Thermodynamic analysis of vacuum decomposition process of molybdenum concentrate. Chin. J. Nonferrous Met. **25**(1), 190–196 (2015)
10. L. Wang, P.M. Guo, J.M. Pang et al., Phase change and kinetics of vacuum decomposition of molybdenite concentrate. Vacuum **116**, 77–81 (2015)
11. D.O. Buker, in *Process for Thermal Dissociation of Molybdenum Disulfide*: US, 3966459. (1976)
12. X.F. Kong, B. Yang, H. Xiong et al., Removal of impurities from crude lead with high impurities by vacuum distillation and its analysis. Vacuum **105**, 17–20 (2014)
13. Y.N. Dai, B. Yang, *Vacuum metallurgy of nonferrous metal* (Metallurgical Industry Press, Beijing, 2009)
14. X.W. Yang, D.F. Qiu, *Hydrometallurgy* (Metallurgical Industry Press, Beijing, 2012)
15. Y.Z. Zhou, Y. Lu, D.C. Liu et al., Thermodynamic analysis and experimental rules of vacuum decomposition of molybdenite concentrate. Vacuum **121**, 166–172 (2015)

Pressure Leaching Behavior of Molybdenum-Nickel Sulfide from Black Shale

Zhigan Deng, Xingbin Li, Chang Wei, Cunxiong Li, Gang Fan and Minting Li

Abstract Molybdenum and nickel in black shale ore exists as amorphous sulfides which has highly active, and easily oxide to sulfuric acid and sulfate at low temperature in the presence of excess oxygen. Take advantage of thc charactcristic to leach molybdenum and nickel from black shale. Oxidation, conversion and dissolution behavior of amorphous colloidal molybdenum and nickel sulfide in pressure leaching process was studied. The effects of stirring speed, temperature, concentration of sulphuric acid, oxygen process and mineral granularity on molybdenum and nickel leaching were investigated. The results showed that dissolution rate increased with increasing temperature and stirring speed, but decreased with increasing concentration of sulphuric acid. And dissolution of molybdenum and nickel from ore only need oxygen without any other reagent.

Keywords Amorphous sulfides · Molybdenum · Nickel · Pressure leaching

Introduction

The molybdenum-nickel sulfide ore is a multi-metal complex and unique mineral resource exists as black shale, which as a sediment-hosted ore layer of the lower Cambrian black shale has been known to local geologists in China. Since 1960s, over the 1600 km belt of Mo-Ni ore has been found successively in some regions of South China, such as Guizhou, Hunan, Zhejiang, Jiangxi and Yunnan Province

Z. Deng (✉) · X. Li · C. Wei (✉) · C. Li · G. Fan · M. Li
Faculty of Metallurgical and Energy Engineering, Kunming University of Science and Technology, No. 253 Xuefu Road, Kunming 650093, Yunnan, China
e-mail: dengzhigan83@163.com

C. Wei
e-mail: weichang502@sina.cn

© The Minerals, Metals & Materials Society 2017

H. Kim et al. (eds.), *Rare Metal Technology 2017*,
The Minerals, Metals & Materials Series,
DOI 10.1007/978-3-319-51085-9_26

[1, 2]. The Mo-Ni ore contains more than 4wt% Mo, at least 2 wt% Ni, up to 2 wt% Zn, 2.5 wt% As and 1–2 g/t of precious metals, primarily Au, Pt, Pd, and Os. In general, the Mo-Ni ore is enriched in nickel, molybdenum and vanadium and is used for extracting nickel and molybdenum raw materials with a gross reserve of 9.37 million tons [3].

The classic process for molybdenum and nickel recovery is roasting, melting and concentration to produce a high impurity nickel-molybdenum alloy. In the process, sulfur dioxide (SO_2) and arsenic trioxide (As_2O_3) release may cause serious environmental pollution. Mineral processing routes have been proposed for the Mo-Ni ore [4–6], however, it is inefficient to enrich nickel and molybdenum through ore-dressing owing to the ore's complex mineralogical characteristics. Research has therefore been initiated into treating nickel-molybdenum ores by direct metallurgical processes.

In recent years, combined process between pyrometallurgy and hydrometallurgy is increasingly applied, such as $NaOH/Na_2CO_3$ leaching under active oxygen conditions [7, 8], direct leaching with $NaOH + Na_2CO_3/NaClO_3$ [9, 10], direct NaOH leaching under active oxygen conditions [11] and direct leaching with H_2SO_4 for nickel recovery only [12]. Moreover, direct bioleaching of molybdenum and nickel from the Mo-Ni ore has been investigated using a molybdenum-resistant thermophilic bacterium sulfolobus metallicus, but the percentage leached is not very high and requires further research [13]. By contrast, oxygen pressure leaching can not only avoid SO_2 or other gases emission, but also can reduce the leaching time dramatically [14, 15].

Molybdenum and nickel in black shale ore exists as amorphous sulfides which has highly active, and easily oxide to sulfuric acid and sulfate at low temperature in the presence of excess oxygen. In this paper, take advantage of the above characteristic of amorphous sulfides to leach molybdenum and nickel from black shale. Oxidation, conversion and dissolution behavior of amorphous colloidal molybdenum and nickel sulfide in pressure leaching process was studied for a direct hydrometallurgical process of extraction molybdenum and nickel from molybdenum-nickel ore by pressure water leaching.

Experimental

Materials

The raw Mo–Ni ore used in the present was obtained from Guizhou Province, China. Analytical grade reagents, including sulfuric acid, were purchased from AR, Chengdu Union Institute Chemical & Reagent.

Experimental Procedure

All the experiments were carried out in a 2 L of acid corrosion resistant titanium autoclave. This autoclave was equipped with a mechanical stirrer with three axial impellers, a heating mantle, a digital temperature controller and an internally mounted cooling coil. Firstly, heated up autoclave with water to a specified temperature in the absence of agitation. Then, the Ni–Mo ore were mixed with the hot water by preset agitator in the autoclave under oxygen pressure. When the desired pressure value was achieved, the agitation and the reaction time were initiated. At selected time intervals, small amounts of slurry were withdrawn and quickly vacuum filtered for analysis.

Analytical Methods

The metal contents of the raw ore sample and leaching residues were analyzed by inductively coupled plasma-atomic emission spectrometry (Spectro Blue, Germany). X-ray powder diffraction (XRD) was carried out using Rigaku D/MAX 2500v diffractometer (Japan).

Nickel and molybdenum in solution were analyzed by atomic absorption spectrometry (WFX-100B, Beijing Beifen-Ruili Analytical Instrument Co., Ltd.). Fe was analyzed by complex titration with potassium bichromate. The potential of solution was measured using a PHSJ-4F digital pH meter by a platinum electrode with a Ag/AgCl electrode used as the reference electrode (Shanghai Leici Instrument Co., Ltd).

Results and Discussion

Characterization of the Materials

The compositions of the molybdenum-nickel sulfide ore from black shale was listed in Table 1. X-ray diffraction of Mo-Ni ore indentified pyrite [FeS_2], quartz [SiO_2], calcite [$CaCO_3$], millerite [NiS] as the main mineral components in Fig. 1. No

Table 1 The main chemical component of materials (wt%)

Ni	Mo	Fe	S	C	P	F	CaO
4.63	5.42	18.24	24.86	9.86	0.47	0.13	14.26
SiO_2	Al_2O_3	MgO	Zn	Cu	TiO_2	Se	V
8.59	2.99	0.74	0.43	0.25	0.23	0.22	0.11

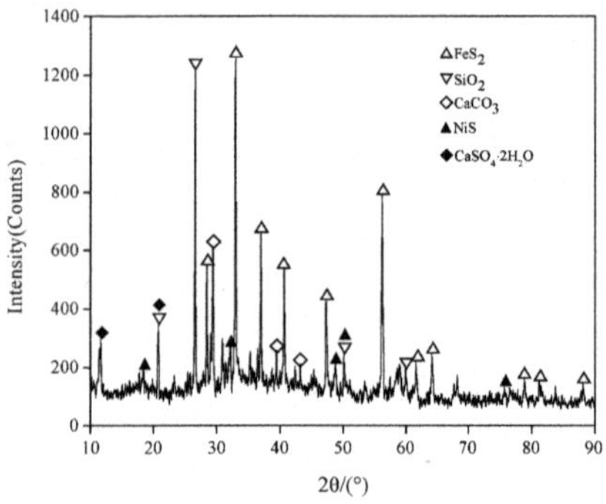

Fig. 1 XRD of the Mo-Ni ore

detected the diffraction peak of the molybdenum in Fig. 1, which occurs mainly due to in the form of amorphous sulfide.

Effect of Agitation Speed

The influence of stirring speed on leaching was studied in the range from 100 to 600 rpm. The results as Fig. 2, which showed that an adequate suspension of the solid particles and adequate distribution of oxygen were observed at 400 rpm. And the nickel leaching was independent of the agitation above this speed. To eliminate the stirring speed as a variable in the kinetics study, a stirring speed of 500 rpm was selected for the subsequent experiments.

Effect of Leaching Temperature

The effect of temperature on molybdenum and nickel extraction was investigated from 105 to 165 °C with distilled water, under an oxygen partial pressure of 0.4–0.5 MPa,

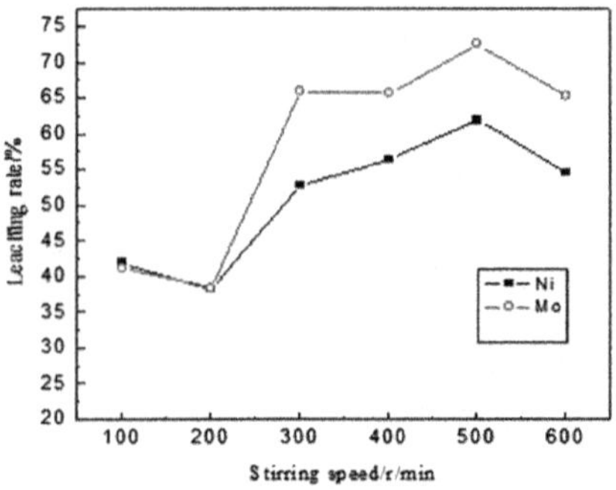

Fig. 2 Effect of agitation speed

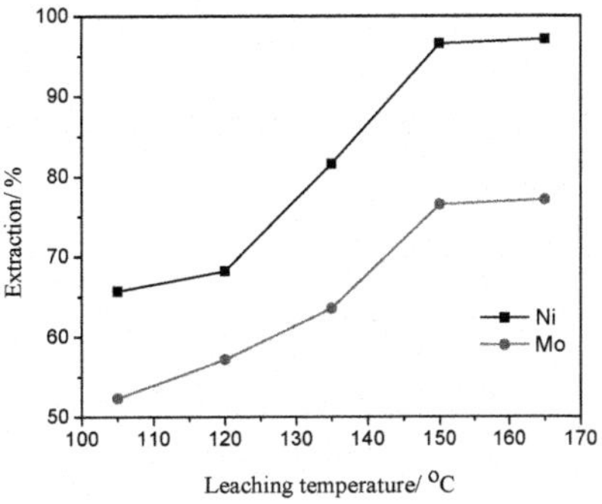

Fig. 3 Effect of leaching temperature

initial sulfuric acid concentration of 0 g/L and a liquid-to-solid(L/S) ratio of 2 mL/g for 5 h. It can be seen from Fig. 3 that nickel extraction increases with increase in leaching temperature from 105 to 150 °C. Beyond 155 °C, the extraction of nickel increases slowly with an increase in temperature. Molybdenum extraction increased with increasing temperature and increased only slightly beyond 150 °C.

Effect of Initial Sulfuric Acid Concentration

To investigate the effect of sulfuric acid concentration on the extraction of molybdenum and nickel, a series of experiments were carried out under the following conditions: initial H_2SO_4 concentration of 0 to 100 g/L, leaching time of 5 h, leaching temperature of 150 °C, oxygen partial pressure of 0.4–0.51.2 MPa and liquid-to-solid ratio of 2 mL/g. As shown in Fig. 4, over 97% of nickel and 76% of molybdenum were leached when the initial sulfuric acid concentration was 25 g/L.

The increase in nickel extraction was insignificant and molybdenum extraction decreased slightly with increase in initial sulfuric acid concentration. Furthermore sulfuric acid consumption increased and produced more impurities. Compared with the experiment where no sulfuric acid was added, the results reveal that the initial

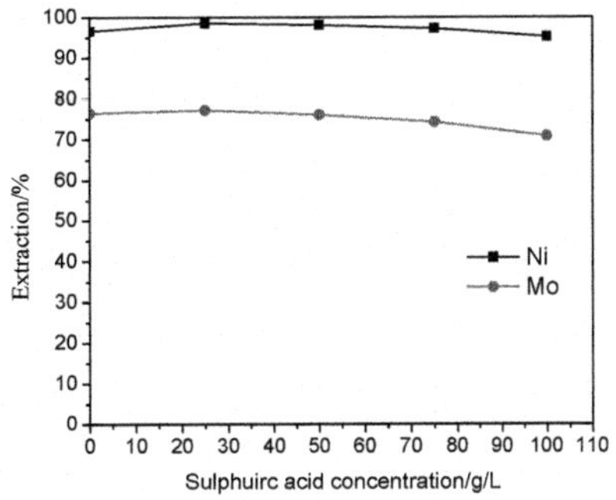

Fig. 4 Effect of initial sulfuric acid concentration

sulfuric acid concentration has little impact on the leaching of nickel and molybdenum from the Mo-Ni ore. Therefore, water leaching without reagent of the molybdenum-nickel ore was optimal under these conditions.

Effect of Partial Pressure of Oxygen

The influence of oxygen partial pressure on the nickel leaching at 150 °C is illustrated in Fig. 5. As seen from this Figure, by increasing the partial pressure of oxygen from 0.1 to 0.7 MPa, there was a gradual increase in the rate of nickel dissolution. More than about 95% of nickel can be dissolved after 180 min under 0.7 MPa of oxygen pressure.

Effect of Particle Size

The effect of particle size on the leaching rate is presented in Fig. 6. As expected, the smaller the size of the particles and the larger contact area, the faster the leaching rate. Nickel extraction is nearly 90% after 120 min of leaching the smallest size.

Fig. 5 Effect of particle pressure on extraction of nickel

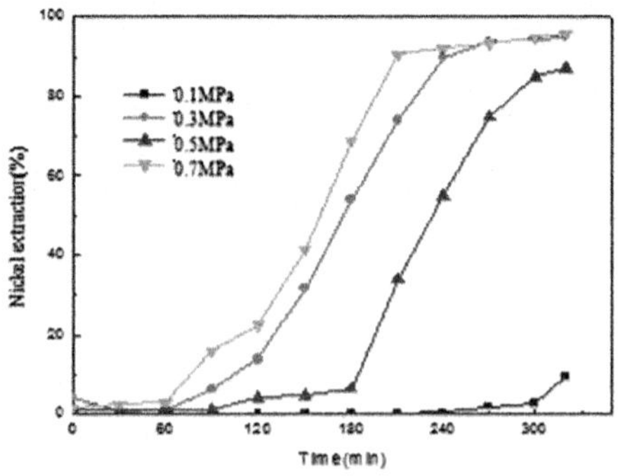

Fig. 6 Effect of particle size on extraction of nickel

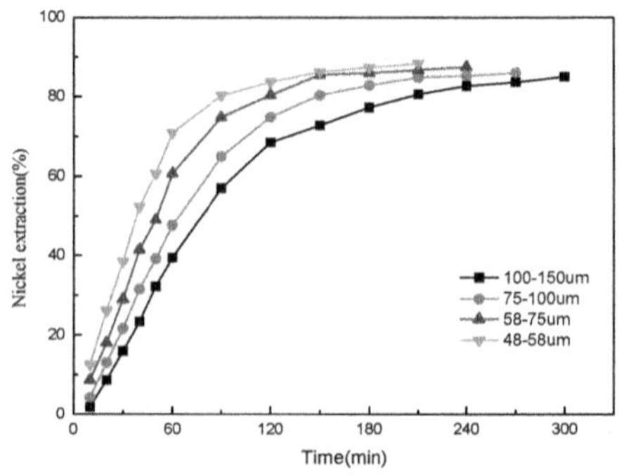

Table 2 Phase changes of sulfide in leaching residue (wt%)

S^{2-}	SO_4^{2-}	S^0	S_T
1.16	11.18	0.78	13.11

Phase Changes of Sulfide

Molybdenum and nickel in black shale ore exists as amorphous sulfides which has highly active, and easily oxide to sulfuric acid and sulfate at low temperature in the presence of excess oxygen. Phase changes of sulfide in leaching residue was given in Table 2. The results showed 95% sulfur in the Mo-Ni ore has been transformed into sulfuric acid, which provided sufficient acid for the leaching of nickel and molybdenum.

Conclusions

Experiments data indicated that dissolution rate of molybdenum and nickel increased with increasing temperature and stirring speed, but decreased with increasing concentration of sulphuric acid. And dissolution of molybdenum and nickel from ore only need oxygen without any other reagent. Under the leaching time for 3–4 h, temperature at 150 °C, liquid-to-solid ratio of 2 mL/g, mineral granularity of 0.074 mm, oxygen partial pressure of 0.7 MPa, molybdenum leaching percentage can be over 80% and nickel leaching percentage can be more than 97%. Furthermore, about 95% sulfur in the Mo-Ni ore has been transformed into sulfuric acid, which provided sufficient acid for the leaching of nickel and molybdenum.

Acknowledgements This work was supported financially by the National Natural Science Foundation of China (No. 51474115) and Yunnan Province Applied Foundation Research Programs of China (No. 2014FB126).

References

1. D.L. Fan, T. Zhang, Y. Jie, J. Paava, B. Kribek, P. Dobes, I. Varrin, K. Zak, Geochemistry and origin of tin-polymetallic sulfide deposits hosted by the Devonian black shale series near Dachang, Guangxi, China. Ore Geol. Rev. **241**, 103–120 (2004)
2. J. Mao, B. Lehmann, D. Andao, G. Zhang, D. Ma, Y. Wang, Re-Os dating of poly-metallic Ni-Mo-PGE-Au mineralization in lower Cambrian black shales of south China and its geological significance. Econ. Geol. **97**, 1051–1061 (2002)
3. Z. BAO, Geological features and metallogenic condition(including Au, Ag mineralization) of Mo-Ni-V ore deposits in the northwest hunan province. Geol. Miner. Resour. Res. **3**, 49–62 (1990)

4. J.M. Juneja, S. Singh, D.K. Bose, Investigations on the extraction of molybdenum and rhenium values from low grade molybdenite. Hydrometallurgy **41**, 201–209 (1996)
5. M. Wang, X. Wang, Extraction of molybdenum and nickel from the Mo-Ni ore by oxidation roasting, sulphation roasting and water leaching. Hydrometallurgy **102**, 50–54 (2010)
6. X. Wang, J. Peng, M. Wang, P. Ye, The role of CaO in the extraction of Ni and Mo from the Mo-Ni ore by calcification roasting, sulphation roasting and water leaching. Int. J. Miner. Process. **100**, 130–135 (2011)
7. M. Wang, X. Wang, W. liu, A novel technology of molybdenum extraction from low grade Mo-Ni ore. Hydrometallurgy **97**, 126–130 (2009)
8. Yu. Juan, Y. Hong-Ying, T. Lin-Lin, Z. Jun, Intensified bioleaching of low-grade molybdenite concentrate by ferrous sulfate and pyrite. Rare Met. **34**, 207–214 (2015)
9. W. Liu, H. Xu, X. Yang, X. Shi, Extraction of molybdenum from low-grade Mo-Ni ore in sodium hypochlorite solution under mechanical activation. Miner. Eng. **24**, 1580–1585 (2011)
10. Z. Zhao, X. Xu, X. Chen, G. Huo, A. Chen, X. Liu, H. Xu, Thermodynamics and kinetics of adsorption of molybdenum blue with D301 ion exchange resin. Trans. Nonferrons Met. Soc. China **22**, 686–692 (2012)
11. Z. Zhao, J. Li, C. Cao, G. Huo, G. Zhang, H. Li, Recovery and purification of molybdenum from Ni–Mo ore by direct air oxidation in alkaline solution. Hydrometallurgy **103**, 68–73 (2010)
12. M. Wang, C. Wei, G. Fan, Z. Deng, S. Wang, Wu Jun, Molybdenum recovery from oxygen pressure water leaching residue of Ni-Mo ore. Rare Met. **32**, 208–212 (2013)
13. J. Chen, C. Gao, Q. Zhang, L. Xiao, Leaching of nickel-molybdenum sulfide ore in membrane biological reactor. Trans. Nonferrons Met. Soc. China **21**, 135–140 (2010)
14. M. Wang, C. Wei, G. Fan, M. Li, Z. Deng, S. Wang, Selective extraction of Mo from a Ni-Mo ore using pressure alkaline leaching. Hydrometallurgy **153**, 6–11 (2015)
15. S. Wang, C. Wei, Z. Deng, C. Li, X. Li, J. Wu, M. Wang, F. Zhang, Extraction of molybdenum and nickel from Ni-Mo ore by pressure acid leaching. Trans. Nonferrous Met. Soc. China. **23**, 3083–3088 (2013)

Selective Recovery of Scandium From Sulfating Roasting Red Mud By Water Leaching

Zhaobo Liu, Hongxu Li and Zihan Zhao

Abstract A completely new process of selectively recovering Sc^{3+} without impurity of Fe^{3+} from Bayer red mud was developed. Prior to water leaching process, red mud needs to be digested and sulfated at a low temperature and subsequently roasted at a higher temperature for a period of time. During the roasting process, the products such as SO_2, SO_3 and H_2SO_4 molecules of thermal decomposition of sulfites including $NaFe(SO_4)_2$ and $NaAl(SO_4)_2$ could be recycled and reused as the raw materials of sulphuric acid. The leaching results indicate that roasting temperature and roasting time are the main constraint on selective leaching performance of scandium. As the sulfated red mud sample is roasted at 750 °C for 40 min and subjected to water leaching at 50 °C for 30 min, 53.0 wt% Sc^{3+}, <1.0 wt% Fe^{3+}, <0.3 wt% Si^{4+} and ~8.9 wt% Al^{3+} could be leaching out. Of note, the solid-liquid separation process is quite easy to be carried out.

Keywords Selective · Ferrum · $NaFe(SO_4)_2$ · $NaAl(SO_4)_2$ · $Na_3Sc(SO_4)_3$ · Red mud

Introduction

Bayer red mud or bauxite residue is a by-product of alumina production and commonly consists of some rare earth elements, such as scandium, gallium and yttrium, especially for the red mud produced from karst minerals [1–3]. Currently,

Z. Liu · H. Li (✉) · Z. Zhao
School of Metallurgical and Ecological Engineering, University of Science and Technology Beijing, Beijing 100083, China
e-mail: lihongxu@ustb.edu.cn

Z. Liu · H. Li · Z. Zhao
Beijing Key Laboratory of Rare and Precious Metals Green Recycling and Extraction, University of Science and Technology Beijing, 100083 Beijing, China

© The Minerals, Metals & Materials Society 2017

H. Kim et al. (eds.), *Rare Metal Technology 2017*,
The Minerals, Metals & Materials Series,
DOI 10.1007/978-3-319-51085-9_27

nearly 3 billion tons of red mud has been accumulated up to now around the world according to the production of metallurgical grade alumina and red mud to Al_2O_3 ratio value of 1–1.5 [3, 4]. While most of them were stockpiled in the red mud residue field or dumped at the sea. These disposal methods not only increase the production cost by occupying acres of storehouses but also leads to environmental problems like ground water pollution and haze [5].

On the one hand, the high content of alkali metal could limit the application of red mud in the field of iron-making and building materials. These oxides could be first reduced to alkali metals by carbon or carbon monoxide in the blast furnace, subsequently volatilized and moved to the upper part of furnace, oxidized by air and then condensed into the bottom of blast furnace circulate, over and over again. These behaviors will definitely increase the fuel consumption and also shorten the lifetime of furnace lining [2]. If use the red mud as building materials directly, the intensity of concrete will be weakened greatly owing to the alkali-aggregate reaction. On the other hand, the importance of recovering the scandium from red mud has been gradually recognized since the scarcity and high market price of scandium [6].

Currently, direct acid leaching, hydro-chemical/hydrothermal process, lime-soda sintering coupled with water leaching could be employed to recover the sodium and scandium from red mud [7–9]. However, the other major elements like Fe, Al, Ca, Ti and Si could be extracted or leaching out simultaneously by these pyro- or hydrometallurgy methods. It is worthwhile to note that the Fe^{3+} and Sc^{3+} ions are difficult to separate from each other during the following extraction and back-stripping process, which are mainly owing to the similar radii between them described by Ochsenkuhn-Petropoulou et al. [8, 10].

In the present study, the sulfation-roasting-leaching process has been wildly and commercially used in the extraction of rare earth elements (not including scandium) from Baotou rare earth concentrates and bastnasite [11]. Besides, this method can also be used in recycling of NdFeB magnets, selectively recovering Ni and Co from iron rich lateritic ores [12, 13]. However, the application of sulfation-roasting-leaching process on Bayer red mud is quite different from the former cases. The primary advantage lies in the fact that the liberation of SO_2 or SO_3 from metal sulfates could be absorbed or recycled by the red mud pulp with a high pH $\sim$12.5 during the roasting process.

The purpose of present work was to investigate the leaching performance of various elements consisting of Ca, Fe, Al, Si, Na, Ti, Sc and Ga from high alkali red mud by sulfation-roasting-leaching process. The effects of roasting conditions on the leaching performance will be explored. Besides, the roasting thermodynamics and phase transitions impacts on leaching mechanism will be analyzed based on the results of DSC, XRD and SEM.

Experimental

Bayer red mud used in the present study was sampled from Chalco in Shangdong province of China. The major chemical composition of red mud was determined by X-ray fluorescence (XRF-1800, Rigaku, Japan), while the minor components of Sc and Ga by Inductively Coupled Plasma-Optical Emission Spectrometer (ICP-OES, Optima 7000DV, Perkin Elmer, USA) after being digested with $HCl–HNO_3–HF–HClO_4$ system. Thermal behavior of sulfated red mud was carried out by Thermogravimetric Analysis-Differential Scanning Calorimetry (TG-DSC, STA409C, Netzch, Germany) in argon atmosphere from room temperature to 1000 °C at a heating rate of 10 °C min^{-1}. Mineral phases were determined by powder X-ray Diffraction (XRD, SmartLab, Rigaku, Japan) with a 2θ scan range of 10–100° at 45 kV and 200 mA with Cu Kα radiation, a speed of 10° min^{-1}. Field Emission Scanning Electron Microscopy (FE-SEM, JSM-6701F, JEOL, Japan) was used to observe the powder morphology. Prior to the SEM observation, the samples need to be scattered over the conductive adhesive with cotton bud and coated with a thin gold film.

Red mud was first mixed with deionized water in a ratio of 1 g:1 ml in a corundum crucible and then with concentrated H_2SO_4 (analytical reagent) in a ratio of 1 g red mud:1 ml H_2SO_4. The extra addition of water is beneficial to make the red mud like newly produced. Well mixed materials were then roasted in a muffle furnace (KZ-1100X-C, MTI Co., China) at a planned temperature with a heating rate of 10 °C min^{-1} and held for a period of time set in advance. The temperature of furnace was adjusted by PID control method with temperature stability of ±1 °C. After the roasting process, samples were cooled inside furnace to less than 250 °C. All the leaching processes were conducted by a magnetic heating stirrer (85-2A KEXI Instrument, China) under leaching temperature 50 °C, leaching time 30 min, liquid to red mud solid ratio (L/S) 1:10 g ml^{-1} and constant stirring speed of ~250 rpm. The separation of leaching liquor and residue was carried out by a vacuum suction filter with medium speed quantitative filter paper.

Results and Discussions

The chemical composition of red mud is listed in Table 1 in the form of element, suggesting Fe, Al, Si, Na in major amounts for more than 54 wt%, Sc and Ga in trace. Figure 1 shows the mineral phases of red mud. It can be seen that this residue

Table 1 The primary chemical composition of Bayer red mud (wt%)

Elements	Fe	Al	Si	Na	Ca	Ti	Ga	Sc
Content	26.74	12.05	8.88	6.54	1.83	1.96	76 g ton^{-1}	921 g ton^{-1}

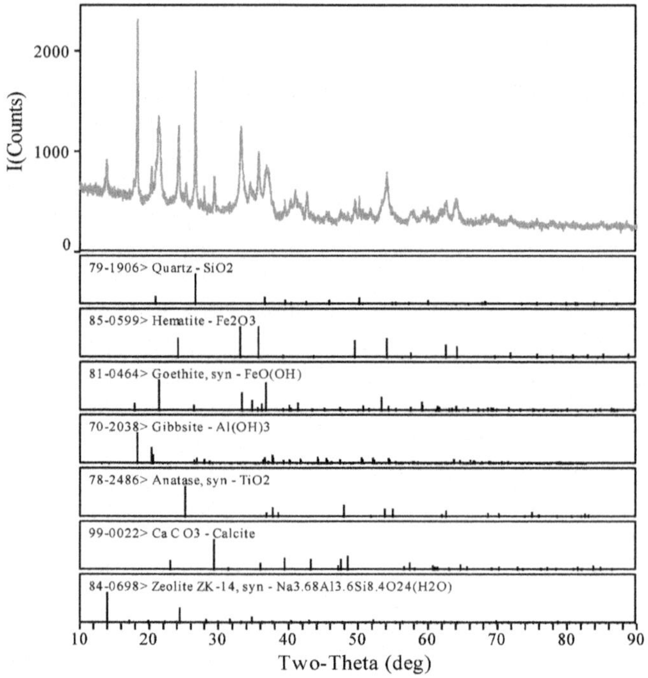

Fig. 1 Mineralogical phases of Bayer red mud

mainly consists of hematite Fe_2O_3, goethite FeO(OH), gibbsite $Al(OH)_3$, quartz SiO_2 and zeolite $Na_{3.68}(Al_{3.}6Si_{8.4}O_{24})(H_2O)_{1.2}$.

Figure 2 displays the sulfated red mud samples roasted at 740 °C for various roasting time. It can be observed in Fig. 2 that the recoveries of Al and Fe decrease rapidly with the increase of roasting time from 40 to 60 min, especially with regard to Al. When the roasting time lasts for 60 min, there are still 5.6 wt% Al can be leached out, while the leaching rate of Fe drops to less than 0.4 wt%. The variation between these two major elements, namely Fe and Al, could be assigned to the various decomposition temperature of Fe- and Al-containing mineral phases. Compared with the theoretical thermal decomposition temperature of 545 °C for $Fe_2(SO_4)_3$ and 524 °C for $Al_2(SO_4)_3$, it can be inferred that the Fe and Al could not be present as the simple metal sulfates [14]. The following XRD results of samples roasted at various roasting temperature could prove that suppose. The recovery of Ca shows a fluctuation around ~30 wt% and an independence of roasting temperature at 740 °C since its high decomposition temperature of 1000 °C. As for the scandium, there is a slight increase in the recovery from 45.5 to 49.6 wt% from 40 to 50 min; then the recovery stables at ~50 wt% as the roasting time is extended to 60 min. The longer roasting time will result to the less soluble of sodium by ~7 wt% compared the recovery of 40 min roasting time sample with 60 min one. Other elements like gallium and titanium are almost insoluble in the leachant. The reason could be first decomposition of Ti-containing sulfates and then gallium-containing

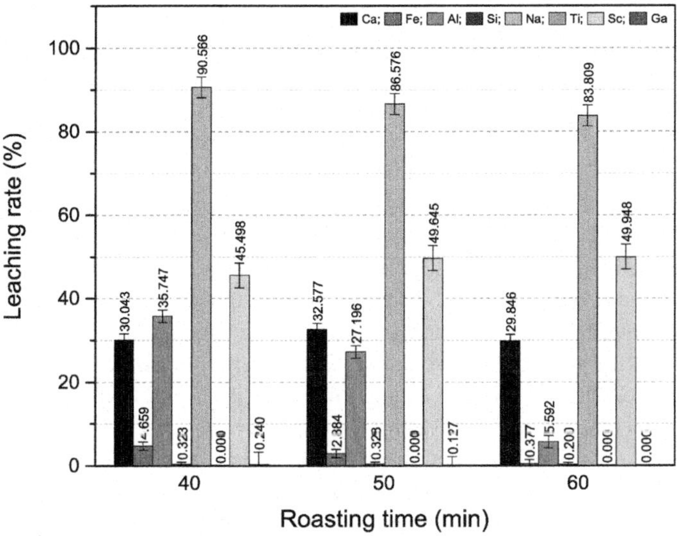

Fig. 2 Leaching performance of sample roasted at 740 °C for various roasting time

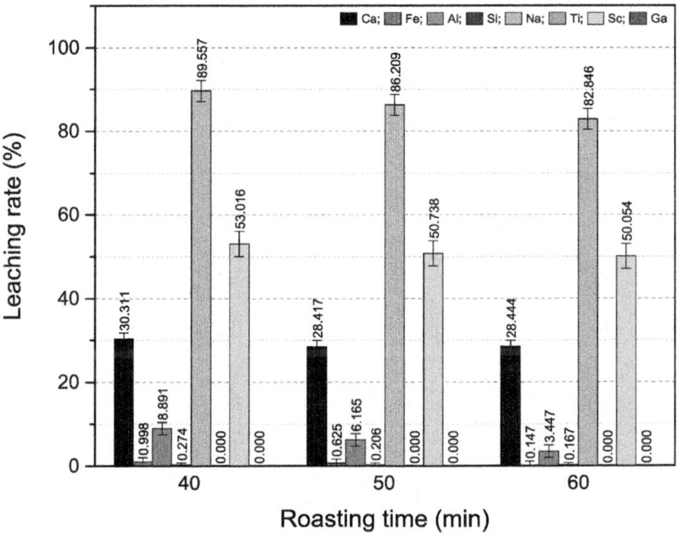

Fig. 3 Leaching performance of sample roasted at 750 °C for various roasting time

ones. In respect of Si recovery, owing to the fact that SiO_2 belongs to acid oxide, all the samples exhibit that only a small quantity of silicon less than 0.4 wt% could be washed out.

As shown in Fig. 3, the maximum leaching rate of scandium could be obtained with sulfated red mud roasted at 750 °C for 40 min. Subsequently, the recovery of scandium begins to decrease slightly to 50.7 and 50.1 wt% corresponding to the roasting time of 50 and 60 min, respectively. Besides, the change of recovery of sodium is similar to the case of scandium. The longer roasting time, the smaller leaching rates of ferrum and aluminum. The leaching rates of Al and Fe have respectively dropped to 3.4 and <0.2 wt% with augmenting roasting time to 60 min. As compared with the samples roasted at 740 °C, these leaching rates of Al and Fe at 750 °C are much smaller than the values of same roasting time. Significantly, both the leaching rates of Al and Fe have decreased to an extremely low concentration except Sc, which makes selectively and effectively extraction of Sc from red mud been achieved. The dissolution of calcium is stable at ~29 wt% and almost unaffected by the increase of roasting time to 60 min. As for the titanium and gallium, both the concentrations in the leachates are less than the detection limit of ICP-OES. The change of silicon is similar to the case of 740 °C. It is noteworthy that the higher roasting temperature will be helpful to promote the agglomeration of crystals and inhibit the dissolution of silicon during the leaching process.

With the further increase of roasting temperature to 760 °C (Fig. 4), leaching rates of some elements, such as Sc and Na exhibit a significant drop with augmenting the roasting time from 40 to 60 min. It is apparent that the drop of leaching rate of Sc is associated with the decomposition state of Al-containing mineral

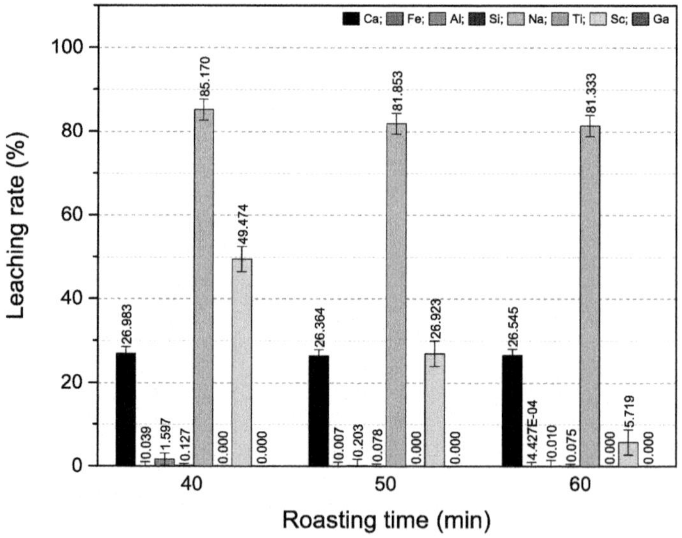

Fig. 4 Leaching performance of sample roasted at 760 °C for various roasting time

phases. Once the decomposition of Al-containing phases has almost been accomplished or the leaching rate of aluminum is less than 1.6 wt%, the extra roasting time will facilitate the aggregation process and the decomposition of Sc-containing mineral phases and last lead to the decrease of leaching rate of scandium. In comparison with the cases of 750 or 740 °C, the dissolution of calcium drops to ~26.5 wt%; ferrum to less than 0.1 wt%; silicon to less than 0.2 wt%; titanium and gallium almost could not be leached out. It can be concluded that the liberation of SO_3 or SO_2 from scandium-containing phases occurs after the decomposition of Al-containing phases. Moreover, the decomposition of Al-containing ones is followed by the Fe-containing sulfates.

Figure 5 illustrates the DSC curve of sulfated red mud at a heating rate of 10 °C min^{-1}. Several endothermal peaks occurring at the temperature range of 25–400 °C could be assigned to the evaporation of adsorbed and crystal water molecules. The strong peak of 740 °C suggests that a dramatic decomposition reaction must occur at that temperature. Compared with the decomposition temperature of $Fe_2(SO_4)_3$ and $Al_2(SO_4)_3$ (Both less than 600 °C), the peak at 740 °C could be attributed to the new generation of Na_2SO_4, $Fe_2(SO_4)_3$ and $Al_2(SO_4)_3$ during the roasting process, which could be supported by the following XRD result. In the present study, experiments of all the roasting conditions were designed based on this temperature point. All the leaching results indicate that the selectively recovery of scandium is sensitively to the roasting temperature and roasting time.

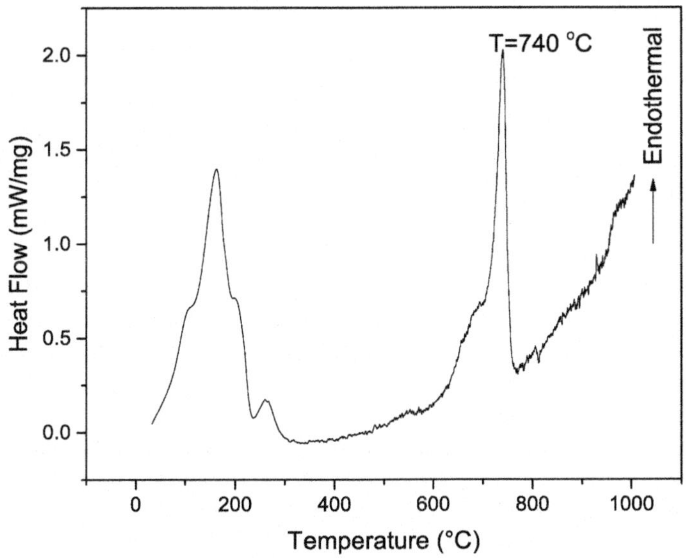

Fig. 5 Thermal behavior of sulfated red mud

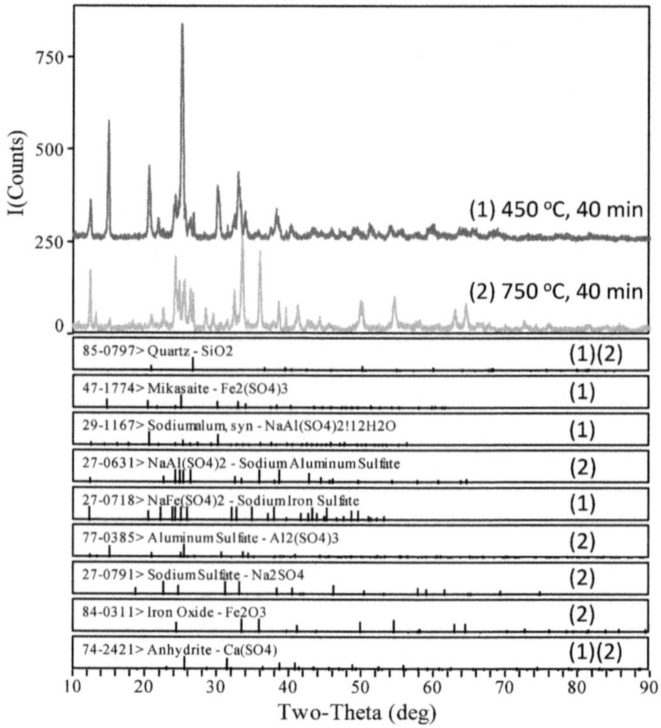

Fig. 6 Mineralogical phases of sulfated red mud samples roasted at 450 and 750 °C

Figure 6 presents the XRD patterns of sulfated red mud samples roasted at different roasting temperatures. The mineral phases of sample roasted at 450 °C for 40 min include $NaAl(SO_4)\cdot 12H_2O$, $NaFe(SO_4)_2$, SiO_2, $Fe_2(SO_4)_3$ and $CaSO_4$. While the sample roasted at 750 °C, most of the metal sulfates have decomposed and appear as corresponding oxides or simple metal sulfates mainly consisting of Fe_2O_3, Na_2SO_4, $Al_2(SO_4)_3$ and SiO_2. When sample is roasted at 740 °C or higher temperature like 750 or 760 °C, $NaFe(SO_4)_2$ will first decomposed to Na_2SO_4 and $Fe_2(SO_4)_3$, then the $Fe_2(SO_4)_3$ phase continues to transform into Fe_2O_3. Simultaneously, the $NaAl(SO_4)_2$ start to break up into Na_2SO_4 and $Al_2(SO_4)_3$. Part of $Al_2(SO_4)_3$ decomposes into Al_2O_3 and the decomposition state primarily depends on the roasting time. On the other side, the Sc-containing phase should not be present as $Sc_2(SO_4)_3$ because the decomposition of $Sc_2(SO_4)_3$ will begin at 500 °C [15]. In analogy to the elements of Al and Fe, scandium is more likely to appear as $Na_3Sc(SO_4)_3$ in the roasting process. The corresponding SEM images were given in Fig. 7, large quantities of particles were aggregated together in sample roasted at 450 °C for 40 min (Fig. 7a); while for the 750 °C roasted sample, slightly more micro-pores could be observed in that sample due to the decomposition of metal sulfates.

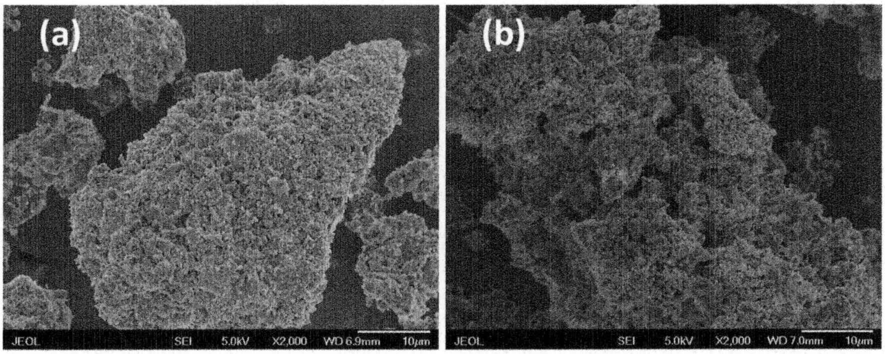

Fig. 7 SEM of sulfated red mud samples roasted various temperatures **a**: 450 °C for 40 min; **b**: 750 °C for 40 min

Fig. 8 A flow diagram of selective recovery of Sc from Bayer red mud

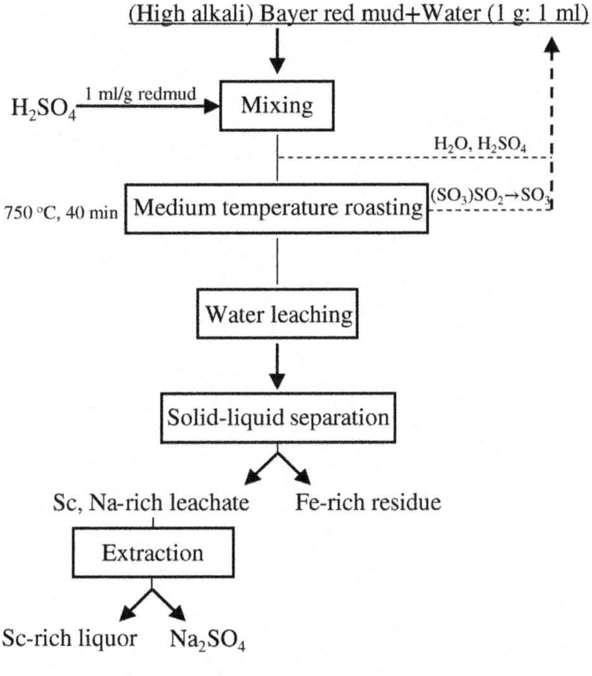

Conclusions

A novel method on selectively recovering Sc^{3+} with few Fe^{3+} and Al^{3+} ions from Bayer red mud was introduced in the present study, as shown in Fig. 8. In this method, most H_2SO_4 could be recycled and reused by gathering the gas products like SO_2, SO_3 and H_2SO_4 molecules from thermal decomposition of sulfites like

$NaFe(SO_4)_2$ and swelling agent $NaAl(SO_4)_2$. The leaching results indicate that the selective performance of scandium is sensitive to the roasting temperature and roasting time. After being subjected to roasting at 750 °C for 40 min and water leaching at 50 °C for 30 min, recovery of ~53.0 wt% Sc^{3+}, <1.0 wt% Fe^{3+}, <0.3 wt% Si^{4+}, ~30.3 wt% Ca^{2+} and ~8.9 wt% Al^{3+} could be obtained; after at 760 °C for 40 min, recovery of ~49.5 wt% Sc^{3+}, <0.1 wt% Fe^{3+}, <0.2 wt% Si^{4+}, ~27.0 wt% Ca^{2+} and ~1.6 wt% Al^{3+} obtained. It is worthwhile to note that scandium is likely present as $Na_3Sc(SO_4)_3$ during the roasting process.

Acknowledgements The authors are grateful for financial support from National Natural Science Foundation of China (grant number 51234008), Beijing Technical Development Project (grant number 00012132), China Postdoctoral Science Foundation (grant number 2016M590046).

References

1. Z. Liu et al., Selectively recovering scandium from high alkali Bayer red mud without impurities of iron, titanium and gallium. J. Rare Earth. (2017)
2. Z. Liu, H. Li, Metallurgical process for valuable elements recovery from red mud—A review. Hydrometallurgy **155**, 29–43 (2015)
3. Y. Liu, R. Naidu, Hidden values in bauxite residue (red mud): recovery of metals. Waste Manage. **34**(12), 2662–2673 (2014)
4. Primary aluminium production, Available from: http://www.world-aluminium.org/statistics/ (cited 22 July 2016)
5. S. Xue et al., Proposal for management and alkalinity transformation of bauxite residue in China. Environ Sci Pollut R, 1–13 (2016)
6. J.A. Ober, Mineral commodity summaries 2016 (US Geological Survey, 2016)
7. L. Zhong, Y. Zhang, Y. Zhang, Extraction of alumina and sodium oxide from red mud by a mild hydro-chemical process. J. Hazard. Mater. **172**(2), 1629–1634 (2009)
8. M.T. Ochsenkühn-Petropoulou et al., Pilot-plant investigation of the leaching process for the recovery of scandium from red mud. Ind. Eng. Chem. Res. **41**(23), 5794–5801 (2002)
9. X. Zhu, W. Li, X. Guan, An active dealkalization of red mud with roasting and water leaching. J. Hazard. Mater. **286**, 85–91 (2015)
10. M. Ochsenkühn-Petropulu et al., Recovery of lanthanides and yttrium from red mud by selective leaching. Anal. Chim. Acta **319**(1), 249–254 (1996)
11. X. Huang et al., Development of rare earth hydrometallurgy technology in China. J Rare Earth **23**(1), 1–4 (2005)
12. X. Guo et al., Leaching behavior of metals from a limonitic nickel laterite using a sulfation–roasting–leaching process. Hydrometallurgy **99**(3–4), 144–150 (2009)
13. M.A.R. Önal et al., Recycling of NdFeB magnets using sulfation, selective roasting, and water leaching. J Sustain Metall **1**(3), 199–215 (2015)
14. H. Tagawa, Thermal decomposition temperatures of metal sulfates. Thermochim. Acta **80**(1), 23–33 (1984)
15. J.G. Li et al., Monodispersed Sc_2O_3 precursor particles via homogeneous precipitation: synthesis, thermal decomposition, and the effects of supporting anions on powder properties. J. Mater. Res. **18**(05), 1149–1156 (2003)

Study for Preparation of Industrial Ammonium Molybdate from Low Grade Molybdenum Concentrate

Qingwei Qin, Zhenwei Liu, Tiejun Chen, Zili Huang, Jianhong Yang and Wei Han

Abstract The preparation of industrial ammonium molybdate from molybdenum concentrate was conducted according to the process of roasting—leaching—purification—precipitation—crystallization. The optimum technological conditions were determined: the roasting temperature was 680 °C, the roasting time was 2.5 h; in the leaching process the alkali excess coefficient was 1.2, the reaction temperature was 80 °C, the liquid-solid ratio was 3:1, the leaching time was 3 h; the amount of $(NH_4)_2S$ added into the solution was 1.1 times of the theoretical amount, the temperature was 80–85 °C, and the terminal pH was 8.5–9 in the purification process; the density of the solution was 1.4–1.6 g/mL, the pH was 1.5–2 in the precipitation process; temperature of dissolution was controlled at 80–90 °C. The recovery of the preparation of ammonium molybdate can reach 85.18%.

Keywords Low grade molybdenum concentrate · Roasting · Leaching · Precipitation

Introduction

One copper mine is rich in molybdenum resources while the molybdenum grade is low (0.020–0.035%) in Hubei province. So it's hard to select molybdenum resources. Copper and molybdenum mixed concentrate was gained by flotation method in copper mine, then was separated into copper and low grade molybdenum

Q. Qin (✉) · Z. Liu · T. Chen · Z. Huang · J. Yang
Key Laboratory for Ferrous Metallurgy and Resources Utilization, Ministry of Education, Wuhan University of Science and Technology, Wuhan 430081, China
e-mail: qingweiqin@126.com

Z. Liu
e-mail: 40315085@qq.com

W. Han
Daye Nonferrous Design and Research Institute Co. Ltd., Huangshi 435005, China
e-mail: 380977134@qq.com

© The Minerals, Metals & Materials Society 2017

H. Kim et al. (eds.), *Rare Metal Technology 2017*,
The Minerals, Metals & Materials Series,
DOI 10.1007/978-3-319-51085-9_28

concentrates (Mo content is about 20–30%) [1, 2]. Ammonium molybdate is produced by standard molybdenum concentrate (Mo $\geq$ 45%) in industry. A flowsheet of producing ammonium molybdate from low grade molybdenum concentrate was explored, which will improve the comprehensive utilization of resources in mining enterprises and economic benefits [3, 4].

Properties of Test Materials

According to certain standard, the molybdenum concentrate, taken from a copper mine in Hubei province, was sent to conduct multi-element analysis after a process of drying and sample preparation. The result was shown in Table 1. It can be known the molybdenum content was 23.80%, with a low content of copper and iron, which is non-standard molybdenum concentrate (molybdenum content of standard molybdenum concentrate is 45% or higher). Main gangue composition were SiO_2, Al_2O_3, CaO, MgO, accounting for 43.63% of the total, so it's a material of a high containing calcium and magnesium. The SEM image of the molybdenum concentrate was shown in Fig. 1.

Table 1 Results of multi-element chemical composition

Element	Mo	Cu	TFe	S	P	Zn
Content/%	23.80	0.23	4.09	19.61	0.28	0.026
Element	SiO_2	Al_2O_3	CaO	MgO	Pb	Re
Content/%	24.03	3.57	5.92	10.11	0.0074	0.014

Fig. 1 SEM image of molybdenum concentrate

The Method of Experiment

According to the composition and dissemination characteristics of molybdenum concentrate in the test (molybdenum mineral was symbiotic with gangue mineral closely. Ca and Mg content was higher while the content of Mo was about 20–30%), its grade was below the standard molybdenum concentrate (Mo $\geq$ 45%). Meanwhile, there's a high cost in producing ammonium molybdate by hydrometallurgy, and a high requirement for equipment at the same time. While the oxidizing roasting method is simple, with a less consumption, lower cost, simpler equipment, etc. This study selected ammonia alkaline leaching process, and the flowchart was shown in Fig. 2. The main advantages of ammonia alkaline leaching process are as follows: (1) compared with molybdenum calcine pretreatment process, there's no corrosion in equipment, and quantity of waste water and calcine is less, so the labor intensity is low. (2) Compared with ammonia leaching-acid precipitating process, the leaching rate of molybdenum is improved by translating traditional process into alliance leaching of ammonia and sodium carbonate.

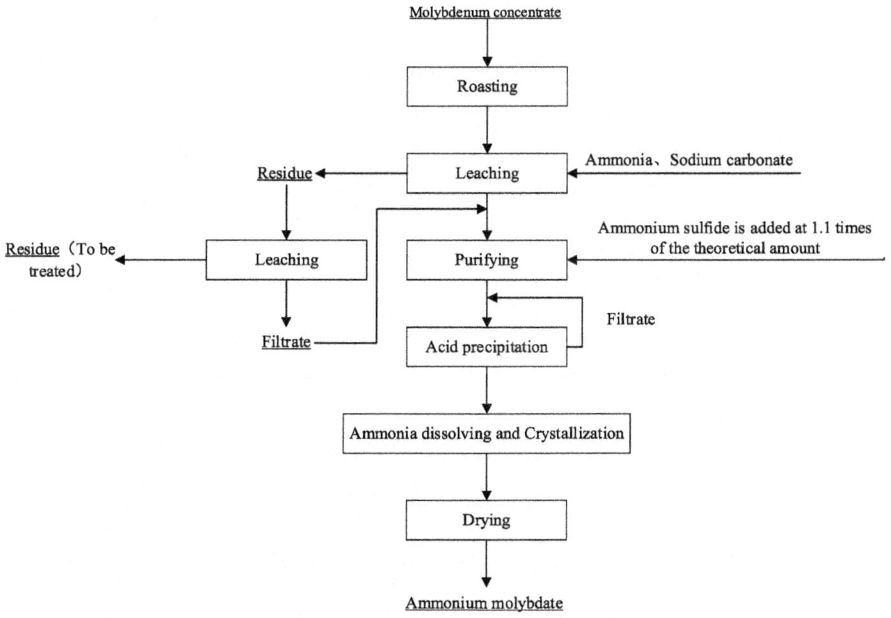

Fig. 2 The process flowing diagram of the preparation of ammonium molybdate

Results and Analysis

The Roasting Process

As the material was low grade molybdenum concentrate, in addition, more talc and calcite were also contained. In order to make the molybdenum concentrate fully roasted and oxidized, MnO_2 was chosen as the oxidant. The influences of roasting temperature, roasting time and the quantity of oxidant were studied in this experiment to explore the optimum roasting conditions. The orthogonal which was designed with three factors and three levels was used in this experiment. The three factors were followed: A-roasting time, B-roasting temperature, C-the quantity of oxidant (MnO_2). The design of orthogonal test was shown in Table 2; the results were shown in Table 3.

Table 2 The orthogonal test table

Factors	Level 1	Level 2	Level 3
A roasting time/h	1.5	2.5	3.5
B roasting temperature/°C	650	700	750
C MnO_2 consumption/g	4	7	10

Table 3 The experimental results

Factors	A time/h	B temperature/°C	C MnO_2 consumption/g	Leaching rate/%
Test 1	1.5	650	4	87.40
Test 2	1.5	700	7	87.64
Test 3	1.5	750	10	86.54
Test 4	2.5	650	7	88.78
Test 5	2.5	700	10	89.82
Test 6	2.5	750	4	84.73
Test 7	3.5	650	10	89.79
Test 8	3.5	700	4	89.13
Test 9	3.5	750	7	84.12
Σ(1)	261.50%	265.97%	261.26%	
Σ(2)	263.34%	266.59%	260.54%	
Σ(3)	263.03%	255.31%	266.06%	
Σ(1)/3	87.17%	88.66%	87.09%	
Σ(2)/3	87.78%	88.86%	86.85%	
Σ(3)/3	87.68%	85.10%	88.69%	
R	0.61%	3.76%	1.84%	
Optimal level	A2	B2	C3	

The order of the influencing factors: B > C > A

It can be known from the results that the order of the influence of each factor on roasting was the temperature, the consumption of oxidant, the roasting time, consecutively; the appropriate process parameters of roasting were as follows: roasting temperature of 700 °C, the oxidant consumption of 10 g, the roasting time of 2.5 h.

The Ammonia-Alkali Leaching

MoO_3 is easily dissolved in alkaline solution and could produce sodium molybdate and ammonium molybdate solution, so molybdenum can be recovered. The orthogonal test of four factors and three levels was conducted in order to study the effect of temperatures, reaction time, excess coefficient of alkali, liquid to solid ratio on the roasting sample by ammonia-alkali leaching. The four factors were as follows: A-temperature, B-reaction time, C-the excess coefficient of Na_2CO_3, D-liquid-solid ratio. The orthogonal test was shown in Table 4; the results were shown in Table 5.

It showed that the order of the influence of each factors on the molybdenum leaching rate was: temperature, liquid-solid ratio, reaction time, alkali excess coefficient. The optimal conditions of the ammonia-alkali leaching process were as follows: the alkali excess coefficient of 1.2–1.4, the reaction temperature of 70–80 °C, liquid to solid ratio of 3:1, the reaction time of 2.5–3 h.

The Purification of the Leaching Solution

The Influence of the Terminal pH on the Purification

The purpose of the purifying process of the leaching solution was to remove the metal ions impurities such as Cu, Fe, etc. The density of leaching solution was 1.14–1.16 g/ml and would be injected slowly according to the amount that was 1.05 times of the theoretical amount of the $(NH_4)_2S$ solution under the condition of stirring. The container was placed in the thermostat water bath for 20–30 min. By this way, the metal ions could react with the sulfide ions. The influence of terminal pH on purification was studied and the results were shown in Table 6.

Table 4 The orthogonal test table

Factors	Level 1	Level 2	Level 3
A leaching temperature/°C	50	70	90
B leaching time/h	1.5	2.5	3.5
C Na_2CO_3 excess coefficient	1.2	1.4	1.6
D liquid-solid ratio	2:1	3:1	4:1

Table 5 The experimental results and analysis

Factors	A leaching temperature/°C	B leaching time/h	C alkali excess coefficient	D liquid to solid ratio	Leaching rate/%
Test 1	50	1.5	1.2	2:1	87.25
Test 2	50	2.5	1.4	3:1	87.01
Test 3	50	3.5	1.6	4:1	87.88
Test 4	70	1.5	1.4	4:1	89.82
Test 5	70	2.5	1.6	2:1	87.50
Test 6	70	3.5	1.2	3:1	90.35
Test 7	90	1.5	1.6	3:1	87.69
Test 8	90	2.5	1.2	4:1	87.23
Test 9	90	3.5	1.4	2:1	86.68
Σ(1)	262.13%	264.76%	264.82%	261.43%	
Σ(2)	267.66%	261.74%	263.50%	265.91%	
Σ(3)	261.59%	264.90%	263.07%	264.92%	
Σ(1)/3	87.38%	88.25%	88.27%	87.14%	
Σ(2)/3	89.22%	87.25%	87.83%	88.64%	
Σ(3)/3	87.20%	88.30%	87.69%	88.31%	
R	2.02%	1.05%	0.58%	1.5%	
Optimal level	A2	B3	C1	D2	

The order of the influencing factors: A > D > B > C

Table 6 The influence of the terminal pH on the effect of purification

No.	Terminal pH	The concentration of Cu^{2+} in purifying liquid	The concentration of Fe^{2+} in purifying liquid	Color of the solution
1	7	1.25 g/L	0.86 g/L	Blue
2	8	0.36 g/L	0.23 g/L	Light blue
3	9	0.003 g/L	0.013 g/L	Transparent
4	10	Minor	Minor	Light yellow

As was shown in Table 6, with the terminal pH increasing, the content of Cu^{2+} and Fe^{2+} in the solution reducing; while the color of the solution turned to yellow after the terminal pH was 10, which showed the fact that the content of $(NH_4)_2S$ was excessive. So the purifying terminal pH should be controlled in the range of 8.5–9, so it would meet the requirement of removing the impurities. Besides, excessive $(NH_4)_2S$ is harmful of the subsequent operation, which would make the existence of impurity S^{2-} in the filtrate.

Table 7 The color change of the solution corresponding to the addition of $(NH_4)_2S$

No.	Multiple of theoretical value	Actual amount of $(NH_4)_2S$/ml	Cu^{2+} content in the filtrate/g L^{-1}	Fe^{2+} content in the filtrate/g L^{-1}	Color of the solution
1	0.9	0.45	1.08	0.56	Light blue
2	1	0.5	0.083	0.15	Semitransparent
3	1.1	0.6	0.002	0.023	Transparent
4	1.2	0.7	0.002	0.008	Transparent and slightly turbid
5	1.3	0.8	Micro scale	Micro scale	Light yellow

Test of the Condition of the Addition Contents of $(NH_4)_2S$ Solution

The purpose of adding $(NH_4)_2S$ solution was to remove the metal impurities such as Cu and Fe effectively, excessive $(NH_4)_2S$ was not convenient for subsequent operation because of the S^{2-} [5]. Therefore, the control of adding $(NH_4)_2S$ is very important. The results of test were shown in Table 7. The actual addition of $(NH_4)_2S$ was a fluctuation value; some adjustments must be made according to the pH value and the change of the color in practice.

From Table 7, the color of the solution become shallow and the content of Cu and Fe in the filtrate were decreased while $(NH_4)_2S$ content increased. When the addition of the $(NH_4)_2S$ reached 1.2 times of the theoretical value, the color of the solution started to mix, then was yellow as the amount of $(NH_4)_2S$ increased. Observing the color change of the solution is also an effective way to judge whether the addition of $(NH_4)_2S$ can remove the impurity or not. The color of $(NH_4)_2S$ solution is yellow, so the color of the solution is yellow when $(NH_4)_2S$ was excessive, therefore, the amount of $(NH_4)_2S$ added into the solution controlled at 1.1–1.2 times of the theoretical value was appropriate.

Effect of the Temperature on Purification

100 ml leaching solution was taken, the density was 1.14 g/ml, the terminal pH was at 8.5–9, after the arrival of the terminal pH value, holding the temperature for 20–30 min. To investigate the effect of temperature on the purification, the results of the test were shown in Table 8.

The results showed that, with the increase of the temperature of the water, the content of Cu and Fe ions in filtrate decreased, the color of the solution become shallow. At a low temperature, the contents of metal ions such as Fe and Cu were

Table 8 Effect of temperature on purification

No.	Temperature of water/°	Cu^{2+} content in the filtrate/g L^{-1}	Fe^{2+} content in the filtrate/g L^{-1}	Color of the solution
1	25	5.43	6.32	Pale blue
2	60	2.35	0.98	Blue
3	80	0.005	0.021	Transparent
4	90	0.004	0.015	Transparent

Table 9 The effect of pH on acid precipitation process

No.	pH	Amount of concentration nitric acid/ml	Shape of precipitate	Color of precipitate
1	3.5	5	Irregular shape	White
2	2.5	12	Granular material	White
3	1.5	14	Granular material	White
4	1.0	16	Non forming material	Light yellow

higher, the speed of the reaction of Fe, Cu and other metal ions with S^{2-} was slower, and there were some metal ions failed to precipitate, and remaining in the solution. It was suitable to control the purification temperature at 80–85 °C after comprehensively considered.

The Effect of pH on Precipitation Process

Precipitated by acid is a process of ammonium molybdate crystallization. Ammonium molybdate solution was heated to boiled, the partial water in the solution evaporated and ammonia volatilized, so the volume of the solution was reduced, the concentration of molybdenum in the solution was increased, and the alkalinity of the solution was reduced so the yield of acid precipitation was increased. Under the condition of stirring, the temperature of acid precipitation was control at 50 °C. The shape of the deposit changed as the pH of the solution decreased.

As can be seen in Table 9, when the pH was 1, the deposit was not formed and the color of the product was yellow, therefore, it was appropriate to control pH at 1.5–2 in practice. When the solid and liquid separate after the acid precipitation, adding ammonia of which the density is 0.9 g/cm^3 and an appropriate amount of deionized water in the deposit, then stirring and gradually heat up to 90–80 °C, when the solution density was 1.5–1.6 g/cm^3, the solution was saturated, the ammonium molybdate can be obtained after cooling and crystallization.

Optimal Process Conditions and Product Inspection

According to the experimental study, the optimum technological conditions were determined: the roasting temperature of 680 °C, the roasting time of 2.5 h; in the leaching process the alkali excess coefficient of 1.2, the reaction temperature of 80 °C, the liquid-solid ratio of 3:1, the leaching time of 3 h; the amount of $(NH_4)_2S$ added into the solution was controlled at 1.1 times of the theoretical value, the temperature was 80–85 °C, and the terminal pH was 9–8.5 in the purification process; the density of the solution was 1.4–1.6 g/mL, the pH was 2–1.5 in the acid precipitation process; temperature of ammonia dissolution was controlled at 80–90 °C, with ammonium molybdate dissolved in ammonia water to saturation. The results of test were shown in Table 10, under the optimized conditions, the ammonium molybdate recovery could reach 85.18% after the leaching of low grade molybdenum. The analysis of ammonium molybdate was carried out, and the results were as follows: it contained 53.56% molybdenum, impurity content 0.0023% copper, and iron of 0.0032%. The SEM image and XRD pattern of the product were shown in Figs. 3 and 4.

Table 10 The results of optimum condition

Project	Filtrate volume/ml	Mo content in filtrate/g l^{-1}	Mo content	Recovery of Mo/%
Roasting				96.43
Leaching	730	59.17	45.20	94.81
Purification	732	56.97	42.71	96.55
Precipitation	80	18.25	41.25	96.50
				85.18

(a)250× (b)1000×

Fig. 3 SEM image of ammonium molybdate. **a** 250×, **b** 1000×

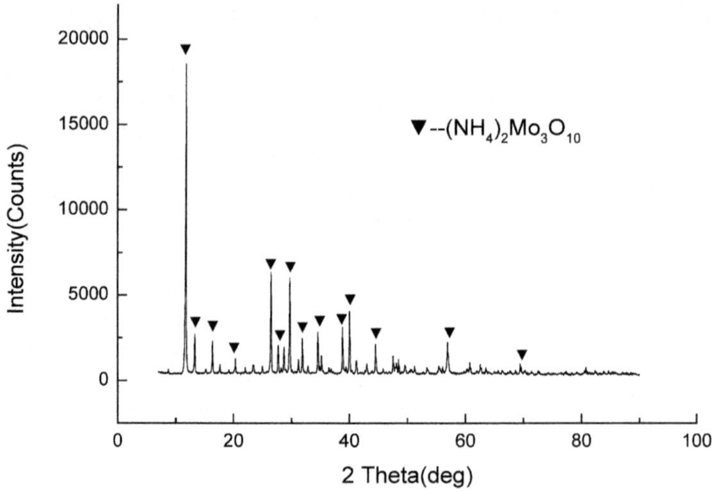

Fig. 4 XRD pattern of the product ammonium trimolybdate

Conclusion

Through the analysis of low grade molybdenum ore, the process of roasting, leaching, purification and precipitation was determined, the feasible process of the preparation of ammonium molybdate was recommended.

The optimum technological conditions were determined: the roasting temperature was 680 °C, the roasting time was 2.5 h; in the leaching process the alkali excess coefficient was 1.2, the reaction temperature was 80 °C, the liquid-solid ratio was 3:1, the leaching time was 3 h; the amount of $(NH_4)_2S$ added into the solution was controlled at 1.1 times of the theoretical value, the temperature was 80–85 °C, and the terminal pH was 9–8.5 in the purification process; the density of the solution was 1.4–1.6 g/mL, the pH was 2–1.5 in the acid precipitation process; temperature of ammonia dissolution was controlled at 80–90 °C. Under the optimized conditions, the recovery of industrial ammonium molybdate could reach 85.18%. It contained 53.56% molybdenum, 0.0023% copper, and 0.0032% iron. The product should be purified further to meet customers' different requirements.

References

1. P. Huiqing, H. Wei, Experimental study and production practice of copper molybdenum separation in a copper mine. Nonferrous Met. (Min. Process. Sect.) **1**, 23–26 (2007)
2. H. Wei, W. Daijun, Improve the selecting of copper recovery of Tongshankou Copper Mine. Min. Metall. **2**, 17–19 (2006)

3. Z. Dechang, Development and application of four ammonium molybdate. Hydrometall. China **1**, 9–12 (1994)
4. L. Guojian, Z. Hong et al., Experimental study on producing ammonium molybdate molybdenum concentrate. Hydrometall. China **3**, 19–23 (2005)
5. X. Tiegen, *Extractive Metallurgy of Molybdenum* (Central South University Press, Changsha, 2002), pp. 35–127

Study of a Synergistic Solvent Extracting System to Separate Yttrium and Heavy Rare Earths: A Deep Investigation on System Behavior

Alessandro Blasi, Corradino Sposato, Assunta Romanelli, Giacobbe Braccio and Massimo Morgana

Abstract Yttrium is a strategical material for its several usages in a wide range of industrial production. For this reason, improving the recovery of Yttrium has become a crucial focus for international research. In this scientific activity, the utilization of two extractants, 2-ethylhexyl-mono 2-ethylhexyl ester phosphonic acid (P-507) and sec-octylphenoxy acetic acid (CA-12), working in a synergistic extracting system, has been tested. A series of trials have been carried out to optimize the experimental conditions in order to separate yttrium from a mix of heavy rare earths in chloride media. Tests with different relative concentrations of extractants and a phase modifier (TPB, trybutyl phosphate), with different pre-saponification rate, have been performed to evaluate the behavior of the synergistic system. Results are very promising to recover and separate yttrium with a high purity.

Keywords CA-12 · P-507 · Solvent extraction · Rare earths · Synergistic extraction

---

A. Blasi (✉) · C. Sposato · A. Romanelli · G. Braccio · M. Morgana
ENEA, Italian National Agency for New Technologies, Energy and Sustainable Economic Development, Research Center of Trisaia, S.S. 106 Ionica, Km 419+500, 75026 Rotondella, MT, Italy
e-mail: alessandro.blasi@enea.it

C. Sposato
e-mail: corradino.sposato@enea.it

A. Romanelli
e-mail: assunta.romanelli@enea.it

G. Braccio
e-mail: giacobbe.braccio@enea.it

M. Morgana
e-mail: massimo.morgana@enea.it

© The Minerals, Metals & Materials Society 2017
H. Kim et al. (eds.), *Rare Metal Technology 2017*,
The Minerals, Metals & Materials Series,
DOI 10.1007/978-3-319-51085-9_29

Introduction

With the increasing demand for rare earth elements and their compounds, the separation and purification of these elements has gained considerable importance in recent years. Lanthanides have several applications in many areas: life sciences, electronics, catalysis, green technologies and so on.

Yttrium is widely used for fluorescent lamps, phosphors, display and solvent extraction is the common method used for its recovery [1–3].

As a branch of solvent extraction, the synergistic extraction has become a common method for the separation of metal ions. It can not only improve the extraction efficiency [4–6] and the extraction selectivity, but also enhance the stability of the extracted complexes: the extracted complexes are more soluble in the organic phase, eliminate emulsification and the formation of the third phase, and increase the extraction reaction rate [5–7].

Xiaobo Sun et al. [8] investigated a new synergistic extraction system using CA-12) and Cyanex272, resulting that adding Cyanex272 to CA-12, the extractability for Er, Tm, Yb and Lu increases much more than that for Yttrium, and thus, results in the increased separation factors between Y and HRE.

Wang et al. [9] studied solvent extraction and separation of yttrium from other rare earths in chloride medium using the mixture of sec-octylphenoxy acetic acid (CA-12, HA) and three different co-extractants (HB) such as bis(2,4, 4-trimethylpentyl) phosphinic acid (Cyanex272), P-204 and P-507. The results of these tests report that double solvent (HAB) system using CA-12) (HA) as main extractant and Cyanex272 (HB) as co-extractant is quite efficient in separating Y from mixed rare earths. Not only the HAB extraction system keeps advantages of high separation efficiency between Y and LRE, but also it overcomes drawbacks of low separation factor between Y and HRE.

This means that co-extractants promote the extraction of HRE significantly in HAB system. Cyanex272 is selected to constitute the HAB system in this work, considering the stripping acidity of rare earths inCyanex272 system is lower than in P204 and P507 systems [10], and low acid consumptions in the stripping stages are expected.

Wu et al. [11] studied the extraction and separation of trivalent Ho, Y, and Er with the mixtures of Cyanex 302 and another extractant, such as P204, P507, P229, CA-100, Cyanex 925, TBP, P350, or N1923. They concluded that Cyanex 302 individually and combined with others show same extraction efficiency.

By the analysis of the status of art about Y separation it's possible to note that synergistic separation is a very promising solution to separate yttrium from heavy rare earth, so in this paper the utilization of of two extractants, 2-ethylhexyl-mono 2-ethylhexyl ester phosphonic acid (P-507) and sec-octylphenoxy acetic acid (CA-12)), working in a synergistic extracting system, have been investigated.

Experimental

Apparatus and Reagents

Rare earths, in aqueous media, were analyzed by using an Inductively Coupled Plasma Optical Emission Spectrometry (ICP-OES) Optima 8300 model by Perkin Elmer. Digital pH meter (Metrohm 780) and an automatic titrator (Metrohm 905 Titrando), equipped by an iSolvotrode probe, were used for pH measurements and to determinate the molar concentration of extractants. Treibacher Industrie kindly provided extractants and rare earths mix in aqueous solution. CA-12) and P507 were titrated by using standard NaOH ([CA-12] = 3.60 M and [P507] = 3.20). These extractants, used without any further purification, were diluted with kerosene. The kerosene was provided by oleotecnica S.p.A. and it's a selected fraction of linear hydrocarbons compounds (C10–C13, aromatic compounds <2%). Pure Sodium hydroxide ACS-ISO was provided by Carlo Erba and it was dissolved in demineralized water and used for the pre-saponification of the extractants. TBP >99% (provided by Sigma-Aldrich) was used as phase modifier. Stock rare earths solution, used for all tests, was checked by using ICP-OES; the total concentration of all metals, in aqueous solution, is 2 M and their relative abundance is reported in weight percentage in Table 1. All other reagents were of analytical grade.

Methods

The experiments were carried out shaking equal volumes of aqueous solution (feed) and extractants mix diluted with kerosene for 40 min with the support of a mechanical shaker at room temperature (298 ± 1 K) to ensure complete equilibration. NaOH was used for the pre-saponification of the extractants in order to reach the required pH at the end of reaction test.

The extractant mixture was prepared by mixing CA-12), P507 and TBP in order to reach the desired value of $X_{CA\text{-}12}$ defined as:

$$X_{CA-12} = \frac{mol_{CA-12}}{mol_{CA-12} + mol_{P507}}$$

The amount of TBP used is the 15 vol% respect to CA-12. The resulting extractant mixture is diluted with kerosene 1/1 in volume. The ingredients in the aqueous phase were analyzed by ICP-OES and the concentration of rare earths in

Table 1 Feed relative concentration of rare earth elements

	Dy	Er	Eu	Gd	Ho	Lu	Sm	Tb	Tm	Y	Yb
Feed (wt%)	9.3	5.4	1.5	11.7	2.3	0.5	4.1	1.7	0.8	58	4.7

organic phase was determined by mass balance. Distribution factor was obtained as $D = [RE]o/[RE]a$, where 'a' and 'o' denote aqueous and organic phase.

Results and Discussion

Several tests at [feed] = 0.5 M were performed. At this feed concentration the extraction system has a poor selectivity due to its high loading capacity. Figure 1 shows that, with higher values of pH, the extractant is able to extract all rare earths contained in the feed.

For this reason, in this work [feed] = 1 M was used.

In Fig. 2 the amount of rare earths extracted versus pH at different $X_{CA\text{-}12}$ is reported. The results show that the extraction of rare earths increase enhancing the pH of aqueous phase. At $X_{CA\text{-}12} \geq 0.7$, reachable values of pH are between 0.5 and 4; pH over than 4.2–4.4 can be considered the upper limit in order to avoid the degradation of extractants. For the highest values of pH, reported in Fig. 2, we observe an amount of rare earths extracted equals to 50–60%. At $X_{CA\text{-}12} = 0.5$ reachable values of pH are between 0 and 1 without degradation of organic phase.

Extraction efficiency is defined as $E = (C_0 - C_f)/C_0$ 100: where C_0 represents the concentration of single rare earth in the feed and C_f represents the concentration of single rare earth in aqueous phase after extraction reaction.

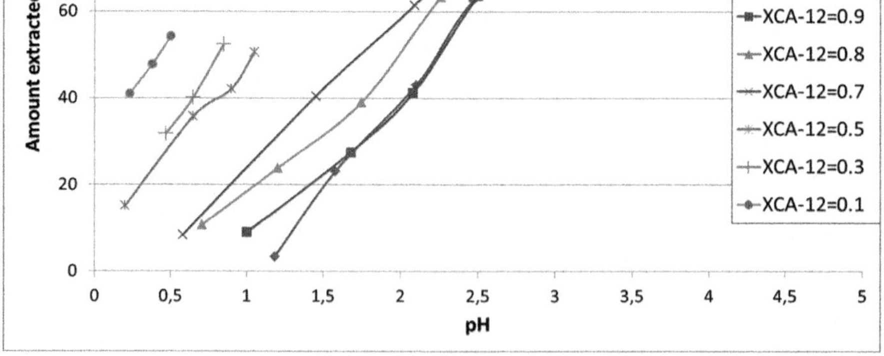

Fig. 1 Total amount of rare earths extracted versus pH at different $X_{CA\text{-}12}$ [feed] = 0.5 M

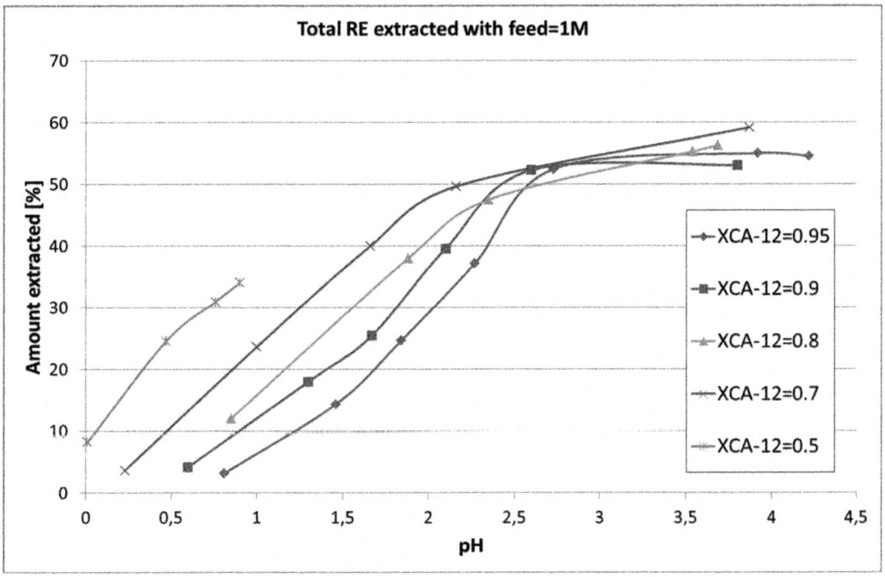

Fig. 2 Total amount of rare earths extracted versus pH at different $X_{CA\text{-}12}$ [feed] = 1 M

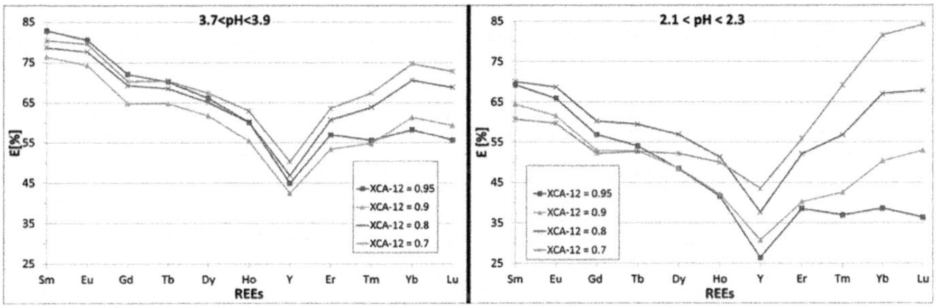

Fig. 3 Extraction efficiency for each rare earths at different $X_{CA\text{-}12}$ in two pH ranges

The optimal pH working range is between 2 and 4 (Fig. 3) and in this pH range the system permit the separation of yttrium from the other metals. Yttrium efficiency values are lower than all other rare earths and it is preferentially leaved in aqueous phase.

Increasing $X_{CA\text{-}12}$ from 0.7 to 0.95 we observe a decrease in efficiency values for heavier rare earths such as Lu, Yb, Tm and Er while for lighter rare earth the extraction efficiency variation is not so relevant.

In 3.7–3.9 pH range we observe an higher difference in efficency extraction between yttrium and lighter rare earths than in 2.1–2.3 pH range; for heavier rare earths the behaviour is the opposite. In a feed with a large amount of lighter rare

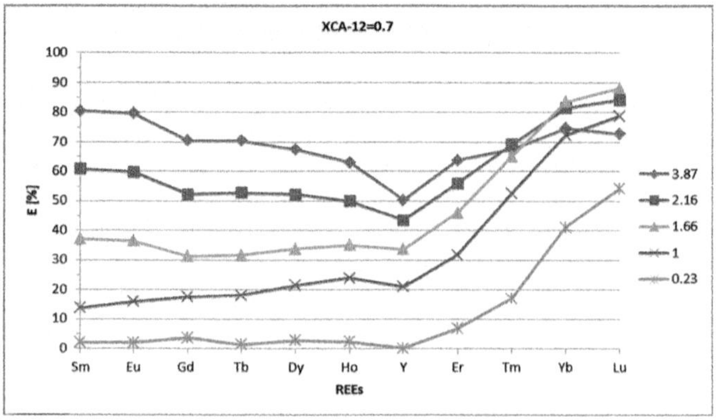

Fig. 4 Extraction efficiency for each rare earth at different pH with XCA-12 = 0.7

earths, as the used one, in which the amount of Sm, Gd, Tb, Dy and Ho is about three times more than Er, Tm, Yb and Lu, we can say that the higher pH range is the best choice for a separation process.

When the pH of aqueous phase falls below the value of 2, with $X_{CA\text{-}12} \geq 0.7$, the separation of yttrium and lighter rare earths became more difficult; this aspect is highlighted in Fig. 4 in which we can observe a drastic decrease in extraction efficiency for Y, Sm, Eu, Gd, Tb, Dy, and Ho.

At $X_{CA\text{-}12} \leq 0.5$ the extraction efficiencies are higher than 30% only for heavier rare earth, such as Lu, Yb, Tm and Er, so the CA-12) + P507 system don't allow an easy separation of lighter rare earth from yttrium (Fig. 5 and Table 2). At these conditions, for the high presence of P507, is impossible to reach pH values greater than 1 without the degradation of organic phase.

In Table 2 are reported the separation factors (β) of rare earth elements versus yttrium, calculated as: distribution factor of single element divided by distribution factor of yttrium. Data shows that for lighter rare earths (Sm, Eu, Gd, Tb and Dy) the system CA-12 + P507 presents the better selectivity with the increase of CA-12) concentration; the best values of separation factor are obtained for $X_{CA\text{-}12} = 1$ (CA-12 alone).

However for $0.7 \leq X_{CA\text{-}12} \leq 0.9$ the separation factor values of Sm, Eu, Gd, Tb and Dy are very good ($2 \leq \beta \leq 4.4$).

Holmium β factor is quite costant changing $X_{CA\text{-}12}$ from 1 to 0.7.

With $X_{CA\text{-}12} \leq 0.5$ the β values are too low to permit a suitable separation of Sm, Eu, Gd, Tb, Dy and Ho from yttrium. For heavier rare earths (Lu, Yb and Tm) the extraction system shows the best selectivity with $X_{CA\text{-}12}$ equals to 0.1 and 0.3. In the range of $0.7 \leq X_{CA\text{-}12} \leq 0.9$ Lu, Yb and Tm presents values of separation factors between 2 and 3 (making separation possible).

Erbium don't show a great difference in separation factors changing $X_{CA\text{-}12.3}$.

By the analysis of reported data yttrium separation can be better performed working at a pH ≈ 3.8 with a $X_{CA\text{-}12}$ equal to 0.7–0.8.

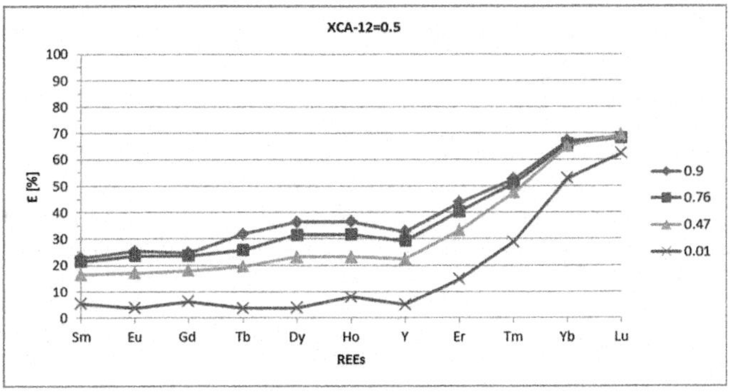

Fig. 5 Extraction efficiency for each rare earth at different pH with XCA-12 = 0.5

Table 2 Separation factors (β) of mixture rare earths elements at different $X_{CA\text{-}12}$

$X_{CA\text{-}12}$	1	0.95	0.9	0.8	0.7	0.5	0.3	0.1	0
Gd	3.8	3.2	2.6	2.4	2.3	0.7	0.2	0.2	0.3
Dy	2.3	2.4	2.2	2.0	2.0	1.1	0.5	0.4	0.5
Er	1.4	1.7	1.6	1.8	1.7	1.7	1.8	1.5	0.9
Sm	7.4	5.6	4.4	3.8	4.1	0.7	0.3	0.2	0.4
Yb	1.2	1.7	2.1	2.9	2.9	4.5	15.5	15.0	7.2
Ho	1.6	1.9	1.7	1.7	1.7	1.1	0.6	0.5	0.6
Tb	3.2	3.0	2.6	2.3	2.4	0.9	0.3	0.3	0.5
Eu	6.3	4.8	3.9	3.5	3.9	0.7	0.1	0.2	0.3
Tm	1.2	1.7	1.6	2.3	2.0	2.6	5.0	4.9	2.8
Lu	1.0	1.6	2.1	2.8	2.6	4.9	24.7	26.8	12.9

Conclusions

Data showed that synergistic effect made possible to tune this extraction system to optimize separation of each rare earths. In the studied case, the separation one-throw of yttrium is achievable and the best conditions are reached with a $X_{CA\text{-}12} = 0.7\text{–}0.8$ at a pH ≈ 3.8. At these conditions the efficiency of extraction for yttrium is around 50% with a β factor value about 4 for europium and samarium and not lower than 1.7 for the other rare earths.

Acknowledgements The authors wish to thank R&D division of Treibacher Industrie, in particular Dr. Stephan Pirker, Dr Christiane Kartusch and Gunther Bierbaumer for support and help during this work.

References

1. L. Xu, Y. Xiao, D.Q. Li, An expert system for solvent extraction of rare earths. J. Chem. Inf. Comput. Sci. **32**, 437–442 (1992)
2. C. Sposato, A. Blasi, G. Devincenzis, P. Garzone, M. Morgana, Comparison among different extractants, As (2-ethylhexyl)-mono (2-ethylhexyl) ester phosphonic acid (P507), secondary-octyl phenoxy acetic acid (CA-12) and BIS(2, 4, 4-trimethylpentyl)phosphinic acid (CYANEX272), in the separation of heavy rare earths via hydrometallurgical processes. Rare Metal Technol. **2014**, 201–206 (2014)
3. A. Blasi, C. Sposato, G. Devincenzis, P. Garzone, M. Morgana, Definition of the process to separate light rare earths by working with (2-Ethylexyl)-Mono(2-Ethylexyl)ester phosphonic acid (P507) in a mixer settler battery. Rare Metal Technol. **2014**, 197–200 (2014)
4. Y. Masuda, Y.W. Zhang, C.H. Yan, B.G. Li, Studies on the extraction and separation of lanthanide ions with a synergistic extraction system combined with 1,4,10,13-tetrathia-7,16-diazacyclooctadecane and lauric acid. Talanta **46**(1), 203–213 (1998)
5. K. Ishimori, H. Imura, K. Ohashi, Effect of 1,10-phenanthroline on the extraction and separation of lithium(I), sodium(I) and potassium(I) with thenoyltrifluoroacetone. Anal. Chim. Acta **454**, 241 (2002)
6. J. Ydberg, T. Sekine, Principles and pratices of solvent extraction, in *Principles and Practices of Solvent Extraction*, ed. by J. Rydberg, C. Musikas, G.R. Choppin (Marcel Dekker, New York, 1992), p. 101
7. Q. Jia, W.P. Liao, D.Q. Li, C.J. Niu, Synergistic extraction of lanthanum (III) from chloride medium by mixtures of 1-phenyl-3-methyl-4-benzoyl-pyrazalone-5 and triisobutylphosphine sulphide. Anal. Chim. Acta **477**, 251–256 (2003)
8. X. Sun, J. Zhao, S. Meng, D. Li, Synergistic extraction and separation of yttrium from heavy rare earths using mixture of sec-octylphenoxy acetic acid andbis(2,4,4-trimethylpentyl) phosphinic acid. Anal. Chim. Acta **533**, 83–88 (2005)
9. Y. Wang, W. Liao, D. Li, A solvent extraction process with mixture of CA-12 and Cyanex272 for the preparation of high purity yttrium oxide from rare earth ores. Sep. Purif. Technol. **82**, 197–201 (2011)
10. J.D. Wang, J.Y. Chen, *Handbook of Solvent Extraction* (Chemical Industry Press, Beijing, 2001), pp. 538–553
11. D.B. Wu, W. Li, D.Q. Li, The extraction and separation of Ho, Y, and Er(III) with the mixtures of Cyanex 302 and another organic extractant. Sep. Sci. Technol. **42**(4), 847–864 (2009)

The Recovery of Bismuth from Bismuthinite Concentrate Through Membrane Electrolysis

Si-yao Peng, Jian-guang Yang, Jian-ying Yang and Lei Jie

Abstract A new membrane electrolysis process for bismuth recovery from bismuthinite concentrate was introduced. Under the conditions of temperature 70 °C, acidity 3 mol/L, solid liquid ratio 1:4 and 1.1 times stoichiometry sodium chlorate dosage for 1.5 h oxidation leaching, more than 98.5% bismuth can be leached from bismuthinite concentrate. The resultant leach solution was subjected to membrane electrolysis. Under the conditions of NaCl concentration 300 g/L and pH 7 in anolyte, bismuth ions concentration 80 g/L and HCl concentration 3 mol/L in catholyte, heteropole distance 6 cm, temperature 40 °C and current density 200 A/m^2, a flat deposited bismuth plate can be obtained after 24 h electrolysis, and the bismuth purity was more than 99%. The current efficiency of cathode was 97.88%. Chemical component analysis results showed that the average convert efficiency of the sodium chlorate was 52.77% and the average oxidizing efficiency of chloride ion was 69.78%, and the resultant anolyte generated in anode compartment could be fully reused as oxidant for leaching bismuthinite concentrate again.

Keywords Bismuthinite concentrate · Chloride leaching · Membrane electrolysis

Introduction

Bismuth is a harmless element among the heavy metals in the periodic table, which has sparked interest in areas varying from metallurgy, alloys and electronics to chemical and medicinal [1, 2]. China is the largest producer of bismuth in the world. Current estimates place the bismuth output of China at more than 60% of the total global output [2]. Bismuth deposits are found in Hunan, Guangxi, and in many

S. Peng · J. Yang (✉) · L. Jie
School of Metallurgy and Environment, Central South University, Changsha 410083, China
e-mail: jianguang_yang@hotmail.com

J. Yang
Jiangxi Environmental Engineering, Vocational College, Ganzhou 342000, China

© The Minerals, Metals & Materials Society 2017

H. Kim et al. (eds.), *Rare Metal Technology 2017*,
The Minerals, Metals & Materials Series,
DOI 10.1007/978-3-319-51085-9_30

other provinces throughout China, of these, Hunan has the richest deposits. In particular, Shizhuyuan, near Chenzhou City, has been celebrated for a long time for the quality and quantity of bismuth ores [4].

The commercial route to extract bismuth from its sulfide mineral bismuth glance at Shizhuyuan Nonferrous Co. Ltd. (Chenzhou, Hunan Province), is a pyrochemical process, which involves reducing the bismuth sulfide to a metallic bismuth in reverberatory furnaces and recovering valuable associated metals such as molybdenum and tungsten from slag via a hydrometallurgy process [5]. However, smelting bismuth under this high smelting temperature causes serious problems of environmental pollution and large energy consumption.

Hydrometallurgical technologies have advantages for processing the low grade and complex ore. Current hydrometallurgical processes for bismuth recovery from bismuthinite concentrate have been studied extensively, such as ferric chloride leaching [6–8], chloride leaching [9, 10], nitric acid leaching, and slurry electrolysis process [11, 12], etc., and the sponge bismuth and bismuth oxychloride and other products can be produced through the electrolysis [13], replacement with iron [14], neutralizing hydrolysis [15].

During the processing of these bismuth minerals, leaching with H^2SO_4 or HCl or H_2SiF_6 is usually involved, and highly acidic solutions containing base metals and bismuth are obtained. Generally, bismuth can be recovered from these spent solutions by hydrolysis, and few authors have reported the separation of bismuth from leach solutions by electrowinning. Zertoubi [16] concluded that the niobium electrode appeared to be suitable for bismuth recovery in acidic media, but the electrochemical re-dissolution of deposited bismuth did not achieve a recovery of more than 50% even in the case of an oxide-free niobium electrode. Reyes-Aguilera [17] have noted a number of other methods for bismuth recovery.

In our previous study [18, 19], a process and related fundamental principle was proposed for the separation and recovery of tin or antimony from a concentrate using a membrane electrolysis process. Further, a new process based on membrane electrolysis for bismuth recovery from a bismuth glance concentrate was proposed here. Thus leaching and membrane electrolysis to recover bismuth was investigated. The current paper demonstrates that not only bismuth can be efficiently recovered from the bismuthinite concentrate, but also $NaClO_3$ can be regenerated from the anode compartment and to be used as leaching agent again.

Experimental

Materials

The low grade bismuthinite flotation concentrates obtained from Hunan Shizhuyuan Nonferrous Metals Co., Ltd. (China). Dry screen analysis of this bismuthinite concentrate showed that almost all concentrate samples were 100 mesh (100% through 100 mesh, 147 μm) and all were leached in this form without further

Table 1 Main chemical composition of bismuthinite concentrates (wt%)

Element	Bi	Fe	S	Ca
Content	25.18	19.33	22.25	3.03
Element	Cu	Mo	Pb	Si
Content	1.99	2.23	2.78	3.065

grinding. Chemical analysis of the bismuthinite concentrate was conducted by chemical titration, except that the Ca, Cu, Mo, Pb were analyzed by atomic absorption spectroscopy, and the results are given in Table 1.

Chemical composition analysis results shows that the bismuthinite concentrates mainly contain Bi, Fe, S, and small amount of Mo, Pb and Cu etc. The X-ray diffraction analysis for mineralogical characterization of this bismuthinite concentrates was performed by using the Rigaku 3014 X-ray diffractometer. The mineralogical characterization analysis result indicated the Bi, Mo and Fe were occurred as bismuthinite (Bi_2S_3), molybdenite (MoS_2) and pyrite (FeS_2).

The hydrochloric acid, sodium chlorate etc., were purchased from Changsha Shenghua, Inc., China, with no further purification introduced.

Method

The procedure for leaching and electrolysis of bismuth from the bismuthinite concentrates was presented as Fig. 1.

Leaching experiments were carried out in a 500 mL conical flask by adding a weighed amount of bismuthinite concentrate, sodium chlorate and sodium chloride to diluted HCl solution at the desired temperature with magnetic stirring at 250 revolutions per minute. The temperature was controlled in a water bath. At the end of each experiment, the insoluble leach residue was filtrated and washed with 1 mol/L HCl and distilled water. The recovery of bismuth was calculated by mass balance using the analysis of the concentrate and the leach residue. The oxidation of sulfur to sulfate ion was not significant under the conditions and duration of these tests. The related chemical reaction was:

$$Bi_2S_3 + NaClO_3 + 6HCl \rightarrow 2BiCl_3 + NaCl + 3H_2O + 3S^0 \quad (1)$$

In the electrowinning experiment, a two-compartment acrylic cell (sizes 90 × 120 × 150 mm) is used, which is separated by a widely used commercial anion exchange membrane (HF-201, Beijing Enling Technology Co., Ltd. China). The cathode was made of titanium plate with an effective area of 45 cm^2 and the anode was a graphite plate with the same surface area. Two electrode were put in the two compartment respectively. And mechanical agitator were used to reduce the concentration polarization.

At the cathode, the electrode reactions were:

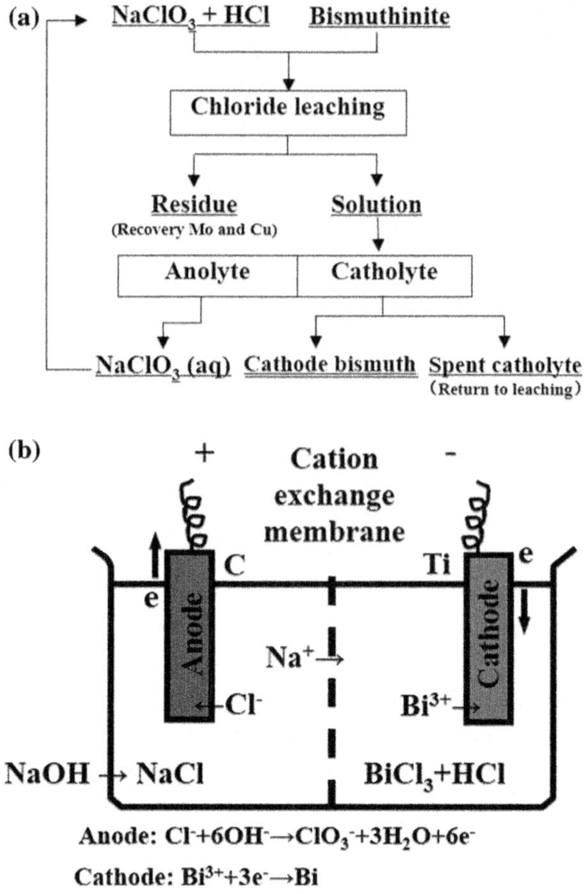

Fig. 1 Schematic diagram of experimental process based on membrane electrolysis

$$Bi^{3+} + 3e^- \rightarrow Bi(\text{main reaction}) \tag{2}$$

$$2H^+ + 2e^- \rightarrow H_2 \uparrow (\text{secondary reaction}) \tag{3}$$

At the anode, anodic reaction was a complex and comprehensive process, including not only the electrochemical reaction but also the chemical reaction [20]. Under the experimental conditions, the chlorine was dissolved in the aqueous solution, then the sodium chlorate was generated by means of a series of electrochemical or chemical reaction.

1. Anode reactions

$$2Cl^- \rightarrow Cl_{2(aq)} + 2e^- \tag{4}$$

$$2H_2O \rightarrow O_2 + 4H^+ + 4e \tag{5}$$

$$6ClO^- + 3H_2O \rightarrow 2ClO_3^- + 4Cl^- + 6H^+ + 3/2O_2 + 6e^- \quad (6)$$

2. Other reactions

$$Cl_{2\,(aq)} + H_2O \rightarrow HClO + Cl^- + H^+ \quad (7)$$

$$HClO \leftrightarrow ClO^- + H^+ \quad (8)$$

$$2HClO + ClO^- \rightarrow ClO_3^- + 2Cl^- + 2H^+ \quad (9)$$

The ideal integrated reaction:

$$Cl^- + 6OH^- - 6e^- \rightarrow ClO_3^- + 3H_2O \quad (10)$$

The cathodic current efficiency (CCE) and anodic current efficiency (ACE) were calculated through the following Eq. (11). Where W_1 represented actual mass of product, W_2 represented the theoretical mass depend on the Faraday's law of electrolysis.

$$\eta = (W_1/W_2) \times 100\% \quad (11)$$

The techniques used to measure the yield of the anodic products ($NaClO_3$, NaCl, and Cl_2) are mainly based on methyl orange spectrophotometric methods [21].

Results and Discussion

Leaching

Through preliminary experiments, various leaching conditions were examined and the most suitable conditions was obtained as follows: temperature 70 °C, HCl 3 mol/L, solid liquid ratio 1:4, 1.1 times sodium chlorate of the stoichiometric quantity as oxidant, reaction time 1.5 h. To verify the reliability of the test, the confirmation experiment were carried out under these conditions. Results were showed in Table 2 and Fig. 2.

The results showed that more than 98.5% of bismuth can be leached from the bismuthinite concentrate, and the metal balance of bismuth, molybdenum, iron and copper reached more than 95%.

The XRD patterns comparison between the concentrates and the leach residue showed that Bi_2S_3 was almost replaced by element sulfur after leaching, while other composition change little, which further confirmed the principle of leaching reaction (Eq. (1)).

Table 2 Results and main metals balance of the confirmation experiment

No.	Leach solution			Residue		Metal balance		
	V (mL)	Element	Composition ($mg \cdot L^{-1}$)	Yield (%)	Composition (%)	Lixivium (%)	Residue (%)	Total (%)
1	489	Bi	48,498	62.25	0.64	97.35	1.64	98.99
		Fe	2623		23.7	8.07	92.85	100.92
		Cu	857		3.64	15.29	82.70	97.99
		Mo	28		3.43	0.61	94.48	95.08
2	495	Bi	48,901	62.05	0.39	99.37	0.99	100.36
		Fe	2510		23.82	7.82	93.02	100.84
		Cu	972		3.46	17.56	78.36	95.91
		Mo	20		3.45	0.44	94.72	95.16

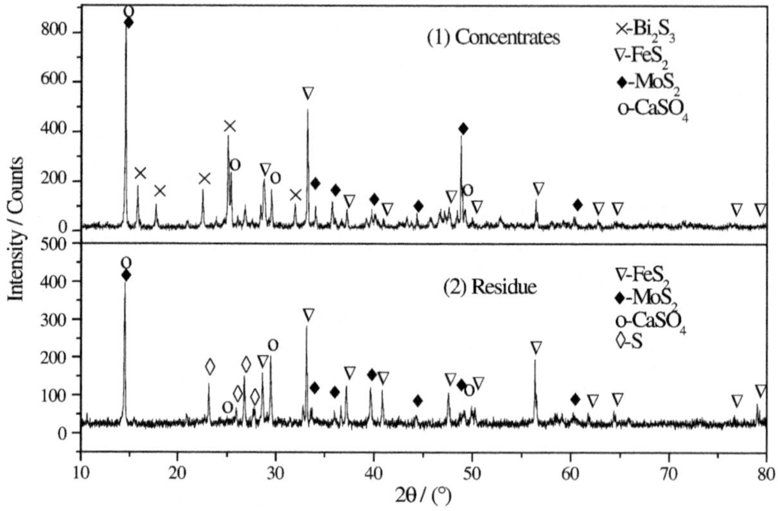

Fig. 2 X-ray diffraction patterns of the concentrates and the leach residue

Electro-Deposition

The membrane electrowinning experiments were carried out in order to study and establish optimum electro-deposition parameters such as the composition of catholyte and anolyte, temperature, cathode current density on the main cell performance parameters such as cathodic current efficiency (CCE), anodic current efficiency (ACE) and cell voltage (U). The appearance of deposited bismuth on the cathode were also determined.

Table 3 Effect of bismuth concentration in catholyte on electro-deposition

Bismuth concentration (g/L)	Cathodic current efficiency (%)	Cell voltage (V)	Cathode surface
30	97.43	1.87	Powdery
40	98.52	1.9	Smooth and compact
50	99.31	1.79	Smooth and compact
60	99.3	1.77	Smooth and compact
70	99.78	1.74	Smooth and compact
80	99.26	1.7	Smooth and compact
90	99.04	1.71	Some tuberculum on the surface

Effect of Bismuth Concentration in Catholyte

The electrowinning of the varying bismuth concentration in catholyte were conducted from 30 to 90 g/L under other cathodic parameters of HCl concentration 3 mol/L, titanium cathode plate, temperature 40 °C, current density 200 A/m^2. The results were shown in Table 3.

The results showed that the powdery metal bismuth was obtained at Bi^{3+} concentration less than 30 g/L, while Bi^{3+} concentration at about 40–60 g/L, the resultant bismuth cathode plate was relatively smooth and compact with some granular deposits, and these particles decreased gradually with increasing bismuth concentration. The bismuth cathode plate was relatively smooth without other tuberculum on the surface at bismuth concentration 80 g/L. The cell voltage gradually decreases with increasing bismuth concentration while the cathodic current efficiency changed little. Considering all these factors, the bismuth concentration 80 g/L was determined as the suitable Bi^{3+} concentration in catholyte.

Effect of HCl Concentration in Catholyte

The effect of HCl concentration in catholyte on electro-deposition were conducted by varying the HCl concentration from 1.5 to 5.0 mol/L under other cathodic parameters of Bi^{3+} concentration 80 g/L, titanium cathode plate, temperature 40 °C, current density 200 A/m^2. The results are shown in Table 4.

Table 4 Effect of HCl concentration on the electrowinning

HCl concentration (mol/L)	Cathodic current efficiency (%)	Cell voltage (V)	Cathode surface
1.5	99.28	1.78	Smooth and compact
2	99.26	1.71	Smooth and compact
3	99.62	1.72	Smooth and compact
4	99.53	1.68	Smooth and compact
5	98.18	1.56	Rough and powdary

Table 5 Effect of different cathode electrode materials on the electrowinning

Cathodic material	η_c (%)	U (V)	Cathode surface
Stainless steel	248.3	2.36	Spongy
Copper	98.94	1.50	Smooth but difficult to strip off
Titanium	99.26	1.71	Smooth and compact

The results showed that catholyte acidity had less impact on the cathodic deposit. Smooth and compact bismuth plates can be obtained under the conducted acidity. As the acidity increased, the cell voltage gradually decreased and the cathodic current efficiency changed little. With the acidity over 4 mol/L, the current efficiency and cell voltage decrease, and the cathode surface become rough, and high acidity resulted in more H_2 evolution. Therefore, 3 mol/L HCl concentration in catholyte was selected for the electro-deposition in this research.

Effect of the Cathode Electrode Material

The electrowinning of different cathode electrode materials were conducted under the cathodic parameters of Bi^{3+} concentration 80 g/L, HCl concentration 3 mol/L, temperature 40 °C and current density 200 A/m^2. The experiment results were shown in Table 5 and Fig. 3.

The results showed that stainless steel was not feasible as the cathode electrode material. Because it was apt to replace bismuth ions from the solution to obtain a spongy bismuth, and the resultant current efficiency was even more than 100% for bismuth replacement and adhered on electrode surface. The compact bismuth plate can be obtained when copper plate as electrode. But it was difficult to be stripped off for close adhesion between bismuth and copper. Titanium electrode performed well in this research and a compact bismuth plate can be obtained, and it was easy to be stripped off. Therefore, titanium is determined as the electrode material in this research.

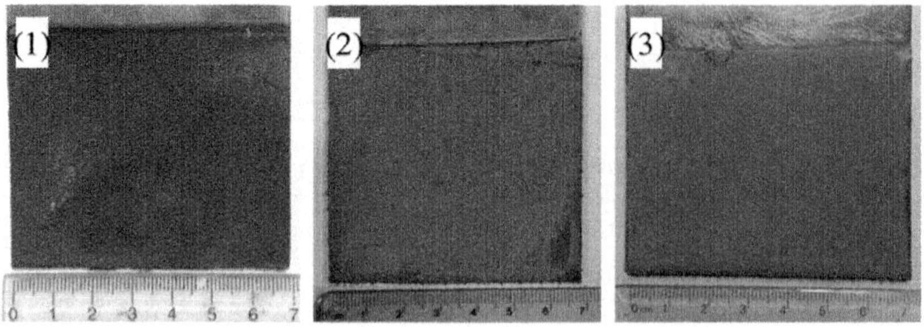

Fig. 3 The optical photos of the electrodeposit at different cathodic materials**a** Stainless steel; **b** Copper; **c** Titanium

Table 6 Effect of electrowinning temperature on the electrowinning

Temperature (°C)	Cathodic current efficiency (%)	Cell voltage (V)	Cathode surface
30	99.7	2.03	Rough and powdary
40	99.34	1.78	Smooth and compact
50	99.36	1.72	Rough and powdary
60	99.36	1.61	Rough and powdary

Effect of Electrowinning Temperature

The electrowinning at different temperatures were conducted from 30 to 60 °C under the cathodic parameters of Bi^{3+} concentration 80 g/L, HCl concentration 3 mol/L and current density 200 A/m^2. The experiment results were shown in Table 6 and Fig. 4.

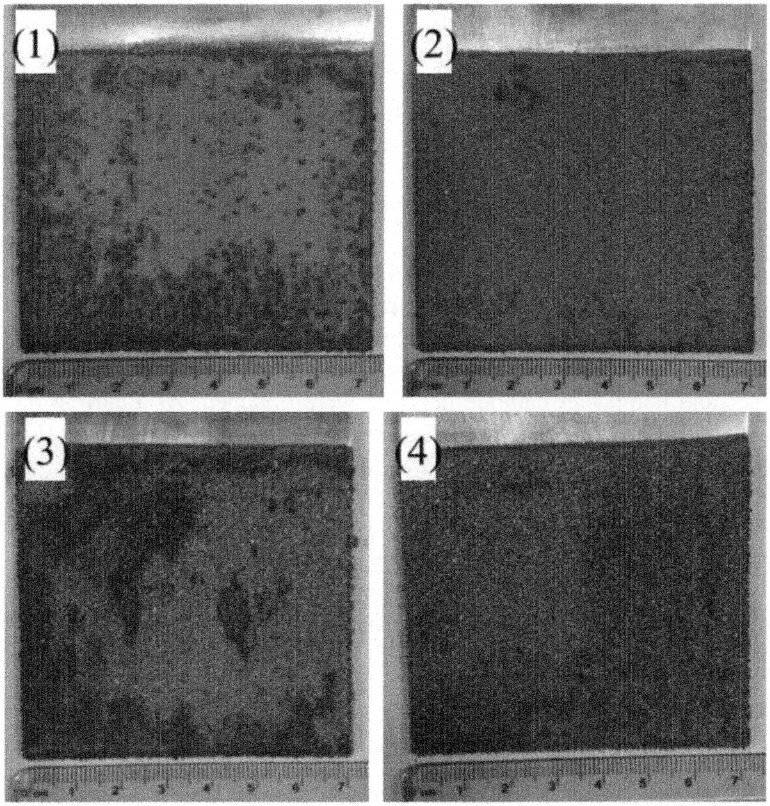

Fig. 4 Optical photos of the electrodeposit at different temperatures**a** 30 °C; **b** 40 °C; **c** 50 °C; **d** 60 °C

Table 7 Effect of cathodic current density on the electrowinning

Current density (A/m^2)	Cathodic current efficiency (CCE) (%)	Cell voltage (V)	Cathode surface
150	99.49	1.52	Smooth and compact
200	99.54	1.73	Smooth and compact
250	99.42	1.86	Rough
300	99.33	2.19	Rough and powdary
350	99.03	2.23	Rough and powdary

Experiment results showed that with the temperature increasing from 30 to 60 °C, all the cathodic current efficiency was higher than 99% while the cell voltage gradually decreased from 2.03 to 1.61 V. The decline of cell voltage may be due to the more active ion mass transfer at higher temperatures. Optical photos of the electrodeposit at different temperatures showed that the bismuth deposit formed at temperatures below 30 °C was rough and powdary with some granular deposits on smooth surface (Fig. 4a, whilst higher temperatures improved both the quality and adherence (Fig. 4b). With further increasing temperature, the bismuth depositing surface turn rough gradually (Fig. 4c d) and the increased temperature also resulted in water and HCl volatilization. Hence a temperature of 40 °C was selected for obtaining bismuth plates of good morphology in this research.

Effect of Cathodic Current Density

The effect of current density were conducted from 150 to 350 A/m^2 under the condition of Bi^{3+} concentration 80 g/L, HCl concentration 3 mol/L and temperature 40 °C. The experiment results were shown in Table 7 and Fig. 5.

The experiment results showed that the lower current density was applied, the smoother and compacter cathode deposits were obtained. At a current density of 150–200 A/m^2, the cathodic bismuth had compact and smooth surface (Fig. 5a b); while at 250–300 A/m^2, the surface was turned rough gradually (Fig. 5c d). When the current density reached 350 A/m^2, the granular deposition on the rough surface became looser (Fig. 5e). On the other hand, with increasing current density, the cell voltage increased steadily, and the cathodic current efficiency remained constant around 99%, while the anodic current efficiency decreased gradually. Considering all these factors, 200 A/m^2 was chosen as the optimum cathode current density in this research.

Effect of the Anolyte pH Value on Anodic Product Yield

According to the theoretical anode reactions (Eq. (4)–(6)), the pH value of anolyte will be reduced gradually during electrowinning. Therefore, some neutralizing

Fig. 5 Optical photos of the electrodeposit at different current density**a** 150 A/m^2; **b** 200 A/m^2; **c** 250 A/m^2; **d** 300 A/m^2; **e** 350 A/m^2

agent must be added to stabilize the pH value, to ensure the electrode reaction stability. Although $NaHCO_3$ was another suitable pH stabilizer from the theoretical point of view, the experiment results showed that a lot of CO_2 gas was released during electrolysis, which inevitably caused more chlorine escaped out of the solution, as well as anodic current efficiency be reduced. Considering all these factors, the NaOH solution instead of the $NaHCO_3$ or Na_2CO_3 solution was selected as anolyte pH stabilizer in this research.

The effect of the pH value to the anodic product yield at different pH value were explored under the conditions of NaCl concentration 300 g/L in anolyte, bismuth ions concentration 80 g/L and HCl concentration 3 mol/L in catholyte, heteropole distance 6 cm, temperature 40 °C and current density 200 A/m^2. The experiment results are shown in Fig. 6.

The results showed that pH value effect the anodic reaction greatly. NaClO yield increases with pH value increasing, and reached a maximum at about pH 9, then decreased gradually after that, which showed that too much acid or alkali weren't benefit for NaClO yield. While, the main product $NaClO_3$ has the highest yield at about pH 7. Hence a anolyte pH of 7 was selected for obtaining highest $NaClO_3$yield in this research, and secondary anodic reaction remain minimal and the total anodic current efficiency can be more than 80% at this pH value.

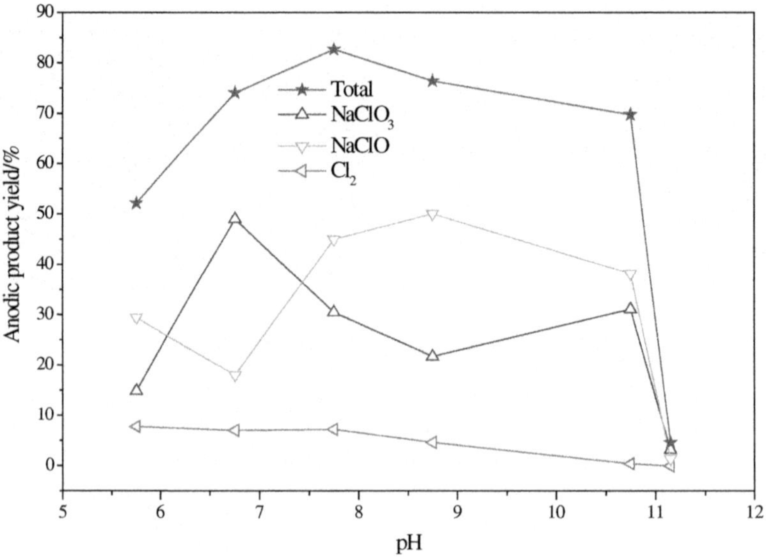

Fig. 6 Anodic product yield at different pH value

Effect of Sodium Chloride Concentration in Anolyte

The effect of sodium chloride concentration in the anolyte were carried out from 100 to 300 g/L, pH 7, and other conditions of bismuth ions concentration 80 g/L, HCl concentration 3 mol/L in catholyte, heteropole distance 6 cm, temperature 40 ° C and current density 200 A/m^2. The experiment results were shown in Fig. 7.

The experiment results showed that the anodic product yield increased with NaCl concentration increasing. The reason can be explained that the chlorine evolution overpotential decreased with increasing NaCl concentration. The higher NaCl concentration was in favor of promoting chlorine evolution and suppressing oxygen evolution. At the same time, more NaCl in anolyte increased solution conductivity and reduced cell voltage. When the NaCl concentration lies in 300 g/L, the cathode surfaces seem better, mostly smooth with metallic luster and anodic product yield reached more than 89%. Since no notable additional benefit was derived in going beyond NaCl concentration of 300 g/L in anolyte, this value was chosen as the optimum in this research.

Effect of Anode Electrode Material

The electrode material should be corrosion-resistant under chloride solution system, and also have the lower chlorine overpotential for promoting the discharge of chloride ions on the anode. Two common electrode material, graphite and DSA (Dimensionally stable anode) Ti -Ru anode, were test in this research.

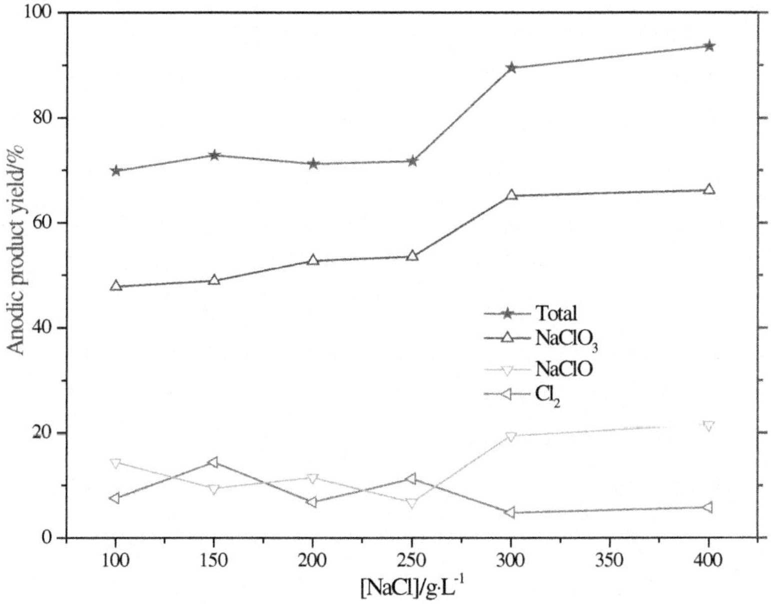

Fig. 7 Anodic product yield for different sodium chloride concentrations

Table 8 Anodic product yield for different anode electrode materials

Material	$NaClO_3$ (%)	Cl_2 (%)	NaClO (%)	Total (%)
Ti -Ru Anode	39.06	4.67	7.14	50.87
Graphite	59.03	7.01	22.06	88.10

The effect of the anode electrode material were tested under the conditions of NaCl concentration 300 g/L, pH 7 in anolyte, bismuth ions concentration 80 g/L and HCl concentration 3 mol/L in catholyte, heteropole distance 6 cm, temperature 40 °C and current density 200 A/m^2. The results were shown in Table 8.

The results showed that graphite plate was better than Ti-Ru Anode, and has fewer secondary reactions. Although Ti-Ru Anode is more widely applied to the chlor-alkali industry than the graphite, the process and parameters in this study were obviously different from the chlorate electrolysis in current chlor-alkali industry. So the economical graphite plate electrode was determined as anode electrode material in this research.

Verification Experiment

Based on the above test, the optimum conditions for the electrolysis of bismuth from a chloride solution and the simultaneous electrolysis synthesizing of the

Table 9 Main chemical composition of the electrolyte ($mg \cdot L^{-1}$)

Element	Bi	Fe	S	Ca
Content	60656	2341	1841	4337
Element	Cu	Mo	Pb	Si
Content	630	13	117	435

sodium chlorate in a membrane cell were as follows: NaCl concentration 300 g/L, pH 7 in anolyte, graphite anode, Bi^{3+} concentration 80 g/L and HCl concentration 3 mol/L in catholyte, titanium cathode, heteropole distance 6 cm, temperature 40 °C and current density 200 A/m^2. It should be mentioned that the above mentioned results were obtained under the relatively short electrolysis time (8 h), and the industrial electrowinning generally have a longer period at least 24 h. Therefore, verification experiment under the optimal conditions was conducted for 24 h to estimate the as resultant cathode and anode product. The chemical composition of the cathode solution was showed in Table 9.

Verification experiment result showed that the cathode bismuth plate has compact surface before 12 h (Fig. 8a), while surface become rough and granular gradually after 12 h (Fig. 8b). SEM image showed the surface bismuth was spherical particles.

From the Table 10, the average cell voltage and cathodic current efficiency were 1.74 V and 97.88% respectively. The average convert efficiency of the sodium

Fig. 8 Optical photos and SEM images of the electrodeposit

Table 10 Anodic product yield after 24 h electrolysis

t (h)	$NaClO_3$ (%)	Cl_2 (%)	NaClO (%)	Total (%)	η_c (%)	U (V)
4	48.41	8.21	15.44	72.06	98.20	1.76
8	56.51	6.07	12.29	74.87		1.74
12	54.62	5.16	10.16	69.94		1.76
16	51.83	3.87	10.49	66.19	97.95	1.72
20	52.48	3.21	10.16	65.86		1.72
24	53.19	2.68	9.30	65.17		1.73
Average	52.77	5.30	11.71	69.78	98.08	1.74

Table 11 Chemical composition of resulting cathodic bismuth plate (%)

Element	Bi	Fe	S	Pb	Cu	Mo
Content	99.886	0.014	0.025	0.005	–	–

chlorate was about 51.17%, and the total oxidizing efficiency of chloride ion decreased from 72.06 to 65.17% because of the decreasing of the convert efficiency of free chlorine and sodium hypochlorite (8.21% → 2.68%, 15.44% → 9.3%). From the result, it was convinced that the concentration of free chlorine and sodium hypochlorite reached a relative stable value and prevented chlorine gas from escaping from the solution. The chemical composition of electrodeposited bismuth plate analysis result showed that the cathode bismuth purity exceeded 99% with little impurity (Table 11; Fig. 4).

Figure 9 showed the XRD patterns of evaporated salt obtained from the resulting anolyte. All the peaks readily indexed to pure sodium chlorate (PDF#05-0610), which mean that the anolyte can be returned to the leaching process as the oxidizing agent.

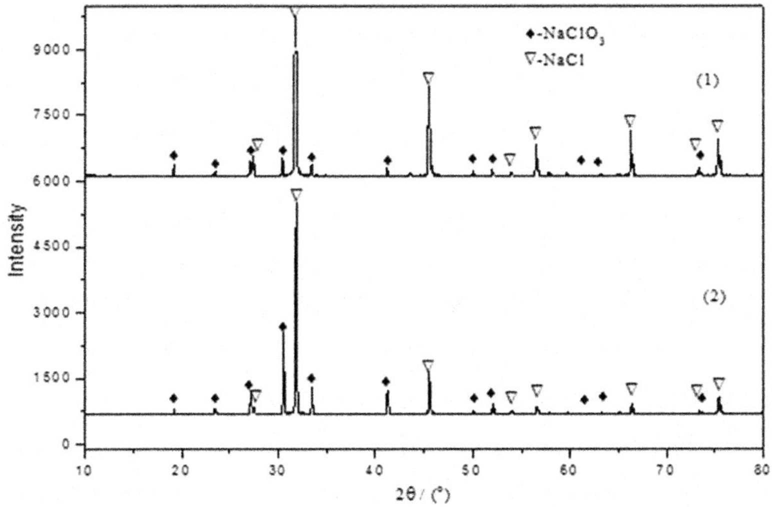

Fig. 9 X–ray diffraction patterns of salt evaporated from anolyte(1) 12 h; (2) 24 h

Conclusion

A new procedure was developed for leaching and electrowinning of bismuth from bismuthinite concentrate containing mainly pyrite and minor quantities of molybdenum, copper, lead and silica. The process has a series of advantages of using common industrial reagents to produce good quality product and to regenerate oxidants in the anode at the same time.

All the process conditions, including leaching and electrowinning of bismuth, were optimized, and the suitable conditions were obtained. Selective leaching of bismuth from bismuthinite concentrate was achieved using chlorate ion as oxidant. By using a membrane electrolytic cell, high purity bismuth plate (more than 99%) was obtained through membrane electrolysis, with 97% cathodic current efficiency and 69.78% anodic current efficiency. Chloride ion in anolyte was oxidized to chlorate ion which may be recycled to the leach again.

Acknowledgements This Project was supported financially by the National Nature Science Foundation of China (51574294) and Innovation Driven Plan projects of Central South University (2015CX001).

References

1. R. Mohan, In your element: green bismuth. Nat. Chem. **2**, 336 (2010)
2. F.K. Ojebuoboh, Bismuth-production, properties, and applications. JOM **44**, 46–49 (1992)
3. C.S. Anderson, Bismuth. U.S. geological survey, mineral commodity summaries 2016; http://minerals.usgs.gov/minerals/pubs/commodity/bismuth/mcs-2016-bismu.pdf
4. H.Q. He, J.C. Wang, Y.C. Jiang, Preliminary analysis on the geological characteristics and genesis of Fe-W-Mo-Bi(-Sn) polymetallic deposit in the southern Huangshaping lead-zinc mine district. Hunan. Mineral Explor. **1**, 323–333 (2010)
5. Wang LG, Bismuth metallurgy, 1st edn. (Metallurgical Industry Press, 1986)
6. J.M. Jin, Research on a molybdenum-bismuth ore processing test. Nonferrous Met. Eng. Res. **33**, 1–3 (2012)
7. B.L. Liu, H.J. Wei, B.W. Wang, Experiment of leaching of stibium and bismuth in the lead anode slime. Multipurpose Utilization Miner. Resour. **2**, 48–50 (2013)
8. Y.P Li, S.H. Wu, Y. Huang, Leaching of bismuth from complex bismuth ores by ferric chloride. Yunnan Chemical Technology **34**, 31–34 (2007)
9. Y. Chen, T. Liao, G.B. Li, B.Z. Chen, X.C. Shi, Recovery of bismuth and arsenic from copper smelter flue dusts after copper and zinc extraction. Miner. Eng. **39**, 23–28 (2012)
10. J. Ficeriová, P Baláz, C.L. Villachica, J. Harvanová, E. Gock, Selective leaching of bismuth from mechanically activated concentrate. Miner. Process. Extr. Metall. **121**, 103–108 (2012)
11. C.Y. Wang, D.F. Qiu, P.H. Jiang, Bismuth hydrometallurgy technology in China. Nonferrous Met. **53**, 15–18 (2001)
12. J.S. Luo, The studying situation and outlook of slurry electrolysis process. Gold Sci. Technol. **11**, 36-43 (2003)
13. H.S. Chen, R.D. Xu, Y. Zhu, S.W. He, H.Q. Hua, Y.W. Li, Recovery of bismuth from lead anode slime by electrolysis. Min. Metall. **24**, 44–49 (2015)

14. J. He, R. Guo, M.Y. Lan, C. Wang, M.T. Tang, J.G. Yang, S.H. Yang, C.B. Tang, J.L. Lu, Replacement of sponge powder bismuth with iron plates under micro-current. Chin. J. Appl. Chem. **30**, 1182–1188 (2013)
15. B. Zhang, Q. Li, W.Q. Shen, X.B. Min, Recovery of bismuth and antimony metals from pressure-leaching slag. Rare Met. **31**, 102–106 (2012)
16. M. Zertoubi, M. Chatelut, O. Vittori, Electrochemical recovery of bismuth in acidic media using a niobium electrode. Hydrometallurgy **34**, 109–118 (1993)
17. J.A. Reyes-Aguilera, M.P. Gonzalez, R. Navarro, T.I. Saucedo, M. Avila-Rodriguez, Supported liquid membranes (SLM) for recovery of bismuth from aqueous solutions. Journal of membrane science **310**, 13–19 (2008)
18. J.G. Yang, J. Lei, S.Y. Peng, Y.L. Lv, W.Q. Shi, A new membrane electro-deposition based process for tin recovery from waste printed circuit boards. J. Hazard. Mater. **304**, 409–416 (2016)
19. J.G. Yang, S.H. Yang, C.B. Tang, The membrane electrowinning separation of antimony from a stibnite concentrate. Metall. Mater. Trans. B **41**, 527–534 (2010)
20. X.D. Fang, Chlorate producing process, 1st edn. (Chemical Industry Press, 1988)
21. B.Fu, Non-ferrous metallurgy analysis manual (Metallurgical Industry Press, Beijing, 2004), pp. 184–188

Author Index

A
Abbasalizadeh, Aida, 87
Akbar Rhamdhani, M., 65
Andrews, Greg, 3
Azimi, Gisele, 47

B
Bai, Chenguang, 193
Blasi, Alessandro, 225, 277
Braccio, Giacobbe, 225, 277

C
Cao, Li, 203, 211
Chamelot, P., 77
Chen, Tiejun, 265
Chorney, Jannette L., 55
Czettl, Christoph, 159

D
Dai, Yongnian, 203
Deng, Zhigan, 247
Diaz, L., 77
Downey, Jerome P., 55
Dreisinger, David, 3
Durandet, Yvonne, 65

F
Fan, Gang, 247
Feng, Naixiang, 167
Firdaus, Muhamad, 65
Friedrich, Bernd, 103

G
Gibilaro, M., 77
Guo, Han-jie, 177

H
Han, Wei, 265
He, Kui, 211
Huang, Qingyun, 193
Huang, Zili, 265
Hu, Long, 137

J
Jiang, Tao, 137, 147
Jiang, Wenlong, 235
Jie, Lei, 285
John Rankin, W., 65
Judge, W.D., 37

K
Kipouros, G.J., 37
Koch, Wolfgang, 103
Konishi, Hirokazu, 93
Konishi, Yasuhiro, 129
Kücher, Gregor, 159

L
Lambert, Adrian, 47
Lee, Jin-Young, 19
Li, Cunxiong, 247
Li, Hongwei, 147
Li, Hongxu, 255
Li, Minting, 247
Li, Qian, 137, 147
Liu, Dachun, 183, 235
Liu, Fansong, 235
Liu, Pei-xiao, 177
Liu, Songli, 203, 211
Liu, Xiaoliang, 147
Liu, Zewei, 235
Liu, Zhaobo, 255
Liu, Zhenwei, 265
Li, Xingbin, 247
Li, Xuepeng, 183
Li, Yan-xiang, 177
Luidold, Stefan, 159
Lv, Xuewei, 193

© The Minerals, Metals & Materials Society 2017

H. Kim et al. (eds.), *Rare Metal Technology 2017*,
The Minerals, Metals & Materials Series,
DOI 10.1007/978-3-319-51085-9

Lyons, Katelyn M., 55

M
Martins, Gerard P., 31
Massot, L., 77
Maurell-Lopez, Ann-Kathrin, 103
Ma, Wenhui, 203
Mcgregor, Kathie, 65
Mishra, Brajendra, 31
Morgana, Massimo, 225, 277

N
Narita, Hirokazu, 219
Nohira, Toshiyuki, 93
Nomura, Toshiyuki, 129

O
Ogata, Takeshi, 219
Oishi, Tetsuo, 93
Okabe, Toru H., 119
Ono, Hideki, 93

P
Peng, Si-yao, 285

Q
Qin, Qingwei, 265

R
Rajesh Kumar, Jyothi, 19
Rao, Xuefei, 137
Romanelli, Assunta, 225, 277

S
Saitoh, Norizoh, 129
Schumacher, Katie J., 55
Seetharaman, Seshadri, 87
Serp, J., 77
Sietsma, Jilt, 87
Sposato, Corradino, 225, 277
Storf, Christian, 159
Strauss, Mark L., 31

T
Takeuchi, Eiichi, 93
Tam, Jason, 47
Tanaka, Mikiya, 219
Taninouchi, Yu-ki, 119

V
Venkatesan, Prakash, 87

W
Webster, Nathan A.S., 65
Wei, Chang, 247

X
Xiang, Junyi, 193
Xiao, Z.W., 37
Xu, Bin, 137, 147

Y
Yang, Chongfang, 235
Yang, Jian-guang, 285
Yang, Jianhong, 265
Yang, Jian-ying, 285
Yang, Yongbin, 137, 147
Yang, Yongxiang, 87

Z
Zhang, Shiju, 203, 211
Zhang, Yan, 147
Zhao, Kun, 167
Zhao, Zihan, 255
Zhou, Yuezhen, 235
Zhu, Kuisong, 203, 211
Zou, Qiang, 137

Subject Index

A
Acid baking, 10
Acid leaching, 238, 240, 244, 245
Acid or alkali washing, 148, 149, 153–155
Acid pressure leaching, 184
Acid treatment, 5
Adsorption, 220–223
Agitation speed, 250
Alloying treatment, 122, 125
Ammonia-alkali leaching, 269
Amorphous sulfides, 248, 250, 253
Anolyte, 294
Arsenic, 183–185, 187, 191
Arsenic sulphuration volatilization, 185
Automotive catalysts, 130, 132, 133, 135

B
Biomineralization, 130, 134
Bismuthinite concentrate, 286, 287, 289, 297
Blast furnace iron, 205

C
CA-12 behavior, 228, 230
Calcination, 13
Calcine, 148–155
Capital and operating cost, 4
Carbothermal reduction-nitridation, 211, 215
Catholyte, 289, 291
Cemented carbide, 160, 164
Ceramic substrate, 130
Chemical leaching, 131
Chloride leaching, 286
Chlorinate, 212, 214, 215
Choice chlorination, 206
Co-deposition, copper, 105
Co-deposition, palladium, 107
Comproportionation reaction, 78, 81, 83, 84
Converter slag, 194, 195, 201
Copper smelting dust, 183–185
Cyanide, 149, 150

D
Direct extraction, 4

E
Earth precipitation, 5
Economic assessment, 14
Electric smelting, 206
Electrochemical cell, 78
Electrochemical extraction, 87
Electrodeposition, 79, 289
Electrolysis, 94, 96, 98–100
Electrowinning, 82, 292
Energy conversion, 38, 40, 44
Energy supply chain, 38
Europium, 31–34, 36
Extraction, 4–7, 13, 138

F
Feed preparation, 9
Ferrochrome, 177, 178, 182
Ferrum, 260, 261
First stage reduction, 169
Foxtrot, 3–5, 7, 9, 14–16

G
Gold, 147–151, 153–155
Gold leaching, 150
Growth conditions, 130

H
Hard metal, 160
Hydrochloric acid leaching, 214
Hydrometallurgy, 23

I
Ilmenite concentrate, 205
Indium, 183, 185, 190, 191

© The Minerals, Metals & Materials Society 2017
H. Kim et al. (eds.), *Rare Metal Technology 2017*,
The Minerals, Metals & Materials Series,
DOI 10.1007/978-3-319-51085-9

Intensifying mechanism, 148, 151, 154, 155
Iono-metallurgy, 25
Iron chloride, 122

L
Leaching, 49, 51, 160–162, 164, 194, 196–201, 267–269, 271, 273, 274
Leaching rate, 188
Leaching temperature, 250
Leaching time, 189
Liquid-liquid extraction, 21–24
Liquid metal, 94, 95, 100
Lithofacies analysis, 180, 181
Lixivium, 138
Low grade molybdenum concentrate, 266, 268

M
Magnet, 40, 42, 44
Mechanical activation, 194, 195
Membrane electrolysis, 286, 297
Microbial cells, 133
Microbial reduction, 132
Microorganism, 130
Microstructure, 66, 67, 70, 71, 73
Microwave, 49
Microwave treatment, 49, 50, 52
Mixed rare earth oxide, 4, 16
Molten chloride, 78, 79
Molten salt, 94, 95
Molybdenite concentrate, 236, 238–241, 245
Molybdenum, 248–253
Molybdenum powder, 236, 238, 244, 245

N
$Na_3Sc(SO_4)_3$, 262, 264
$NaAl(SO_4)_2$, 255, 262, 264
$NaFe(SO_4)_2$, 255, 262, 264
NdFeB magnet, 66, 67, 70, 71, 73
Neodymium, 78–83, 86
Nickel, 248–253
Non-blast furnace iron, 205
Non-titanium extraction, 207

O
Organo-phosphorus, 25
Oxidation, 5, 66, 67, 70, 71, 73

P
P-507, 278
Palladium, 137–145
Palladium and copper deposition, 113
Palladium precipitation, 141, 142, 143
Panzhihua, 204, 206–209
Parameter influence, 104
Partial pressure, 252
Particle size, 252
Permanent magnets, 40
Phase change, 253
Phosphogypsum (PG), 48–52
Phosphor dust, 31, 36
Photovoltaic, 40, 41, 44
Platinum Group Metals (PGM), 119, 129–133, 135
Precipitation, 5, 272–274
Pressure leaching, 185, 248

R
Rare Earth (RE), 3, 4, 6–8, 11, 16, 38, 40–44, 77, 83, 86–88, 91, 94–96, 100, 219, 223, 226–231, 278–283
Rare Earth Elements (REEs), 47–50, 52, 55, 56
Raw materials, 168
Reaction rate, 160–163
Reactive anode, 88, 90
Reagents, 13
Recovery, 48, 52, 220, 222, 223
Recycling, 32, 120–122, 129
Red mud, 255–259, 261, 263
Reduction, 79, 80, 83, 85, 86
Reduction smelting, 212
Releach, 12
Renewable energy, 38, 39, 44
Rich titanium materials, 212
Roasting, 267–269, 273, 274

S
Sample preparation, 68
Scanning Electron Microscopy (SEM), 50
Scanning Transmission Electron Microscopy-Energy-Dispersive X-ray Spectroscopy (STEM-EDS), 50
Sec-octylphenoxy acetic acid (CA-12), 226–231, 278, 279, 282
Second stage reduction, 171
Selective, 256, 260, 261, 263, 264
Semidirect recycling, 160
Separation, 94, 100, 220
Silver electrorefining, 105
SNC lavalin, 8
Sodium sulfide, 141, 143
Solid reduction chromium process, 178, 181, 182
Solvent extraction, 226, 278
Spent catalyst, 138
Status and development, 204, 205, 207, 208
Sulfide precipitation, 138

Sustainability, 38, 42–44
Synergistic extraction, 278, 284
Synthetic rutile, 213

T
Thermodynamic analysis, 178, 182
Thermodynamics, 56, 57, 60, 62, 178
Titanium-bearing blast furnace, 207
Titanium extraction, 207
Titanium iron, 204
Transmission Electron Microscopy (TEM), 50

U
Utilization of Vanadium-Titanium Magnetite, 204, 209

V
Vacuum decomposition, 236–242, 245
Vanadium, 193–196, 198–201
Vapor phase extraction, 62
Vapor treatment, 123
Vulcanizater, 186, 187

W
Waste Fluorescent Lamp, 32
Water leach, 5

X
X-ray diffraction (XRD), 50

Printed by Printforce, the Netherlands